W9-AEV-067
WITHDRAWN
L. R. COLLEGE LIBRARY

Transforming Traditions in American Biology, 1880–1915

Transforming Traditions in American Biology, 1880–1915

Jane Maienschein

CARL A. RUDISILL LIBRARY
LENOIR-RHYNE COLLEGE

THE JOHNS HOPKINS UNIVERSITY PRESS
Baltimore and London

©1991 The Johns Hopkins University Press
All rights reserved
Printed in the United States of America

The Johns Hopkins University Press
701 West 40th Street
Baltimore, Maryland 21211
The Johns Hopkins Press Ltd., London

The paper used in this book meets the minimum requirements of American National Standard for Information Sciences—Permanence of Paper for Printed Library Materials, ANSI Z39.48-1984.

Library of Congress Cataloging-in-Publication Data

Maienschein, Jane.
Transforming traditions in American biology, 1880–1915 / Jane Maienschein.
p. cm.
Includes bibliographical references.
Includes index.
ISBN 0-8018-4126-7 (alk. paper)
1. Biology—United States—History. 2. Biology—United States—Philosophy—History. I. Title.
[DNLM: 1. Biology—history—United States. 2. Science—trends—United States. QH 305.2.U6 M217t]
QH305.2.U6M35 1991
574'.0973—dc20
DNLM/DLC
for Library of Congress 90-15623
QH
305.2
.U6
M35
1991
157742
Jan. 1993

Contents

Illustrations

Acknowledgments

THIS BOOK began to take shape during a visit in 1983–1984 at Harvard. Ernst Mayr arranged office space for Michael Ruse and me in a large storeroom of the Museum of Comparative Zoology. There, surrounded by dusty brachiopods, sponges, and such, I decided that the book I had been planning on Ross Harrison and embryology needed considerable reshaping. During a stormy, snowy week at the Marine Biological Laboratory (MBL) on Cape Cod, it became clear that the lives and careers of four friends from Hopkins intertwined so much that they really formed one story. The collective biography began.

As with any long project, this one has had its share of delights and frustrations. I have certainly incurred even more than the usual set of debts along the way because so many people have contributed in so many ways. For the actual writing and rewriting and re–rewriting, Richard (Chip) Burkhardt, Lindley Darden, and Richard Creath have been especially helpful in reading the whole manuscript carefully. Chip read parts twice and made important suggestions about the shape of the conclusion. Lindley and I continued a tradition that we began nearly a decade ago during an intense week of arguing in Woods Hole. We both finished our books at the same time this past year and continued our now familiar exchange of "I don't know what you mean here" comments at critical points. My husband, Rick, maintains final approval rights over everything I publish and has worked hard with his usual insistence on precision of language and on clarifying ideas. I am, of course, also hopelessly indebted to him for his infinite patience and calm ability to sculpt a phrase, no matter how elusive the idea or how apparently difficult it seemed to get things just right.

Others, including Richard Burian, Fred Churchill, Paul Farber,

Jonathan Harwood, Eli Gerson, Scott Gilbert, Sharon Kingsland, David Magnus, Ernst Mayr, Lynn Nyhart, Phil Pauly, Rudy Raff, and Jennifer Tucker, have read parts of the text, earlier drafts, or related work and have contributed intellectually in innumerable ways. Special friends Keith Benson, Jim Collins, and Ron Rainger have also read and discussed parts of the project at length and have offered encouragement and intelligent critical suggestions at key times. So has Gar Allen, even though my interpretations and approach diverge from his in fundamental ways. The innumerable hours we have spent discussing—and disagreeing about—everything from Morgan to Cuban politics have been fruitful and fun. In addition, my music teacher–Girl Scout director mother, Joyce Kylander Maienschein, and my nuclear physicist father, Fred Maienschein, have maintained faith that this work is really worthwhile and that I would actually finish someday, despite their occasional doubts about why anyone would want to study old science, and especially old biology, anyway.

In understanding the technical details of research methods, I have received help from a number of outstanding biologists. Robert Briggs and Donald Costello each spent long hours, respectively at Indiana and Woods Hole, helping me to reproduce old experiments and observations. They joined in my projects with enthusiasm and understanding of the historical inquiry. Briggs, for example, insisted that I had to keep working with the old tissue culture methods that Harrison would have used and should not resort to modern tricks to preserve asepsis, even though this meant that I kept growing lovely random batches of bacteria rather than nerve fibers. Costello gently instructed me, "This is the way that Conklin (or Morgan) would have done it." He also insisted that I use his collecting net and go down to the same place on the MBL dock that Wilson had gone to gather my *Nereis* worms; back in the lab, he then helped me follow the cleavage stages. Others, including Edgar Boell, Evelyn Hutchinson, Sally Wilens, J. P. Trinkaus, Dick Whittier, and many, many more, offered information and stories about my scientists and their work. Jane Oppenheimer provided an inspiration by making it clear in her important work on the history of embryology that the details of the scientific research matter in a way that too many historians seem to have forgotten.

The families of the four biologists have also been very helpful. Isabel Morgan Mountain has been delightful in helping to clarify details and shed light on her father's life and work. Richard Edes Harrison and Elizabeth Harrison have each supplied information and answered questions as well. Isabel Conklin has recorded reminiscences of early life at Woods Hole and has donated valuable photographs to the MBL Library. Wilson's granddaughter Linda Timmons also generously made available some extremely important unpublished letters and photographs.

Without the help of librarians and archivists, this project would not have been possible. Christiane Groeben, archivist at the Naples Stazione Zoologica, provided copies of a rich collection of letters from each of the four central characters. Jane Fessenden and Ruth Davis of the MBL allowed me free access to the invaluable letters, photographs, reprint collections, and other archival sources there. Judith Schiff and Patricia Bodak of Yale University's Manuscripts and Archives have continued to provide support many times over the years, as has Daniel Meyer at the University of Chicago. Ann Blum was archivist at the Museum of Comparative Zoology and introduced me to the Agassiz Collection there, as well as to institutional records. Caroline Rittenhouse at the Bryn Mawr College Archives has been generous in providing photographs and other materials from the collections there. In addition, I have received help in locating and copying materials from the archivists at the American Museum of Natural History's Department of Vertebrate Paleontology, the American Philosophical Society, the California Institute of Technolgy, the Johns Hopkins University, the Library of Congress, and Princeton University. All passages are quoted with permission of the library or family that holds the materials.

Photographs came from the archives at Bryn Mawr College, the Johns Hopkins University, the MBL, and Yale University, as well as from the private holdings of Linda Timmons. In addition, I also studied hundreds of wonderful photographs largely taken at Columbia University and the MBL by Alfred Francis Huettner and in the private collection of Robert and Millie Huettner. The Huettners also provided valuable insights into the lives and careers of the Columbia biology program during the time when Robert's father was a student there.

Funding support for this project came from the National Science Foundation (grants SES-83-09388 and SES-87-22231). In addition, Arizona State University provided internal support in the form of summer grants and travel minigrants.

Most important in producing this document itself have been the departmental administrative assistant, Nita Dagon, who manages all the paperwork and makes sure that things get done, and Joy Erickson, who has typed, revised, reformated, generated bibliography, and then edited everything at least a million times. Irma Garlick's intelligent and sensitive copyediting helped to make this a better book, as did the work of editors and production staff at the Johns Hopkins University Press. Thanks to everyone who has helped.

Transforming Traditions in American Biology, 1880–1915

Introduction

THIS BOOK is a study of scientific change, specifically change in biological traditions around the turn of this century, with the focus on American studies of development and heredity. At root, the changes involved epistemological shifts concerning what should count as good science, as legitimate problems, as knowledge, as evidence, and related assumptions. The changes took place within institutional settings and were brought about by individual scientists, of course. Hence, this is a story of individuals, and institutions, of scientific commitments, and the way scientists do their work, and of how all these come together in the process of transforming traditions.

Furthermore, this episode of scientific change occurred at a critical time for biology and for American science generally. As historians William Coleman and Garland Allen have shown in their respective textbooks, and as others have agreed in a host of studies, the several decades around the turn of this century proved a pivotal time for biology.[1] In particular, we may legitimately consider it the time when biology became modern and experimental. Despite disagreements of interpretation about just why changes occurred at this particular time and about just who effected the changes and how, historians have widely agreed that biology in 1910 looked very different from biology in 1880: the sorts of papers produced, the organisms used, the questions asked and the problems attacked, the methods and approaches adopted—all changed in at least some important ways.

New techniques and equipment also made it possible to carry out many kinds of work that simply had not been feasible before. As a result of successes in the related medical fields, new sources of funding had emerged to support more researchers in the biological sciences. In addition, the United States had begun eagerly to enter the scientific research arena and to compete for international attention. New institutions, such as the Johns Hopkins University and the Marine Biological Laboratory, were founded as settings aggressively designed to allow students and faculty to carry out biological *research.* Thus, a wider range of biological work became possible.

In addition, more people took up biology as a profession. Individuals who would never have attended college or sought professional research positions before began to enter biology and to establish their own research programs. By the end of the nineteenth century—for the first time in the United States—it had become possible to earn a living doing biology. With an increasingly progressivist, pragmatic, and scientifically sympathetic public, the time was ripe for change. These changes occurred on both sides of the Atlantic, and this book shows that the American developments involved responses to European innovations. Yet the United States experienced a greater expansion and perhaps a unique development of biological science, and it is here that this book focuses most closely.

One of the most significant changes that took place within biology was the move from a developmental to a hereditarian focus. This meant that, whereas biologists had once looked primarily at the way in which organisms develop over time and at all the environmental as well as internal factors effecting that development, they now turned increasingly to the earliest stages, shortly after which the new offspring had inherited something from the parents. The questions became: Precisely what is inherited, how, and to what effect? This brought with it a focus on individual organisms and on fertilization and the very first stages of development.

At the same time, the emphasis began to shift from external factors to factors internal to the organism itself. Whereas the nineteenth century had seen environmental influences as primary in directing development, the twentieth century looked inward. No longer did they ask what the weather was when a mother conceived, for example. Instead, it mattered what internal materials and patterns her offspring inherited from the previous generation.

These changes in research questions, accompanied by changes in techniques, equipment, and institutional settings, were fundamentally epistemological. Perhaps surprisingly, they did not involve any large-scale shifts in theory or any significant metaphysical revisions. Instead, they centered on how to do science. The nineteenth century had maintained the general assumption that one can only do science—that is, study nature effectively—if one does not do violence to the natural subject under study. As a result, each organism must be a vital, interactive whole that cannot be ripped apart for study. Yet the materialistic implications of Darwinian evolution theory and current medical advances had already cut away at this assumption. Studying parts, including cells and tissues, began to make sense in its own right. In addition, manipulative and interventionist experimentation became accepted as a legitimate way of attacking organisms. Consequently, doing biology was already becoming a different enterprise from what it had been at midcentury. By 1900, additional changes were coming in what one sought as the outcome of science, in what counted as appropriate procedures for gathering knowledge, and in how to determine that one had achieved knowledge at all. In other words, what have often been (problematically) labeled respectively as scientific goals, process of discovery, and process of justification had all changed. These changes were clearly epistemological at root, and different commitments clustered as different ways of doing biological science.

These epistemological changes transformed the existing biological traditions. This study focuses on one of those traditions and looks at a second with related concerns. By the 1870s, a morphological tradition that studied structure and pattern in organisms had already embraced evolutionary concerns, which put an emphasis on groups of organisms, rather than on individuals, and on the way that external or environmental factors shape development. A physiological tradition, which studied functions and processes within organisms, had focused on adults. At the end of the nineteenth century, the new techniques, new institutions, new questions, and new individuals moving into biology combined to transform each of those dominant traditions. Some morphologists began to focus more closely on the changing internal structures of individual organisms, and some physiologists began to explore physiological processes of development and heredity. Both traditions thereby underwent transformation and expansion, though each persisted in modified form. The transformation occurred piecemeal and gradually. As a result, this is a story of evolutionary change, not one of revolutionary overthrow or rejection of something old in favor of something new.

Many people contributed to transforming these traditions, of course, and this study focuses on only four. Yet they are not just any four, capriciously or randomly chosen. These four remained at the center of activity throughout the period. More important, each quickly established himself as a highly visible success according to any standard. Each published important work shortly after graduate school, set out a productive research program in a major university setting, attracted able and eager graduate students and institutional funds to support them, worked with the others to found respected journals and summer research facilities, assumed influential roles in all aspects of professional biology, and became known as one of the very few international leaders in the field. There is a good deal of disagreement about details and eventually significant divergence in their ideas and their work; yet the four overlapped and shared important central commitments about their conception of how biological science should work. They remained close friends, their lives and careers intertwined in many ways. They continued to learn from one another and to suggest revisions in one another's work as well. As biologist-historian John Moore puts it, they made up a "remarkable quartet," and all "were basic to the development of embryology, genetics, and cytology in America and to a somewhat lesser degree throughout the world."[2]

Edmund Beecher Wilson was the oldest and, in Moore's words, "one of the noblest minds in the formative years of experimental biology in America."[3] With his textbook, *The Cell* (1896), Wilson established himself as one of two leading cytologists in the world, along with his friend Theodor Boveri. His study of changes in the chromosomal apparatus during development and heredity remains the starting point for many cytological studies today. Edwin Grant Conklin's examination of cell lineage in gastropods and ascidians provides the takeoff point for modern studies of cleavage patterns. Conklin maintained a strong interest in evolutionary questions and in how individual embryological development is shaped by evolution and can inform our understanding of evolutionary processes. Thomas Hunt Morgan made major contributions to understanding the basics of frog development, of regeneration, and of sex determination in individuals before moving on to take up his Nobel Prize–winning work in *Drosophila* genetics. The youngest of the four, Ross Granville Harrison carried out the first successful tissue culture in his work on neurodevelopment. He almost received a Nobel Prize for that work but was denied the honor because of the intervention of politics during World War I. After that, critics acknowledged the technique as one the most fundamentally important of the century but pointed out that Harrison had abandoned it and moved on to other studies of embryology, so that others earned Nobel Prizes while he did not.[4] Clearly, then, each of the four deserves serious historical attention in his own right.

This book goes beyond the individual stories, however, to provide a collective biography of sorts. It is the account of these leading biologists whose careers and lives were interwoven in many ways.

Fig. 1. The young Wilson, about 1875. From the collection of Linda Timmons.

They each moved to graduate work in biology at the Johns Hopkins University, so that this is also a story of Hopkins and of the institutions to which each then proceeded.

It is therefore on these four graduate student friends and their work, beginning at Hopkins, that this study focuses. They did not all attend in the same year, though they overlapped in many years in various professional capacities. Each was profoundly influenced by his educational experience. Each obtained a Ph.D. degree from the Johns Hopkins University with morphologist William Keith Brooks as major advisor and studied with British physiologist Henry Newell Martin as well. They took many of the same courses and read many of the same books. They became and remained friends. They organized journals and societies together and began various lines of research work with one another's help. They even tried at various times to hire each other. Thus, we have four young zoologists who began at roughly the same time and place, with very similar early professional training and with closely intertwined early careers. They all participated in the morphological and physiological traditions at Hopkins.

Yet, as their respective research programs developed during their first decades of work, they began to diverge. By 1915, they were all widely regarded as being among the very few world leaders in zoology. Wilson had become identified as the leading cytologist, Morgan as the foremost geneticist, Conklin as a first-rate evolutionary embryologist, and Harrison as one of the world's two best experimental embryologists (alongside Hans Spemann). True, their fields of interest remained closely related; yet their areas of research, the questions they asked, and the methods they selected diverged in the course of the decade in rather notable ways.

The problems for this study are, first, to establish the shared origins and the subsequent divergence of four biological research programs and, second, to account for that divergence. Hence, this is a historical study of central players in the development of biology. Yet, since it is also a consideration of the nature and causes of scientific change, we must also explore in some detail precisely how and why these four research programs emerged and changed. At one level, the unit of analysis is the individual lines of research and their settings; at another, it is the larger traditions to which those researches belong. I hope by the end of the book to have demonstrated that both units of analysis are basic for understanding the nature of scientific change.

It is crucial to understand the scientific work and its changes through time in order to get at any larger patterns of transformations in traditions. In addition, to explore the epistemological norms, it is critical to examine the way the scientists carried out their work. This emphasis requires a fine-grained descriptive look at the science done before moving on to larger questions. Such detail and description are absolutely essential in order to get the story "right" and to understand the scientific work. Yet presentation of "the facts" clearly also involves interpretation at all stages and therefore depends on an analytical as well as descriptive approach. Because it incorporates biography, institutional history, technical analysis of the scientific work, and analysis of philosophic questions about scientific change, my approach is eclectic rather than falling into any one currently popular historiographic framework. It is more morphological, concentrating on the form and the patterns of change in the traditions and lines of research within them, rather than physiological and focused on the functioning or larger role of the traditions.

The first part considers the common background at the Johns Hopkins University. After introducing the principal players and their paths to graduate work at Hopkins, chapter 1 considers the research ideal that Hopkins offered to would-be biologists at the time: just what was the mission of this institution that attracted Wilson, Conklin, Morgan, and Harrison? Chapter 2 addresses the actual Hopkins experience: What was the biological work like which all four shared there? What was the character of research within the morphological and physiological traditions that they ingested there? The three chapters of part II examine the ways in which the traditions underwent transformation in Europe and the United States and the impact of European innovations on American biology: what factors influenced change more broadly, what forces affected our four, and how did they contribute to the transformation process? Throughout, there is an emphasis on the generation of research programs within changing traditions: what happened, how, for whom, and why? Part III turns to the protagonists: the four chapters detail the establishment and development of each of the diverging research programs. They also explore in greater depth the epistemological commitments underlying the research and show the ways in which shifts in those commitments contributed to the transforming of traditions.

PART I

Four Friends from Hopkins

1

The Hopkins Ideal

EACH OF THE four principals in this story selected the Johns Hopkins University for his graduate degree in biology. Yet each found his path to that premier program in a different way. This chapter details, to the extent that we can know, the background of each and the reasons why each selected Hopkins. In addition, by considering the ideal that Hopkins represented at the time, the chapter explains why it would have been rational to have chosen that particular graduate university and its exceptional biology program.

Edmund Beecher Wilson

Born in 1856, Edmund Beecher Wilson grew up in the small town of Geneva, Illinois, where his father practiced law before eventually going on to serve as chief justice of Illinois. His mother's side of the family boasted descent from a crew member of the *Mayflower*, and Wilson prized his centuries-long American, New England, and Cape Cod ancestry. He also referred proudly to the seafaring side of the family, though his childhood in the Midwest may have exaggerated the exotic attractions of the sea.

Wilson's maternal grandparents lived in Geneva, which presumably explains why his family settled there as well. This put him in close contact with his mother's sister and family, so that at age two and a half young Eddy was reportedly adopted by his aunt, Mrs. Charles Patten. Hence he reported that he "had in effect two homes and four parents between whom I . . . hardly distinguished in point of love and loyalty."[1] The Pattens encouraged his interest in gathering living things such as birds, snakes, and toads, and even gave him a room in their house for his collection. Both his education-minded parents and his relatives encouraged his interests in music as well. Wilson later wrote that "it would not be easy to imagine a happier environment for a boy who somehow managed to combine an early passion for natural history with an almost equal love for music; who grew up in an atmosphere of warm affection and sympathetic understanding at home, and was surrounded by a circle of intelligent and cultivated people."[2]

Wilson did well in school, so that at age sixteen he was persuaded to teach in a "little country district school" where his brother had taught the year before—for thirty dollars a month and board. There he experienced the prototypical prairie life with its frosty early mornings of breaking ice on the wash water and huddling around the wood stove, long walks to school through driving storms and blowing snowdrifts, getting to school early to light the stove, and teaching all grades and all subjects at the same time in one room. He found his one year "a grand experience," which he "would not have missed, as the saying is, for a farm." Yet he resolved not to continue that glorious time, despite the high praise from one pupil's father: "He's young but his age don't hurt him none."[3] Instead, he determined to enter college and devote his life to biology, or science at least. Clearly his cousin Samuel Clarke's influence begins to appear here.

In some sense, Wilson followed Clarke around for nearly a decade. Clarke had attended Antioch College before going on to Yale's Sheffield Scientific School, and Wilson followed him there. He had passed an exam for West Point, even receiving the highest score, but he was too young to enter the military academy and therefore ruled out of that opportunity. So with two hundred dollars in his pocket, young Wilson headed for Antioch. Yet already changes were afoot to call him onward. After one year at Yale, Sam Clarke began sending rave reviews from the Sheffield Scientific School there. He encouraged Wilson to try it also. The fact that "Sheff," as it was called, had much more liberal views about the required classical background, in which Wilson was rather weak, plus the fact that Yale offered opportunities for earning one's way through school, attracted this young man from Illinois. He determined to enter Yale but now needed another year to prepare himself in the sciences instead of the classics. He spent that year in Chicago with his family, getting ready for Yale, then serving as a recorder for a summer survey of Lakes Ontario and Erie in order to earn money. In the fall of 1875 he entered Yale's scientific school.

At Yale, Wilson became excited about biology. He found William Henry Brewer's lectures covering heredity and evolution in agricul-

Fig. 2. Wilson (*standing, left*), Baltimore, 1879. Also, *clockwise from top right:* K. Mitskuri, Samuel F. Clarke, and William T. Sedgwick. From the collection of Linda Timmons.

ture particularly enthralling, as did many undergraduates at the time.[4] Wilson also took courses with Sidney I. Smith and undertook his thesis work for his bachelor's degree. With two papers on the Pycnogonids (sea spiders), he received his Ph.B. degree in 1878.[5] For that work and during the next year, when he remained at Yale as an assistant, he learned about the meticulous studies of cell division produced by Professor Edward Laurens Mark at Harvard. Smith reportedly gave Wilson one of Mark's papers and remarked that the man had written a very lengthy study of snail development and had "only got as far as the 2-cell stage." Wilson then "wondered what the author could find to fill two hundred pages on the subject. I looked over the paper and saw my first picture of karyokinesis. Then and there was born my determination to find out something about cells, protoplasm, cell division, fertilization, and development." He further noted that, "from that determination I have never swerved, although it often seems to me that cell structure and cell life seem in their essentials as mysterious today as they did fifty years ago."[6] Pursuing his interests in biology, Wilson spent the summer of 1877 under Spencer Fullerton Baird dredging for marine life for the United States Fish Commission. For the next year, Yale offered Wilson a further position as assistant and graduate student that would have kept him there. But cousin Sam made his mark again.

Sam Clarke had gone to pursue graduate work in biology at the Johns Hopkins University. He sent wonderful reports about the opportunities there, opportunities that Wilson probably only vaguely appreciated before he arrived. Yet the prospects were sufficiently enticing that both Wilson and his friend William Sedgwick, who was an assistant at Yale, applied to attend Hopkins themselves. They both received fellowships, and they both joined Clarke there in 1878. At Hopkins, they studied both physiology and morphology, respectively with Henry Newell Martin and William Keith Brooks, though Wilson worked most closely with Brooks.[7]

Edwin Grant Conklin

Like Wilson, Edwin Grant Conklin was a midwesterner; he was born in 1863 in the small town of Waldo, Ohio. His background was less illustrious, but just as highly valued by this young son of a country physician. He particularly enjoyed having been born in the year that Lincoln signed the Emancipation Proclamation and coming from the central pioneering tradition of the United States. He had ridden about Ohio in a covered wagon, lived and worked on a farm, and attended a one-room country school with one teacher—all sources of considerable pride for him.[8] Yet his father's professional background likely pointed him toward gaining more education than a midwestern farm boy might otherwise have sought.

Conklin graduated from high school in nearby Delaware, Ohio, then attended college at Ohio Wesleyan University. Brought up in a religious family, he considered the ministry as a career. As one biographer put it, with the combined religious influences from home and his sectarian college, "he barely escaped being a Methodist minister."[9] He did study the King James version of the Bible and passed an exam on it to receive a "local Preacher's license." Fortunately for the history of biology, during his third year of college Conklin needed money. He therefore decided to try teaching, though he retained a lifelong concern with showing the compatibility of religion and science. For part of one year, Conklin, like Wilson a decade earlier, taught in a one-room school, serving as teacher, janitor, and disciplinarian—all for thirty-five dollars per month. Like Wilson, Conklin found one year of that enough but later regarded the experience as one of the most valuable of his life, since it gave him self-confidence.[10]

As Wilson had, Conklin became interested in natural history and the outdoors fairly early. Life on a farm, reinforced and directed by fossil collecting trips at college, undoubtedly prepared him for a greater interest in science.[11] Yet it was in college that his natural history interests really began to develop. Conklin presumably attended Ohio Wesleyan because it was the local Methodist school. Though not noted for an outstanding scientific or research orientation, the university had at least one special teacher, Edward Nelson. Nelson held one of the first Ph.D. degrees from Yale and introduced Conklin to biology as he took his classes on field trips to study fossils, botany, and local fauna in the rivers and ponds.[12] He then encouraged him to pursue that field professionally himself. Conklin graduated with a B.S. degree in 1885 and a B.A. degree in 1886.

Immediately upon graduating, Conklin once again needed to earn some money. Presumably, he also felt the call to carry out some missionary work. Thus, he spent the years 1885–88 at Rust University in Mississippi, a black college run by the Freedman's Aid Institution of the Methodist Church. He taught Latin, Greek, and all the science courses during his three years there. He then determined to enter graduate school in biology because he realized the lack of opportunity for promotion at Rust and his lack of preparation for a professorship in biology.[13] He selected the Johns Hopkins University and began in 1888, ten years after Wilson had arrived.

Once again, we have only clues about why this young would-be biologist selected Hopkins. Presumably he, too, had heard rumors of the educational experiment there and resolved to enter the promising biology program. Perhaps he, too, was attracted by the unusual fellowship program that would make it possible to attend school and earn a living at the same time. Whatever the reasons, at Hopkins he joined Wilson and Morgan in immediately aligning himself with Brooks's morphological work, though he also took courses in physiology with Henry Newell Martin. Conklin loved the Johns Hopkins University, reporting that: "I cannot begin to describe adequately the stimulus for scholarly work and research which I received at Johns Hopkins. It was as if I had entered a new world with new outlooks on nature, new respect for exact science, new determination to contribute to the best of my ability to 'the increase and diffusion of knowledge among men.'"[14]

Thomas Hunt Morgan

Wilson's maternal ancestors had arrived on the *Mayflower*, and Morgan's came to this country only shortly thereafter. In 1636, his paternal forebears sailed for America. Arriving in Boston, one branch of the family eventually made its way southward, to Kentucky. While Wilson prided himself on his lengthy New England heritage and Conklin on his pioneer background, Morgan had a similar longtime history in the American South. And while Wilson's family featured a seafaring mate on the *Mayflower*, Morgan's boasted a Civil War general—on the Confederate side. Further, Thomas's father was a diplomat who then entered the Confederate army himself. After a stay in the Ohio Penitentiary for his wartime actions, Morgan's father married a woman from an established family in Baltimore, herself related to Francis Scott Key, to a Revolutionary army colonel, and to a former governor of Maryland. Thus, Thomas came from a mixed but richly cultured and educated background.

Thomas Hunt Morgan was born ten years after Wilson and three years after Conklin, in 1866, in Lexington, Kentucky. Like Wilson, he also felt the early attraction of living nature. He enjoyed rambling about the countryside collecting living birds, as well as fossils and other objects. Reportedly, he even organized collecting expeditions of small boys throughout the surrounding country.[15] In summers when he visited the Baltimore relatives in the mountains of western Maryland, he wandered the country there. In particular, the areas where railroad lines were being cut through afforded excellent observing and collecting grounds.

At age sixteen, Wilson had taken his first job teaching school on that chilly prairie, and Conklin was finishing high school. At age sixteen, Morgan entered the Preparatory Department of the State College of Kentucky, now the University of Kentucky, as one of its first students. This program, designed to prepare the poorly schooled students for college, served in effect as a college prep course to provide what high school should have given. Upon enrolling in college proper, Morgan took many of the science courses taught and received his B.A. degree in 1886 with highest honors. During those undergraduate years, he pursued his interest in geology and natural history and spent several summers carrying out geological and biological field work with the U.S. Geological Survey.[16]

From undergraduate days and exploration of natural history in the hills of Maryland and Kentucky, Morgan moved to the seashore. In 1886, he attended the summer session at the marine biology school at Annisquam, Massachusetts, a school organized by naturalist Alpheus Hyatt and the Boston Society of Natural History.[17] Morgan praised the Annisquam experience: "I continually congratulate myself that I am here . . . without doubt the work will be of the greatest assistance to me next winter."[18] For the next winter he would be working as a graduate student in biology at the Johns Hopkins University.

The Johns Hopkins University attracted Morgan for graduate work as it had Wilson and Conklin. Yet the reasons remain rather vague in all three cases. Wilson had heard reports from his cousin. Conklin probably was attracted by the fellowships. Morgan may have heard of the uniqueness of Hopkins while at Kentucky, perhaps through a student Joseph Kastle who had preceded him from Kentucky to Hopkins. Perhaps the fact that his mother's family lived in Baltimore made him more aware of the existence of the new university, which had begun only ten years earlier in 1876. Yet Morgan later wrote, "Little did I know then how little I appreciated that a great university had started in their midst, and I think this was typical of most of the old families of that delightful city." Perhaps also his southern background inclined him to a southern school, since Hopkins was considered a "southern university in a southern city."[19] More likely, it was the "rather vague rumor that reached us undergraduate students in far distant colleges" that had its effect.[20] Like Wilson and Conklin, Morgan did not have clear ideas about what Hopkins offered, but he had an excitement about natural history and the study of living things that directed him toward natural sciences. And he had at least heard rumors that Hopkins was the place to go for graduate work in such areas. There Morgan also began to work with Brooks, though he eventually regarded his philosophical inclinations as unattractive. He found physiologist Martin stimulating as well and especially admired his rigorous experimental work.

Ross Granville Harrison

The youngest of the four protagonists in this story, Ross Granville Harrison was born in Germantown (essentially Philadelphia), Pennsylvania, in 1870. His father was a mechanical engineer from the area and sent his young son to Mrs. Head's school to begin his education. Mrs. Head evidently believed in introducing her charges to natural history, during this time of "nature study" popularity, for the group took many field trips into the countryside. As Harrison's primary biographer put it, "the teaching of the children included real trips to study nature and animals in their natural surroundings."[21] Weekend bicycle trips with school chums also made up an important part of young Ross's life, and on one occasion he reportedly had the exciting experience of saving another boy from drowning.[22]

After three years at Mrs. Head's in Germantown, Ross moved with his family to Baltimore, where other family members already lived. There he attended public schools and spent one year at Marston's University School, apparently taking a course of study to prepare him for university work. During this time he gained a solid grounding in the classical study of Latin and Greek. He also continued his interest in natural things, exploring the countryside through walks and cycling trips, often with map in hand. Only the study of birds failed to interest him, because he felt that it had suffered from too much amateur interest and smacked of mere stamp collecting rather than of serious science.[23]

In Baltimore, he gained an avid curiosity for information about living creatures, no doubt encouraged by his particular surroundings there. He lived at least periodically with his aunt, whose home has been described enthusiastically by a friend as a veritable wilderness of natural wonder:

> This was a place with a grotto and fish pond and formal terraces, hemlock-bordered gardens and trellises for grapes, and greenhouses and hothouses of many kinds. There was a cool house of circular construction, for camellias, a warm greenhouse from which we once stole a ripe pineapple, and hothouses where were gardenias and orchids. William Herrise, the Alsatian gardener who presided over all these extrava-

> gances, was choleric and marvelously bearded, a man who seemed to have stepped right out of a German fairy tale. He bred strange breeds of fowls, of French and German origins, sultans and Houdans, golden Hamburgs and silkies.[24]

Presumably because he was in Baltimore, and possibly influenced by his father's scientific and technical training, Harrison decided to enter the local Johns Hopkins University for his undergraduate work. While at Hopkins, he belonged to the Tramp Club, which tramped energetically around the countryside. He took long walks and bicycle trips with whatever friends he could persuade to join him, or on his own, and generally acquired a great affection for the area.[25] Harrison entered Hopkins intending to take the premedical course of study in preparation for a medical career, a course that included biology and especially physiology with Martin. After graduating with an A.B. degree in 1889, however, he decided not to go straight into graduate work in medicine, perhaps in part because the Hopkins medical school had not yet opened (it began in 1892). Instead, he would enter graduate school in biology. And what better place than the research-oriented, local Johns Hopkins University.

At Hopkins, the nineteen-year-old Harrison joined Morgan, who had arrived three years earlier at age twenty, and Conklin, who had entered the program the year before at age twenty-five. Wilson had entered Hopkins in 1878 at age twenty-one and had left in 1881. Though Harrison was the youngest and the newest arrival, he nevertheless had the most experience with the Hopkins approach to doing biology, with its emphasis on research and its combination of physiological and morphological work in the same department. Despite his undergraduate attraction to and his continued respect for Martin's research, Harrison joined Wilson, Conklin, and Morgan in choosing to carry out his Ph.D. work in morphology, specifically the morphology of development (or embryology) with Brooks.

Harrison, with his background medical interests, may have gained most from the exposure to practical work and from the intense interaction among the group of active and enthusiastic students sharing problems and approaches. Yet surely all three of the contemporaries gained from the existence of their excited and highly motivated group. Harrison, Conklin, and Morgan became close friends during their Hopkins days and remained so through life, despite their respective wanderings and changing research emphases. They also each came to know Wilson and to work with him in various capacities facilitated by the Hopkins connection.

Other Students

Of course, these four were not the only graduate students in biology at Hopkins. Martin had his coterie of followers, and so did Brooks.[26] A full roster of those who graduated with doctorates in zoology during Hopkins' first twenty years includes:[27] Ethan Allen Andrews, Robert Paine Bigelow, Adam T. Bruce, Samuel F. Clarke, Edwin Grant Conklin, Herbert W. Conn, Henry T. Fernald, George Wilton Field, Ross Granville Harrison, James L. Kellogg, Henry M. Knower, George Lefevre, James P. McMurrich, Maynard M. Metcalf, K. Mitsukuri, Thomas Hunt Morgan, Julius Nelson, H. Leslie Osborn, Shosaburo Watase, Edmund Beecher Wilson, and Henry van Peters Wilson. The Ph.D. graduates in physiology under Martin included: Henry Gustav Beyer, Henry Herbert Donaldson, E. A. Hartwell, William Henry Howell, George Theophilus Kemp, Frederic Schiller Lee, William Thompson Sedgwick, Henry Sewall, and C. Sihler.

Of these, this study looks closely only at the four friends introduced above, who shared many experiences, maintained close professional connections, achieved wide recognition as biological leaders in the early twentieth century, and by 1910 had advanced to major leadership positions in biology.[28]

Why There? The Hopkins Mystique

Given the "vague rumors" that attracted students, what was it about Hopkins that excited such rumors and enthusiasm? Most likely it was the mystique of the institution itself rather than the reputation of the biological program per se. For the Johns Hopkins University was from its inception extraordinarily well funded, with special fellowship programs to attract students to a serious research career. It was also uniquely research-oriented, stressing the importance of creative laboratory science that went far beyond the amateur nature study movement popular in American society in the last quarter of the nineteenth century. It had a commitment to fund research laboratories and to offer experimentally oriented courses of study as no other American university of the time did. The Johns Hopkins University had a special mission, inspired by Johns Hopkins himself.

A wealthy Baltimore businessman and a Quaker, Johns Hopkins never married because the local meeting of the Religious Society of Friends forbade him to marry his first cousin, whom he loved. With no family, he looked to worthy causes for the vast wealth he had accumulated through railroad investments. Persuaded that a modern hospital with an associated medical school and research university would best serve Baltimore, Hopkins incorporated two groups of trustees for the hospital and for the university in 1867. When he died in 1874, the provisions of his will took effect. His trustees were left largely free to determine how best to proceed, for Hopkins's will outlined few specifics. He really only emphasized two things: the erecting of a research community of good men rather than fancy buildings, and the desire for a group of "free fellowships" to attract the best young men from southern states to study at the university without charge.[29]

By the 1870s, Baltimore already had ten colleges. What distinguished the new entry envisioned by Johns Hopkins was its role as a research university and its endowment. With an endowment of seven million dollars for the university and a new science-based hospital, the yearly income was over four hundred thousand, more than the full endowment of most colleges.[30] This fund proved sufficient to create the pioneering research university that Johns Hopkins and his trustees sought, though the hospital required additional support and opened only seventeen years later.

The university trustees began by appointing a president. For that position they chose Daniel Coit Gilman, who had served as administrator in various capacities at Yale University and at the University of California.[31] Gilman proved to be an excellent organizer and efficient worker, immediately acting on his first mandate of 1 May 1875, to build a "literary and scientific department" for training young scholars, or in other words a graduate school. He promptly began to gather the best faculty he could find.

By the mid nineteenth century, the idea of a research university was "in the air," Gilman wrote. The infusion of funds that the Morrill Land Grant Act had brought to education after 1862 had stimulated development of more advanced educational programs in state schools as well as private. Yet the ideal of a university for training scholars remained just in the air rather than actually put into effect. The university founded by Johns Hopkins was intended to overcome that lack of action. Though Johns Hopkins had not himself set out exactly what he meant by a university, it was clear that he expected it to have several faculties that would work in parallel to offer advanced training and produce scholars. With no American precedent, achieving such a goal required careful study of European models. Historians have made much of the impact of the German ideal on the Johns Hopkins University and other American schools built later, and there clearly were some such influences. Others have pointed to the English model as more important. But Gilman stressed that "we did not undertake to establish a German university, or an English university, but an American university, based upon and applied to the existing institutions of this country."[32] This would be a uniquely American experiment in higher education, he said.

He worked carefully but effectively to bring together his first faculty. "'Evolution,'" Gilman wrote, "was then beginning to be the note of the times, and our best advisers urged upon us 'Development.' 'Be slow,' they said, 'plant good seeds and see what they yield.'"[33] Thus, Gilman traveled through parts of the United States. He then urged a young physicist at Rensselaer Polytechnic Institute, Henry A. Rowland, to accompany him to Europe, and the two spent the summer of 1875 in Dublin, London, and on the continent. There Rowland learned of the current European work in physics, and Gilman recruited faculty members. As a result, by fall 1876, Gilman had chosen a first group of six men. Three were British (James Joseph Sylvester, mathematics; Henry Newell Martin, biology; Charles D'Urban Morris, classics), two Americans trained in Germany (Ira Remsen, chemistry; Basil Gildersleeve, Greek), and the sixth an American who received his higher degrees only later because his Jewish heritage and isolation held him back (Henry A. Rowland, physics).[34] These six ruled as the only full professors on the faculty for seven years.

In those early days, enthusiasm ran high, and the faculty and students shared a spirit of adventure.[35] They were all part of a brave American experiment in university education. As philosopher Josiah Royce reported with excitement Hopkins at the time experienced "a dawn wherein 'twas bliss to be alive." This spirit echoed elsewhere throughout the new university. As Royce stressed from his perspective as one of the early fellows there, the university experienced an unusual spirit of cooperation: cooperation with other universities through passing on the benefits of experience, and cooperation among the different types of study and hence among departments. Royce felt that "the lesson of the hour" from Hopkins was cooperation and congeniality.[36]

In the early decades, faculty and students fraternized relatively easily. The pragmatic philosopher Charles Sanders Peirce arrived in 1879 and started the Metaphysical Club. He gathered groups of students and faculty for long discussions around his open fire, some of which the physiologist Martin joined. Both biology instructors, Martin and Brooks, also regularly scheduled scientific meetings that combined entertainment and scholarly pursuits.[37] After 1880, Martin also presided over the Naturalists Field Club, a group that met on

Saturdays at two o'clock to wander about the countryside in order "to study the fauna, flora, and geology of the neighborhood of Baltimore."[38] Others joined the Tramp Club, which hiked throughout the area and logged impressive mileage. Scholars in one field listened to their colleagues lecture in other fields through such meetings as those of the Scientific Association. The enthusiasm and sense of community helped to make Hopkins a vital university in those early years. This spirit in turn helped attract eager young American scholars. The situation changed and the fervor cooled as the university expanded into the twentieth century, with increasing specialization and fragmentation, and as it encountered new competition from such schools as Columbia University and the University of Chicago. But for the first decades, hopes ran high and results followed.

From the beginning, the intended close connection of the university with the medical school and hospital made it clear that the biology program would be important and that it "should receive a large amount of attention, more than ever before in America." Gilman said in his inaugural address, "In our scheme of a university, great prominence should be given to the studies which bear upon Life,—the group now called Biological Sciences."[39] Historian Philip Pauly has pointed out that the promised medical school, because it was so many years in actually appearing, stimulated the development of biology but decreased the usual medical constraints on what the program would be like. Thus, biology and the other sciences at Hopkins could develop more effectively than elsewhere.[40] In hiring his first biologist, a physiologist intended to build bridges to the medical school, Gilman could find no American to fill the position. This should not have been surprising, since there was really only one professional and nonmedical American physiologist teaching in the United States, Henry Bowditch at Harvard University.[41] So Gilman turned to England, to that leading biological educator Thomas Henry Huxley for a recommendation. Huxley strongly endorsed Henry Newell Martin. Accordingly, Gilman offered Martin twenty-five hundred dollars "to organize a laboratory & school of biology, on a plan similar to that of Prof. Huxley at So[uth] Kensington."[42] Martin declined the offer but responded that he would accept four thousand dollars for the first year and $5,000 thereafter, with assurances of a good laboratory and the title and authority of a professorship. Gilman agreed. Martin sailed with Huxley (who was on his way to help inaugurate the new university) to New York in 1876 and officially took up the positions of chair of biology and director of the biological laboratories when they opened in 1883.

Martin, born in Ireland in 1848, worked as an apprentice to a

Fig. 3. Henry Newell Martin, probably around 1876. From the Ferdinand Hamburger, Jr., Archives, Johns Hopkins University.

physician and also entered medical school at University College, London, where he became attracted to the physiological work of Michael Foster. When Foster moved to Cambridge, Martin followed with a scholarship and as demonstrator for him. He then served as Huxley's assistant at the Royal College of Science in South Kensington. There, under Huxley's instruction, he compiled *A Course of Prac-*

tical Instruction in Elementary Biology. In 1875, Martin received the first D. Sci. degree in physiology from Cambridge. As biographer Rosenberg has put it, "Still in his twenties, Martin was clearly one of England's most promising young physiologists." He reportedly hesitated over the Hopkins offer partly because of his serious commitment to research and his concern that the laboratory in this unknown place and untested new institution would not prove adequate.[43] He nonetheless left for Baltimore.

Upon arriving at Hopkins, Martin declared his conviction that researchers must also be active teachers, and in the long run Martin's institutional role was "almost certainly" more significant than his scientific contributions.[44] In his inaugural address, delivered as he embarked on his teaching career at Hopkins, Martin stressed the importance of pursuing biological knowledge without regard for practical results; those would follow from the advance of knowledge. This pursuit of knowledge involved two duties, he said: "We have to make provision for the advancement of knowledge, and for its diffusion." Biological research and biological teaching must advance side by side.[45]

Biological teaching should rely heavily on practical instruction, with lectures secondary and laboratory work primary. Elementary students should gain a broad general background, then learn to dissect and to use the microscope. Basic laboratory courses should provide familiarity with facts and methods. Some advanced lectures would supplement that grounding. Then, on the way to becoming specialized independent researchers at the advanced level, students would go through a stage of repeating recent experimental work. They would then be expected to discuss critically the work and its interpretations. As Martin pointed out, this stage served to eliminate the "triflers," who included those "with a burning desire to undertake forthwith a complicated research, though they hardly know an ordinary physiological instrument when they see it; much less know to handle it. But they cannot wait; they must begin the next morning, believing, I presume, that laboratories are stocked with automatic apparatus,—some sort of physiological sausage machines, in which you put an animal at one end, turn the handle, and get out a valuable discovery at the other." Finally, advanced students should begin independent investigation, with an occasional seminar to discuss recent research and methods.[46] The laboratory was clearly central to this ideal teaching approach.

Martin also committed himself to teaching in other ways. In the fall of 1877, he first offered a class in physiology for teachers. This Saturday class was to provide practical laboratory experience, and the result was evidently a great success. Martin had persuaded the administration to offer a free microscope as a prize to the top student; he believed that this should entice better teachers to enroll. At the year's end, he reported that several of the teachers had done excellent work and that several deserved the microscope. The inducement to learning would not be necessary for the second year, he decided, and was in principle even a bad idea.[47] Martin reported the experimental course a great success. A similar course in zoology followed the next year, and others followed thereafter, underlining Hopkins's general commitment to teaching as well as to training professional researchers.

Martin's course for teachers did raise one problem, however: women. He allowed one of the women teachers, Emily Nunn, to enroll in the regular biology laboratory as well as in the Saturday course, apparently through a misunderstanding.[48] The trustees disallowed her enrollment, but a few summers later, she was quite purposefully allowed to join the session at the Chesapeake Zoological Laboratory under Brooks. Around Miss Nunn and others ran a controversy about whether women should be allowed at Hopkins and in biology in particular. Gilman wrote, "I should be sorry to have this institution be discourteous to any one seeking knowledge but the Biological Laboratory where experiments in respect to animal life are in progress is not well adapted, in the opinion of some at least of our trustees, to the co-education of young women and young men."[49] Elsewhere Gilman said that the question of women's enrollment remained difficult but added, "Of this I am certain, that they are not among the wise, who depreciate the intellectual capacity of women, and they are not among the prudent, who would deny to women the best opportunities for education and culture."[50]

In addition to teaching teachers, Martin also organized a course of lectures for employees of the Baltimore and Ohio Railroad. Johns Hopkins's bequest to the university had included a preponderance of railroad stocks, so the university and railroad had rather close ties. The president of the railroad asked Martin to organize a series of popular lectures in 1882. The result included slide illustrations and occurred before an audience of several hundred. Martin, Brooks, and two associates from the biology department gave lectures titled respectively "How Skulls and Bones Are Built," "Some Curious Kinds of Animal Locomotion," "How We Move," and "On Fermentation." They achieved such success that the railroad printed the lectures for free distribution to its employees.[51] Clearly, in Martin, Gilman had found an effective choice for the advancement of biology in the new university.

The second member of the Biology Department faculty also arrived in the fall of 1876, though the initial intention had been for him to enter Hopkins on a fellowship as an advanced student. William Keith Brooks was born in 1848, the same year as Martin. Unlike Martin, he had followed a path typical for those interested in broader questions of animal morphology.[52] His brother reported that young Will loved "tromping" through the countryside, collecting things, and then organizing them into a museum at home. Impressed by reading the works of philosopher George Berkeley while an undergraduate at Hobart College, Brooks retained his love for philosophy when he transferred to Williams College after his sophomore year. There he developed further his interests both in philosophy and in natural history. He impressed his friends with his own microscope, a very rare possession even for a committed naturalist. And he especially enjoyed participating in the active local natural history society, the Lyceum of Natural History of Williams. After graduating from Williams College, Brooks worked for one unenthralling year in his father's business, then turned to teaching as his career. In 1873, he entered Harvard University for graduate study under the famous naturalist Louis Agassiz, who died late that year. Brooks received his Ph.D. degree in 1875, under Louis's son Alexander Agassiz. Brooks felt the influence of the charismatic elder Agassiz, even though he only knew him for one summer session at the Penikese Island laboratory and into the following fall. The impression nonetheless sufficiently stimulated Brooks that he modeled his biological station at Hopkins and some of his own work after Louis Agassiz's.[53]

In 1876, Brooks applied for one of the new Johns Hopkins University fellowships. His successful application abruptly established his career. As Gilman wrote, for these fellowships he wanted "men of mark, who show that they are likely to advance the sciences they profess." The first year, 156 applied; 20 received fellowships for a total cost to the university of ten thousand dollars. These attracted first-rate students and later provided a source of young, well-trained faculty for the university. As John French put it in his history of the university, "Probably no expenditure of ten thousand dollars in American education has ever had so large and so enduring a return from the investment."[54] Brooks gained one of those first fellowships but was immediately promoted instead to "Associate." This meant that he had joined the faculty and began teaching right away. He was to supplement Martin's work in physiology with courses in morphology. In 1883, he gained the title of associate professor of morphology, and in 1889, that of professor.

Fig. 4. William Keith Brooks. From the Marine Biological Laboratory Archives.

The Biological Ideal at Hopkins

Appreciating what biology meant at the Johns Hopkins University is particularly important for assessing the role Hopkins played in the students' development. In particular, the emphasis on morphology *and* physiology as the two parallel sides of biology became central.[55]

This division owes its origins in part to Huxley. When President Gilman had visited England to gather ideas for building Hopkins, he had been strongly influenced by Huxley's opinions about biology as well as his recommendations about personnel. He then decided that he wanted a laboratory and biology department like Huxley's.[56] One of Huxley's pronouncements on the nature of biology centered on its traditional division into problems of structure and function, or morphological and physiological problems. Indeed, Huxley's address to the university shortly after it opened underlined that bipartite nature of biology.

Gilman believed that it was Huxley who had first introduced the term *biology* into English usage and that it replaced *natural history* to describe the study of life.[57] In fact, neither claim was quite true since there were other contenders besides Huxley and the two terms were not strictly synonymous. But Huxley's use of *biology* was adopted by the Johns Hopkins University, and that usage gave the word a highly visible place in American science. Huxley's famous lecture at Hopkins hints at his views.

On 12 September 1876, Huxley addressed a general audience on the subject of university education. Whether this lecture accompanied the formal opening of the university or was "simply" a well-attended public lecture seems a matter of some dispute but no obvious consequence.[58] The invitation to Huxley occasioned considerable concern among Baltimore citizens because they feared the advent of evolutionary thinking and rampant atheism on their new university campus. Many cited the absence of an opening prayer as a further signal of such antireligious sentiment. Yet Gilman insisted that he had not anticipated any hostility, that he had not imagined that biology would wave a red flag to so many in the religious community.[59] Despite the concerns, Huxley nonetheless presented his lecture, and he spoke about biology, as well as educational matters more generally. Huxley suggested that a foundation of biological knowledge required for preparation of a medical student would demand familiarity with "the great truths of morphology and physiology, with his hands trained to dissect and his eyes taught to see."[60] He elaborated further on the nature of biology in another lecture delivered shortly after he returned to London.

In his London lecture of 1876, Huxley stressed that the term *biology* was not "simply a new-fangled denomination, a neologism in short, for what used to be known under the title of 'Natural History.'" Rather, the term had come into use over the preceding half-century and included the study of living beings, both animals and plants.[61] Study of structure through dissections and anatomical examinations was matched with study of the way structures function. Huxley made clear in his lectures that proper study of life should be called *biology* but should include both morphology and physiology. A Johns Hopkins University professor of life science, then, should properly be called a professor of biology rather than of zoology, natural history, comparative anatomy, physiology, or other such specialized subcategories. Biology was the proper category for liberal learning. Any biologist should have gained familiarity with all aspects of biology.[62] But, realistically, each professor would specialize. Thus the division of studies into physiology, which came first because of its close relation to medicine, and then morphology.[63] Students would study some of both but choose one or the other for specialized independent research.

The commitment to Huxley's version of the combination is very important for understanding the Hopkins students' convictions about what constitutes good biological work. Physiology and morphology, Martin and Brooks, belonged to different traditions, following quite different approaches with different problems of interest. The four students under consideration here accepted Brooks's general problems as most interesting but followed an approach that was closer to Martin's experimental emphasis. It seems clear that the exposure to elements of two dominant traditions, unique in the United States at the time and perhaps unique *to* the United States, strongly influenced the development of the four successful American research programs in biology on which this study focuses. The style of research which emerged reflected the dual, and occasionally schizophrenic, commitment.

In fact, the exact relations between physiology and morphology, between Martin and Brooks, are difficult to assess. The courses remained quite separate. The reading lists remained distinct.[64] The types of problems addressed and methods employed by Martin and Brooks remained radically different. There are hints of jealousies and increasing divisiveness between the two principals and the two subject areas.[65] Biographer Rosenberg has detailed Martin's conviction, for example: "Although of necessity he taught general biology and animal morphology, Martin thought consistently in disciplinary terms; he never lost sight of his identity as a physiologist, and he was deeply committed to establishing the independence of physiology from the needs and attitudes of clinical medicine."[66] As well as independence from morphology, it seems. Brooks equally strongly stressed as central the significance of evolutionary problems, which Martin largely ignored and which were widely acknowledged as morphological; Brooks thus urged his own disciplinary independence.

What Hopkins offered the students was what Martin had called for in his inaugural address: access to both physiology and morphology.

At least one critic attacked this Hopkins program as a "sham" that merely pretended to teach biology. Without botany, what the program offered could only be considered zoology, he urged, and not biology as a whole.[67] In fact, one Hopkins graduate acknowledged that there had been plenty of lobster in the curriculum, but scarcely enough vegetables to make a decent salad.[68] Indeed, botany failed to gain support time after time, perhaps because it did not offer the direct connections with medicine that the trustees and president sought to develop.[69] Yet at least one Hopkins student did manage to overcome the weakness, evidently, for Shosaburo Watase received a job offer from Chicago in 1899 precisely because he did both plant and animal, and especially both morphological and physiological work.[70]

Despite the general absence of botanical study, the Hopkins program did nevertheless offer something unique, an ideal for what was "biological." It did embrace those problems and methods elsewhere divided into different departments (anatomy and physiology, zoology and anatomy, for example) or into different traditions (physiology and morphology). It did produce a generation of students who embraced biology in a larger sense, influenced by morphological questions but with awareness of physiological concerns and incorporating physiological methods. To understand that combination, central to the new American biology at Hopkins and elsewhere later, we must examine more closely the doing of biology at Hopkins and the traditions in which Martin and Brooks participated.

2

The Hopkins Experience in Biology

AS EACH STUDENT entered the Hopkins biology program, he—recall that only men were admitted—was exposed to the research ideal and joined an expanding group of prospective biologists. But each individual had to make a choice, and the groups divided into two interconnected but largely distinct traditions. As the yearbook, the *Hopkins Medley*, reported in 1890, the Biological Department had fifty-eight degree candidates: forty undergraduate and eighteen graduate students. The overall perception was that "the postgraduate weighs the respective advantages of physiology and morphology before selecting the field for his life's work. He who chooses the first is privileged to ride up and down on the elevator and to breathe the iodoform-laden breezes of the kymograph room."[1] The second group was more likely to roam the countryside or head for the seashore before returning to the microscopes and staining materials in the laboratory.

For Hopkins physiologists, live animals entered the new laboratory building inconspicuously, through the back alley and then up in the elevator to the top floor. There the students of physiology stood over benches and cut into the living, anesthetized mammals. They sought with such interventionist experimental methods to learn what normally invisible processes were going on inside the living animal. The specialty of the laboratory, isolating the mammalian heart to study its action, brought with it a bloody mess, the pungent smell of anesthetics and antiseptics, and the mechanical accompaniment of measuring and recording devices. The researcher was surrounded by electrical motors and batteries, surgical equipment, chemical apparatus, and high-powered microscopic equipment as well.[2] He had to keep things moving along quickly before his animal died. As a result, the whole enterprise took place in a very vital and intense atmosphere.

Fig. 5. The Biological Laboratory at the old campus, about 1890. From the Ferdinand Hamburger, Jr., Archives of the Johns Hopkins University.

In contrast, morphology at the Hopkins laboratory had a more static and certainly less dramatic flavor. The morphologist sat with his low-powered Zeiss microscope and observed and then described the embryo specimens he had collected, perhaps during the previous summer's group research session at the seashore. Probably he had at that time applied a fixative to preserve the dead material in as close to a natural state as possible. Then, when he had a chance to sit down and spend some time in detailed observation, he proceeded to look at his embryo specimens. He might go farther and embed them in paraffin or some such substance, then slice them up with a knife or microtome and look at the pieces with his microscope. He was surrounded by shreds of paraffin, small specimen bottles, and various stains, for example. This was the morphologist's approach to learning about what goes on in the invisible interior of an organism, in this case of an embryo rather than an adult. Of course, the process was somewhat more active during the collecting months, especially since the researcher also had to make observations of living organisms and of those that would not keep. The approach remained the same, only busier.

Clearly, choosing between physiology and morphology represented making a commitment to one of two alternative traditions, with different approaches, organisms, problems, and methods. Yet one great strength of the Hopkins program remained the opportunity to learn from both and to import aspects of one tradition into the other even while concentrating on one. Once a student selected one side or the other (and all did choose), he had bought into an emphasis on a set of courses, researches, and readings.

Students took a full complement of courses from both Henry Newell Martin and William Keith Brooks, as well as from any other faculty associates. In his first year, Thomas Hunt Morgan, for example, took two physiology courses, general biology, mammalian anatomy, and osteology. With more physiology and anatomy in his remaining three years, he began to concentrate more and more on morphology, so that his last year of courses consisted of physiology with Martin and morphology readings and seminar with Brooks. Ross Granville Harrison's program included much more mathematics and general science in his first year, because of his scientifically oriented undergraduate career and his continued medical interest; after that, he pursued a curriculum similar to Morgan's. Edmund Beecher Wilson also took physiology, morphology, osteology, anatomy, and general biology, as well as a practical histology course.[3] Though the archival records do not include registrar's reports for Edwin Grant Conklin, the courses available in the catalog were virtually the same, and the departmental philosophy remained the same. Therefore, the educational experience seems to have remained substantially constant from the time Wilson arrived in 1878 past the time Harrison left in 1894. In large part, the books read for seminars also remained constant, with additions to keep up to date.

The Hopkins experience therefore consisted of a shared set of courses and readings, and also of a shared laboratory experience. Especially the youngest three of our group, who arrived after the new biological laboratory was constructed in 1883–84, were exposed to the full experimental setting of Martin's vivisection research. In addition, they learned field work and embryological methods from Brooks, augmented and reinforced through the sessions at the Chesapeake Zoological Laboratory in the summers.

Working with Martin

Martin saw himself as a physiologist rather than as a zoologist or biologist more generally. Thus, he concentrated on understanding the "play of forces in living organisms."[4] Mechanics, experimental physics, and chemistry all provide important background information for the physiologist. Beyond that, he should be prepared to work on living beings instead of dead bodies in order to observe living, functioning processes.

At Hopkins, Martin followed up on some of his teacher Michael Foster's work, examining the action of the mammalian heart. He succeeded in isolating the heart and sustaining its beating even after removing it from the animal. This isolation allowed very useful examination of the effects of artificially—that is experimentally—altered conditions such as temperature, arterial pressure, or ethyl alcohol concentration on heartbeat. Another series of experiments explored the character of respiration. The results of his research, including his fifteen scientific papers and six more general lectures, appeared together in 1895.[5]

Though his work on living mammals brought the ire of some antivivisectionists, particularly in England but also in the United States, Martin publicly justified the invasive approach. He pointed out that he had always been quite open about exactly what was going on in his laboratory. He even invited critics to visit, though there is no evidence that any did. The university continued to back Martin and his right to pursue his scientific work with his chosen methods, even when popular antivivisectionist sentiment grew stronger.[6]

In his work on the mammalian heart, Martin pursued an approach typical of the medical physiological tradition.[7] In part, that tradition involved the use of what we would today call model systems, though they were not so labeled then. Instead of asking how the rabbit's, dog's, or human being's heart functions, the physiologist would ask how the mammalian heart functions. He would then proceed to study rabbits or dogs or whatever served as a useful, accessible, available system for study. If that system yielded answers about the questions at hand, it was considered to be generalized information applicable to other animals with similar body parts. In other words, since mammalian hearts are all basically alike, they must function alike. Similarly for respiration, digestion, reproduction, and such. The individual was of interest only insofar as it shed light on the general phenomena.

To get at those general processes, the physiologist began by developing a technique to study the functioning heart in isolation from the rest of the body. In one paper, for example, Martin began by asking what effect pressure, both arterial and venous, has on the heart, or pulse, rate. To test the effect of pressure alone, the heart must be removed from (and hence the researcher must control) other influences. Then he changed the pressure in various ways and observed the effect on pulse rate. He began, that is, with a well-defined "what if" question: if the pressure is altered, what effect will the alteration have on pulse rate? He performed manipulative experiments to produce the desired conditions and ended with a series of observations about what actually happened in those cases, obtaining definite results from his clearly formulated, relatively narrowly defined questions. His presentation of results was clear; he referred to well-established experimental results with such comments as, "On this point I think that I have been able to add a little to our positive knowledge."[8] With respect to the implications for more general theories, he concluded with such phrases as "the results confirm such-and-such a position" or "the evidence supports such-and-such a conclusion." He did not orient his experimentation toward testing between two distinct general hypotheses, nor did he go on to draw more general theories with the belief that he had definitively proven anything beyond a controlled observation. Rather, he showed how the experimentally derived data conformed to some available "working hypothesis" or other. Only the results, the definite results of experimental work, bring positive knowledge in this approach. Theorizing alone cannot do so.[9]

Martin's emphasis on "positive knowledge" was important for the Hopkins students but remains complex and difficult to assess critically. This is not the place to enter a full-scale discussion of positivism or pragmatism. But Martin does seem to have absorbed and passed on bits of both intellectual movements, or at least ideas that look quite like parts of those movements. Hints of positivism that echo August Comte and may well have been passed through Huxley to Martin appear in the images Martin drew. He sketches armies of heroic soldiers marching resolutely toward understanding and knowledge through science. This bold march in pursuit of science is not for everyone, but neither is it restricted to a select few. Martin explained that

> assuredly not every one of her [Science's] followers will become a Linnaeus, or a Cuvier, or an Agassiz. It may not be given to any of us to make some brilliant discovery, or to first expound some illuminating generalization; but we can, each and all, if we will, do good and valuable work in elucidating the details of various branches of knowledge. All that is needed for such work, besides some leisure, intelligence and common-sense (and the more of each the better), is undaunted perseverance and absolute truthfulness incapable of the least perversion (either by way of omission or commission) in the description of an observation or of an experiment, or of the least reluctance to acknowledge an error once it is found to have been made.

But this perseverance will be met by failure at times, so that constant probing must continue. "This love of truth must extend to a constant searching and inquisition of the mind, with the perpetual endeavor to keep inferences from observation or experiment unbiased, so far as may be, by natural predilections or favorite theories. Perfect success in such an endeavor is, perhaps, unattainable, but the scientific worker must ever strive after it; theories are necessary to guide and systematize his work, and to lead to its prosecution in new directions, but they must be servants, and not masters."[10]

The march of science, for Martin, does not lead inexorably toward the goal of some absolute external truth or even some proof of any one true theory. The physiologist asks answerable, specific questions and then performs experimental manipulations to gather information about living processes; hypotheses guide the work, but the goal is to achieve positive or definitive information about the way force or energy performs bodily functions.[11] This search rejects the postulation of hypothetical abstract "somethings" since physiologists, as Martin maintained in 1884, "had given up hunting essences and absolutes and had decided that their business was to study the phenomena exhibited by living things, and leave the noumena, if there were such, to amuse metaphysicians."[12]

In much of his commentary on how science should work, Martin may have been influenced, or perhaps redirected, by Charles Sanders Peirce. Peirce taught philosophy, specifically logic, at the Johns Hopkins University from 1879 to 1884, until he was rather unceremoniously and unhappily fired. Peirce had two close friends at Hopkins in addition to his scattering of student followers, according to reports: physicist Henry Rowland and Martin.[13] Peirce and Martin agreed on a number of points about how science should work since Peirce also held a very practical goal for science: science develops opinions, through working hypotheses and experimental tests. The opinions fated to achieve general agreement among scientists are those that may be considered true, yet there is no external truth somewhere outside science toward which science progresses. This view, which Peirce clearly held and Martin seems to have echoed, was not generally shared by that apparent majority whose rhetoric stressed science as seeking and disseminating some noble, absolute, and universal truth.[14]

A very similar limited, more cautious view of what science offers came to play a central role in the work of the Hopkins students under consideration here. Perhaps the views of science that Martin and Peirce held gained general acceptance or were in the air. Perhaps—and I think this more helpful than suggesting an unverifiable direct link to Peirce—the students listened to Peirce's friend Martin, remembered what he said, and were thus sensitized to the experimental and focused scientific approach of the physiologist.

In addition, Martin designed the new laboratory building and oversaw every detail of its construction. Every day he was there, checking to guarantee that even the smallest features came out right.[15] Laboratories had to have proper lighting, properly sized rooms to accommodate animal experiments, as well as study rooms and lecture halls. Floors must be sturdy enough to support equipment, and elevators and other facilities must be provided to facilitate moving the experimental animals around. Animal rooms, and especially those for vivisection work, should be separated from more public and introductory instructional areas. Martin worked through all such considerations. The result was a very handsome laboratory building, placing physiological experimental work on the highest level. In his address at the opening of the building, Martin stressed the centrality of physiology to the development of modern research medicine and the importance of the opportunities the new facility would bring for physiology.[16]

Yet despite his successes and the example he set, Martin actually had few students complete their doctorates during his term at Hopkins from 1876 to 1893.[17] As his associate in physiology

William Henry Howell said, Martin was excellent and enthusiastic with beginning students and "thoroughly enjoyed introducing young students to the beauties and marvels of living structures and their adaptations." But with advanced students, "certainly he could show an extraordinary amount of apparent indifference toward some poor fellow floundering in the difficulties of his first research. . . . Martin was capable of letting them drift in an altogether heartless manner."[18]

Martin was reportedly a very poor lecturer and apparently nearly lost his job in 1878 as a result.[19] Yet a former undergraduate student, Allen Kerr Bond, later wrote that "Professor Martin had a fascination for his students. He had a poor delivery, but the most intelligent blue eyes I have ever seen, and when he was stirred up over some scientific recital they fairly danced with excitement."[20] This, from a student who was manifestly not excited by some of his teachers, speaks highly of Martin's power to inspire students personally even if not to lecture well. In addition, Martin was a "warm and successful graduate teacher and colleague."[21]

One student whom Martin deeply impressed as both an undergraduate and graduate student was Ross Harrison.[22] Although Harrison ultimately did not accept Martin's physiological interests as primary, he recalled in his Croonian Lecture before the Royal Society of London in 1935 that he had attended Martin's general biology course as an undergraduate in biology at Hopkins and "there received the first inspiration to follow my chosen calling. The awakening of my serious interest in biology was thus due to . . . one whom I shall always remember with gratitude and respect."[23] Similarly, Martin greatly influenced Morgan, as did Howell. "From them he learned to appreciate the value of physiological approaches to biology; and I think he was inclined to turn to them rather than to Brooks at times," one biographer noted.[24] Martin's physiological approach to well-defined problems and the eventual production of definitive, positive knowledge—but not necessarily truth—held considerable attraction as a young generation of Americans sought to become biologists and to achieve results and recognition in an increasingly manipulative experimental environment. Thus, Martin particularly exerted influence through his direction of the research laboratory, through teaching, and through his particular laboratory-oriented philosophy of science. Students learned from his approach and his epistemological norms even if they failed to accept his particular research problems and choice of organisms.

Working with Brooks

Yet critics of Martin's emphasis and approach quickly emerged. One lamented in the *American Naturalist,* the physiological facility deserved praise, but more commitment needed to be made to biology in its own right rather than as an aid to medicine:

> While we welcome the auspicious opening of this department of the young university at Baltimore, we cannot escape a feeling of regret that the professor of biology, instead of aiming first at a secure position in medical physiology, even at the sacrifice of much else in biology, should not have fixed upon a higher ideal more worthy of a great university. Morphology, botany, general physiology and biology are all to wait, content to be the handmaidens of medical physiology. The magnificent opportunity to accomplish the immediate development of a school of scientific biology has been deliberately renounced. We deplore what we consider a serious mistake, and are unable to justify the postponement of proper university work in order to favor one class of professional men.[25]

In fact, despite the lament, the morphological side of biology was in reasonably good hands and was progressing well, thanks to Brooks.

In contrast to Martin, Brooks has been portrayed as one who did not experiment but who recorded nature and drew suggestive conclusions instead.[26] As his student Andrews suggested, Brooks "was not an experimenter, but an observer of natural processes, from which he endeavored to interpret logically. He saw too many factors to be long satisfied with the sharp cut result that seemed to follow from experimentally severing some portion of the phenomenon from the rest."[27] Brooks, in short, rejected Martin's particular sort of isolated and narrowly focused experimental emphasis, as well as his physiological concern with obtaining definitive results that would represent sure contributions to positive knowledge. Yet, as Keith Benson has shown, Brooks was not really the blind and passive observer that the standard view suggests.[28] And he certainly did not reject experimentation. At one point, he noted, for example, that "Embryologists are rapidly adding, by experimental methods, to our knowledge of the mechanics of development."[29] At another, he wrote to President Daniel Coit Gilman about his plans for experimentation using hybridization in plants to determine the significance of sexual reproduction.[30]

There are really two sides to Brooks: the more philosophical and nonexperimental Brooks who appears in his theoretical papers and books, and the more practical and empirical researcher revealed in his research papers. The former, exemplified in papers of 1882–83 titled "Speculative Zoology," remained abstract and failed to exert as much direct influence on many of his students as his research, though the indirect effect was clearly strong.

Conklin, perhaps the most sympathetic to Brooks's commitment to philosophical discourse, recalled of that work that Brooks had exhorted his students to read carefully and then to reread so that "what may seem obscure, may, on review, be found consistent and intelligible." Conklin noted of Brooks:

> Much that he has written still seems to me obscure, although I have read it more than once, but I bear in mind his parting request, and in the meantime profit by that which I do understand, and am charmed by the classical and almost poetical diction in which it is written. Whatever one may be inclined to say of his conclusions and theories, it cannot be denied that in an age when biological investigators have been content with discovering phenomena, he attempted to go back of phenomena to their real meaning and significance and to point out the relationship of these newly-discovered phenomena to the great current of philosophy which has flowed down to us from the remote past.[31]

Wilson recorded even more positive assessments of the importance of Brooks's theoretical side when he wrote that from Brooks he had learned "how closely biological problems are bound up with philosophical considerations. He taught me to read Aristotle, Bacon, Hume, Berkeley, Huxley; to think about the phenomena of life instead of merely trying to record and classify them."[32]

Brooks's students generally found discussion of the philosophical views interesting, but perhaps as much because they brought the group together in consideration of fundamental assumptions about science and about nature as because of the particular ideas. Wilson wrote, "It was through informal talks and discussions in the laboratory, at his house, and later at the summer laboratories by the sea that I absorbed new ideas, new problems, points of view, etc."[33] Even if they did not always appreciate the points Brooks tried to make about the existence of psychic and physical realities or about the order of nature and its creator, for example; even if they did not themselves see discussion of free will versus determinism as a pressing problem for their own materialistic scientific exercises, the students still gained an appreciation of the problems and controversies that had occupied scientists. They also learned to think about what biology is and what it does. They were, as a result, far more reflective about what counts as legitimate science than one might expect of men who became leading experimental biologists in the twentieth century.

In these discussions and in his theoretical writings, Brooks's convictions emerged. Biology, like physics, seeks to provide order and to uncover the existing order of nature. Like physics, too, it seeks proof. Yet, since biologists often have only circumstantial evidence with which to work, they must draw conclusions from that evidence. Some would say, "Stick to the facts," but then scientists would "become mere observing machines." This is undesirable and science does not demand it. Yet it is "hopeless for us to attempt actual proofs of the more deep-seated phylogenetic relationships."[34] Science has its limits.

Since Brooks assumed that the traditional morphological problem of reconstructing phylogenetic relationships represents a central goal for biological science, given that evolution is a fundamental assumption, whatever methods proved necessary to address that problem would be appropriate for biology. He thereby justified morphological theorizing as necessary and desirable. Morphologists must go farther beyond the facts than physiologists would allow in order to reach the desired generalizations and laws. So, "from this point of view it is plain that good may result from honest but erroneous attempts at morphological speculation, for the logical restrictions of sound reasoning are often studied to the best advantage in the error of acute thinkers."[35] Proof proving elusive, Brooks felt that the morphologist must turn to guiding hypotheses, which he recognized may be only tentative and temporary.

Yet in another work published the same year, Brooks appeared to shift this emphasis. In *The Law of Heredity*, he outlined his task as follows:

> I shall give first, an outline of the chief hypotheses which have been proposed, from time to time, as an explanation of heredity, with reasons for rejecting them. I shall then present briefly, in outline, a statement of what I believe to be the true explanation. I shall then try to show that this theory furnishes a basis or foundation for the theory of natural selection, and removes the most serious difficulties which have been urged against the latter theory. I shall then show that there is no *a priori* reason for rejecting the theory of heredity; and that it furnishes an explanation of many well-known factors which cannot without it be seen in their true relations. I shall then attempt to show that it is supported by direct proof, and finally I shall give a statement of the theory in a more extended form.[36]

This emphasis on absolute proof and truth also sets a different tone than his later words in the same book that one should put forth a hypothesis and not expect to achieve "truth," but rather should "test it by applying it to the various observed phenomena of heredity in order to see how far it explains and interprets them."[37] His readers may well have wondered precisely what relation evidence does hold to hypotheses and which way one had better proceed—by going to the full theory and offering final proofs to get at truth or by testing a suggestive guiding and eminently revisable hypothesis. Fortunately, his students had the opportunity to talk through such apparent inconsistencies as they gathered at Brooks's house, in journal club meetings, or around the dinner table or the laboratory during summer sessions at the seashore.

Yet Brooks was not so inconsistent as it might seem. Despite apparent contradictions in approach, the important goal of science throughout remained the putting forth of an explanatory theory that would allow data to fall into place. Having a successful theory in that sense was ultimately as important as achieving any definite "results" in a particular experiment or making such smaller-scale contributions to "positive knowledge." Brooks as morphologist had a different emphasis from Martin as physiologist.

Brooks felt that the morphologist held larger-scale goals for science similar to those of the physiologist. He also agreed with the physiologist in holding to a materialistic and physicalist ontology. But he differed in epistemological emphasis. For the morphologist, the best approach involved detailed study of individuals, or of individual kinds of forms. Thus he would begin by describing developmental details for the gastropods, or for crustacea, or perhaps for a particular sort of polychaete worm. He would then move to comparing those details with developmental and adult structural details of other organisms. Then, and only then, he could develop a hypothesis about how the various organisms were evolutionarily related and about what the homologies or dishomologies therefore meant. Developmental and structural phenomena were assumed to be particularized to the individual type of organism until they were shown to be general. This was in contrast to the physiologist, who had by the nineteenth century long held basic processes as general across different animal groups. The morphologist had a different approach with a different relative emphasis on observation and on the development of a hypothesis.

Brooks thus agreed with Martin about the validity of solid, careful observation but gave theory a more important place than Martin did: "Exact observations are permanent additions to our stock of knowledge, and although they may be supplemented they cannot be superseded; while any theoretical views which are reached in the present imperfect state of our knowledge of zoölogy are liable to be entirely set aside by the discovery of new facts, the history of our knowledge of [the oyster] Salpa shows that a theoretical interpretation may be of the greatest utility, and yet be entirely false."[38] In particular, Brooks wanted a theory to explain the striking parallels in development among different organisms, and to outline the historical evolutionary relationships, or phylogenies, among diverse organisms.[39] In part, this emphasis on developing theoretical explanations was a product of the morphological legacy as transmitted through Louis Agassiz, and the particular form that Brooks's research took reflected the American morphological tradition Agassiz had spawned.

Agassiz's Legacy: The Morphological Tradition from Harvard to Hopkins

Until his death in 1873, Louis Agassiz remained the dominant force in American academic morphology. As head of the Museum of Comparative Zoology at Harvard University and as organizer of the short-lived but influential Anderson School of Natural History on Penikese Island, he directly influenced the next generation of American morphologists.[40] Brooks had been a student of Agassiz's, attended the Penikese summer school, and worked for a short time as an assistant (essentially a curator) at the museum.

While steadfastly denying that evolution occurs, Agassiz nonetheless sought to discover underlying unities of organic form just as the evolutionists did. Instead of seeking phylogenetic relations, however, he looked for anatomical adherence of organisms to one of the four basic types that morphologists had found to exist: the radiates, molluscs, articulates, and vertebrates.[41] Organs, for Agassiz, appear in the course of embryonic development in the order of their importance rather than reflecting any ancestral, phylogenetic past. Structural homologies proved useful as clues for establishing systematic relations, just as they did for the proponents of evolution and phylogeny. For Agassiz, embryology also played a useful role for determining the relative rankings among animals, though his were stable, essentialistic, and hierarchical rather than evolutionary rankings. Marine animals were particularly useful since they were relatively abundant, simple, and diverse.

Agassiz's method of teaching, at least at the Penikese school, centered on collecting, observing life histories of organisms, and tracing embryonic stages of development; classification followed. Agassiz also stressed the careful description of differences as prior to identifying underlying samenesses. The procedure was slow and did not always produce exciting results, but the deliberate process produced compelling descriptions that could serve as "positive" information. It also gave the students respect for the organisms they studied and for the process of collecting, observing, and accurately describing what they saw. In retrospect, one student recalled that he had found Agassiz's approach frustrating at first but had come to appreciate its value.[42] Agassiz thus taught an approach as well as ideas in morphology, so that even those students who rejected details of the latter could retain the former. Even students who endorsed the evolutionary ideas that Agassiz insistently rejected could accept Agassiz's basic morphological approach.

When Louis died in 1873, his son Alexander did his best to take over his father's responsibilities. Despite the unexpected deaths of both his father and his wife within a very short time, Alexander managed to keep the museum under control and to begin the second session of the Penikese school. Poor health kept him from finishing directorship of the latter or from persisting past the second year, but the effort was important nonetheless. At first, Alexander found it difficult to resume work, but with the help of his stepmother (Elizabeth Cary Agassiz, who had been the first president of Radcliffe College), he recovered enough to carry out exploring trips, to head the museum, and to keep a laboratory running at his home in Newport, Rhode Island. Though he never achieved a status or developed a driving research approach as strong as his father's, Alexander made considerable efforts to keep alive his contacts with the expanding morphological and zoological community.[43] For example, he worked to obtain an academic position for Brooks and in 1876 helped to upgrade Brooks's new Hopkins position.

The younger Agassiz had already helped to obtain a job for Brooks at the University of Cincinnati.[44] Presumably the pressure of that offer prodded the Johns Hopkins University into upgrading Brooks's appointment, even before he arrived in Baltimore, from fellow to associate. In 1877, Brooks accepted the Hopkins position; the next year he was promoted to assistant professor, and he spent the rest of his life at Hopkins despite several further job offers elsewhere.

With the death of Louis Agassiz and the opening of Hopkins, the dominant morphological tradition in America passed from Harvard to Brooks, and later to Charles Otis Whitman (first director of the Marine Biological Laboratory [MBL] in Woods Hole, Massachusetts, and first chairman of the biology program at the new University of Chicago) and to others. Brooks largely adopted Agassiz's approach, though he also incorporated other emphases. In fact, Brooks remained firmly within the dominant mainstream morphological tradition, influenced by a predominant concern with the developing structures of embryonic germ layers, with life histories, homologies, unities, embryonic recapitulation of the phylogenetic past, and phylogenies.[45] He did not devote his energies to identifying the ancestor of us all, as so many other morphologists had done (including Ernst Haeckel, Carl Gegenbaur, Adam Sedgwick, Anton Dohrn, and William Bateson, for example), since he felt that by the 1880s so many theories had appeared that virtually all eligible candidates for the one ancestral organism had already been nominated.[46] While not endorsing the extreme reaction to such uncertainty by his student Shosabura Watase, who said, "I am done with this whole phylogeny business," Brooks did not himself continue and did not demand that his students work to identify the elusive original ancestral form.[47]

Nonetheless, studying ancestral lineages or phylogenies of individual organic forms did remain productive, Brooks held, in agreement with Haeckel, since individual development repeats at least parts of its ancestral past. Each organism has a history, understanding of which helps to illuminate the structure and origin of that organism and its relation to other organic forms. Identifying homologies in parts of different organisms should also help to reveal those histories and those relationships, he felt. Studying life histories from germ layers to reproducing adult stages would also help to illuminate those relationships. Homologies and life histories thus loomed large in Brooks's work, as they certainly did not in Martin's.

Brooks's students began with similar interests and considered life histories, early development, and homologies in a wide variety of marine invertebrates. They embraced the fundamental morphological collecting, preparing, observing, recording, and describing techniques as well. They accepted the problems and basic assumptions of the morphological tradition. Yet, as we shall see, they came increasingly to embrace an epistemology closer to Martin's than to Brooks's, just as the morphological tradition itself moved in the direction of narrower, clear-cut questions and definitive results as a proper immediate goal for science. Even Brooks, in his own research at the Chesapeake Laboratory, showed such tendencies to focus on the problem at hand rather than on the distant theoretical explanation.

The Chesapeake Zoological Laboratory

What Brooks tried to do in his own research was to produce concrete and indisputable facts that could then, in turn, form the basis for constructing phylogenies and explanatory theories. Most of Brooks's research during his years at Hopkins resulted directly from his summer sessions with the Chesapeake Zoological Laboratory or related trips to marine research stations.[48] Therefore his research focus remained on marine animals.

The experiences at the Chesapeake Zoological Laboratory each summer reinforced the camaraderie and exposed the students, as the daily laboratory work alone could not, to the way Brooks worked and to the problems that he felt were important. The *Hopkins Medley* reported that the student on the morphological side of biology "takes his exercise out of doors, and on these balmy spring mornings deems it absolutely necessary to roam over the beautiful hills of Baltimore in search of specimens."[49] How much more appropriate even was the image of Brooks's morphology students at their seaside marine stations, joining a community of investigators pursuing similar problems with related organisms and roaming about exotic locales collecting, observing, and recording.

Brooks was unquestionably a careful observer of nature, and where better to observe nature than on the coast, especially since, as he wrote, "nearly every one of the great generalizations of morphology is based upon the study of marine animals, and most of the problems which are now awaiting solution must be answered in the same way."[50] The Chesapeake Zoological Laboratory, designed along the pattern of Louis Agassiz's Penikese Island laboratory, clearly played a major role in stimulating Brooks's students, as it did Brooks himself. It is no accident that his students each attended at least one summer session of the laboratory or went to the United States Fish Commission laboratory in Woods Hole, Massachusetts.[51]

Brooks described how the students worked together, overcoming mosquitoes, heat, primitive supplies, and other problems.[52] As Andrews later wrote of one summer session, "Not the least of the good result of this Expedition was the fact that it formed part of the life experience of Brooks's men who later became outstanding as teachers and as investigators."[53] Many of the students published results of their summer work in papers that reflect an exploratory descriptive approach to questions of animal form. Embryological concerns remained central, as they did for Brooks, and some consideration of evolutionary, phylogenetic relationships between organisms appeared often. But most of the work traced development and life histories of a variety of organisms. Though not what Brooks's students became known for later, this careful descriptive work, accompanied by detailed drawings and thus meticulous observation, formed a vital part of their training and their convictions about what matters in science. And some published major contributions to the body of such descriptive and phylogenetic studies.[54]

What Brooks and his coterie of students found at their American seashore was a host of new, unexamined organisms. No one had studied the rich southern fauna of North Carolina or of Jamaica or Bermuda before. Every new organism brought with it new information about its particular life history and its reproductive habits, as well as about its place in the evolutionary tree of life. Time and again, Brooks and his students cited the wealth of new animals and the intense activity involved in trying to collect and study as many of them as possible.

The leader sent President Gilman reports from their summer location each season, keeping him informed and justifying the expense of sending a group of graduate students off to the beach. On one early trip, Brooks exulted at finding eggs of the crustacean decapod *Lucifer* in great supply at Beaufort: such a piece of luck to find eggs that reflect the unmodified ancestral form of development, unlike most where modification and adaptations have destroyed the record of past stages. *Lucifer* has a life history that is "perfectly unmodified, or ancestral," he wrote. It retains developmental stages omitted from other crustacea because of pressures on them. The new discovery casts "a flood of light" on the history of crustacea and on their embryology and "is like the discovery of a key to an unknown literature." Several years later, he wrote from the Bahamas, "Everything is so new and strange that we find exploring and collecting much more exciting than laboratory work," though they were, of course, pursuing laboratory work as well.[55]

At one point, Brooks sought a permanent setting for the summer laboratory. In the search, a group of young men from the Johns Hopkins University's biology program set out in the summer of 1891 for Jamaica's Blue Mountain, with its now famous coffee plantations. They spent the first night in Kingston, in a house where the rats scampered up the bedposts on their way to the roof. They continued the next day by coach, then took to horseback as the road became too steep and winding. Finally, they completed the trip (to the summit at 7,360 feet elevation) on foot, arriving "in time to cook our supper on an outdoor fire and then sleep, well wrapped in the blankets we had brought and found so necessary. The air there was cool enough to make one's breath visible," even in the midst of summer. They recorded in the guest book that "never had we been so near to Heaven."[56]

In effect, this was a biological reconnaissance mission. The participants in 1891 in the Chesapeake Zoological Laboratory, directed

Fig. 6. Harrison (*second from left*) and Morgan (*second from right*) on the way to Blue Mountain, Jamaica, with the Chesapeake Zoological Laboratory group, 1891. From the Marine Biological Laboratory Archives.

by Hopkins morphologist William Keith Brooks, were scouting Jamaica to locate a prime site for a possible internationally sponsored Columbian Marine Laboratory. The Blue Mountain area offered intriguing possibilities for botanical work, while the numerous attractive harbors would support a range of embryological study.

More important, though, the trip served to cement even more strongly the friendships and collegiality of the Hopkins group that took part. Fifteen men, plus Brooks's wife and two children, made up the party that year, and there were more or fewer in the other summer sessions. The group, throughout the late 1870s, the 1880s, and into the 1890s, included many of those destined to become the top biologists in the United States. They worked closely together, generally rooming in a large and inelegant bunkhouse building that also served as a communal laboratory, sometimes with crabs crawling through the doors and along the walls. As one participant recalled later, the arrangements could produce unanticipated hazards, as when the crabs knocked over all the pure alcohol bottles, at a time

Fig. 7. At the Chesapeake Zoological Laboratory, Jamaica, 1891. From the Marine Biological Laboratory Archives.

when there was no more of the essential preservative available.[57] Sometimes the peripatetic laboratory convened in Beaufort, North Carolina, or in Bermuda, before it moved to Jamaica. Brooks discontinued the highly successful summer expeditions only in 1897, after one of the participants died during one of the island's recurrent yellow fever epidemics and another died after returning to Massachusetts General Hospital in Boston.[58] After that, Brooks sent his students instead to the apparently safer United States Fish Commission laboratory in Woods Hole, Massachusetts.

The laboratory sessions at the seashore provided a wealth of diverse material far greater than that available in the colder waters of Baltimore: the environment offered a summer of collecting and drawing exotic organisms. The sessions afforded opportunities for both careful observation and creative thinking that went beyond anything possible back home in the landlocked indoor laboratory during a winter in Baltimore.

The opportunity provided by new materials also helped determine the research emphasis of the group. Before undertaking detailed comparisons or formulating sophisticated theories about the causes of development, for example, there was much work to be done simply in describing in detail the developmental stages of all these new, diverse organisms. Since the sea was thought to contain the most primitive organisms and therefore hold the keys to basic relationships among animals and to early evolutionary history, study of these marine organisms would eventually advance toward the basic goals of morphological science. In the meantime, the group stressed the careful observation and description of embryological development, thus focusing on that particular part of morphology where form first comes into being, just as Martin attacked the small corner of medical physiology that looked at function of the heart.

Brooks worked with tunicates, crustaceans, molluscs, and coelenterates, in each case striving first to identify the new forms: specifically, where did they fit in relation to other known similar species and genera? Second, he asked how the different parts of organs develop, according to what pattern and in what order: does the organism reproduce sexually or asexually, and precisely how and when; with what are the parts or the developmental patterns homologous, and thus to what other organisms is this one most closely related? Third, what do we learn from current embryological relations about ancestral development? With hydromedusae, for example, he learned about the origin of alternation of generations, while coelenterates raised questions about the phenomena of asexual budding. He had occasional practical successes as well. In particular, his work on oyster development beginning in 1878 showed that new oysters develop outside rather than within the mother's shell. This meant that artificial production of oysters could be possible, a suggestion that various fisheries departments pursued, with Brooks's assistance. Yet such practical results remained secondary.[59]

The morphological marine study began with collecting eggs. One had to get enough to follow them through all the developmental stages and to preserve and observe each stage. Furthermore, they had to be staggered over time; those species that bred quickly and whose eggs passed through the same stages simultaneously caused trouble since it was hard to get enough specimens or to observe them fast enough. As Brooks wrote to Gilman during his second summer season at Beaufort, collecting could be difficult. Either the weather was calm and excellent for surface collecting but fraught with mosquitoes and intense heat, or else the conditions were tolerable for people and only moderately good for collecting. Brooks reported Wilson's troubles with collecting enough worms, which lived in burrows several feet deep: Wilson "planned a buried torpedo, to throw up the mud." This would cause a sort of "scientific fourth of July celebration," but Brooks was not sure whether it would work to dislodge the rather stubborn mud and the deep-lying worms.[60] Even with sufficient eggs collected, the rest could prove trying, as when the interesting stages were transparent and hard to watch in the glass dishes.

Yet Brooks insured that his students had the full range of necessary basic equipment and helped them learn to collect and observe their chosen organisms. Generally, as he carried out his own research, the students learned from his example. They read his papers and discussed techniques. Since Brooks often adopted a policy of benign neglect toward his students at Hopkins, the intense summer experience was crucial to becoming a biologist. In fact, two months at the summer laboratory were required of all Hopkins graduate students in morphology.[61] The group in 1891 wanted a major trip so much that they lobbied Gilman in order to be allowed to go, and many recalled the central importance of that and other years' experiences for their subsequent biological careers.

Brooks and the Morphological Tradition

Once Brooks had introduced his students to the basic techniques of selecting and collecting species, and once he had provided them with high-quality low-power microscopic equipment, he let them figure out for themselves how best to proceed, perhaps using his *Handbook of Invertebrate Zoology* as a guide. He encouraged them to try different embedding materials from soap to paraffin. He urged them to learn hand sectioning as well as more modern microtoming techniques, including the preparation of transverse or longitudinal sections. For staining, he left them to discover what worked, often the hard way. He recommended to one student that he try the traditional Beale's Carmine stain, for example. "After waiting in vain to get some response, he ventured to ask whether Doctor Brooks had ever used the method. Yes, he had. 'What did you think of it?' 'Twasn't worth a damn.'"[62]

The students who stayed with him did learn. Indeed, they thrived. Each year, Brooks reported the successes in his reports to Gilman and in the *Johns Hopkins University Circulars.* In the earliest years, Wilson dredged the polyp *Renilla* and traced life histories of several annelids, so that by 1881, Brooks was discussing Wilson's publication of results. By the next year, Wilson had received international recognition in the form of an invitation to work for three years at the Naples Zoological Station to carry out comparative work on *Renilla* species. Howell explored marine animals' physiology of blood. Others worked on siphonophore development or on sexual development of oysters, for example, while still a few others pursued botanical research.

We see the students who worked with Brooks through seminars and reading courses in morphology, and especially those who attended the summer Chesapeake Zoological Laboratory sessions and had sustained exposure to Brooks's work, writing very similar sorts of papers. At least, they did while they remained at Hopkins. Wilson's work on *Renilla*, Conklin's earliest study of *Crepidula*, and Morgan's on sea spiders, as well as unpublished works by Morgan and Harrison, closely paralleled Brooks's own research papers, though not his philosophical writings. They presented detailed descriptive life histories and examined evolutionary relationships.[63] They participated in a morphological tradition that focused on such problems.

Yet by the time the last of our four finished his degree, Brooks had become less active in that tradition himself. He performed fewer studies. He sent students to the United States Fish Commision laboratory in Woods Hole instead of going with them or taking them to sessions of the Chesapeake Zoological Laboratory. He spent less time with students and turned more to philosophy and to his beloved Bishop Berkeley, whose philosophy he found particularly interesting. He had been a good teacher and had served as an American exemplar for doing patient descriptive embryological study. He had laid out standards for careful work and had stimulated students to think about the "larger problems" of embryology and heredity. Brooks's chief contributions to the development of American biology probably lay in his somewhat enigmatic ability to inspire outstanding students and in his direction of the Chesapeake Zoological Laboratory, just as Martin had inspired physiologists, provided a fine research laboratory, then withdrawn from research.[64] Each introduced his students to his own tradition, and together they offered a modified Hopkins tradition in biology.

The many eulogies about Brooks by his students, some of whom gained great success in biology, occasionally sound a bit odd for that sort of tribute. They contain the usual adulations but also have a protective flavor as well. Conklin wrote fondly that "There was a sort of helplessness or lack of worldly wisdom on his part which made his students feel responsible for him, and which increased their affection for him."[65] Somewhat shy, not aggressive, Brooks was evidently an effective teacher in at least some fundamental ways. Yet his contribution as a teacher remains difficult to evaluate.

"His lectures were often vivid and picturesque, as well as clear and logical," Conklin recalled.[66] When Brooks lectured, others listened, because of his clarity and ability to entrance listeners despite his modest appearance and reserved delivery. As a lecturer, then, Brooks succeeded where Martin did not. But Brooks neglected his students and reportedly had absolutely no interest in their daily work. He often appeared quite unsympathetic to their research, claiming no knowledge or interest. He left advanced students to flail about on their own to define their research problems and to discover for themselves how to pursue whatever problems they had defined. He concerned himself little with specifics of laboratory techniques or practical problems. As a day-to-day guide to laboratory research, then, Brooks offered even less than Martin did.

To what extent did Brooks really stimulate his many outstanding students? Dennis McCullough has argued that, more or less, Brooks was in the right place at the right time. He gave students freedom in their research, letting them follow their own paths. The community of shared investigations in the Hopkins laboratories in the 1880s and early 1890s, where students and associates worked together even when Brooks did not join them, helped produce successful and active students. McCullough's assessment is probably accurate at least in part. If only because Brooks's students also studied with Martin and thereby gained considerable practical knowledge of laboratory techniques, Brooks probably does not deserve full credit for his successful scientific progeny. Still, a number of students, including Wilson, Morgan, Conklin, and Harrison, did follow Brooks's emphasis on embryological problems, evolutionary perspective, and morphological outlook. They did write favorably of Brooks's important influence on their work. Students worked with Brooks rather than Martin for a reason that reflects more than accident.

As Martin's associate in physiology, William Howell, reported of Brooks, "from him [his students] attained the stimulus to real thinking."[67] His student Ethan Allen Andrews emphasized Brooks's "creative, philosophical thought" and stressed that "his philosophical mind left its impress upon their ways of thought in whatever part of zoology they labored. The old problems of heredity are now attacked by new methods, but some of the foremost investigators are bound to Professor Brooks, more or less intimately, by nurture got when he was a stimulating if not also a formative part of their environment." Brooks was an inspiring teacher and wise friend, Andrews concluded affectionately.[68]

The close affection Brooks felt for his students and they for him, the informal summer interaction, and his conviction that he was confronting the large and truly important problems of biology undoubtedly all made Brooks an attractive teacher. Several of his students referred to his kindness and concern for the students' well-being. His informal style and affection for his students appears early, as when he wrote fondly to Gilman that the locals "have a profound admiration for Wilson's legs, and speak of them in the warmest terms. Should I say that Wilson makes a spectacle of himself every day digging for worms, at low tide, in a short bathing suit."[69] Conklin has recorded, for example, that Brooks was very deeply affected, even more than would be expected, by two deaths because of yellow fever during the 1897 summer expedition to Jamaica.[70] And Brooks's reports of the results from the Chesapeake Zoological Laboratory manifest his pride in his students' productive output.

McCullough reported that Brooks's preoccupation with phylogenies, descriptive morphology, and philosophizing prompted adverse reactions from his students. Garland Allen has reiterated this conviction.[71] No doubt. No teacher ever entirely avoids some such criticism from the next generation. But the evidence McCullough and Allen offer seems very weak. True, Brooks's students did not follow along for long, continuing to do what their teacher did; they moved off in different directions. Yet the emphasis in biology changed significantly through the 1890s and into the twentieth century, so this move should not surprise us. Their failure to follow Brooks forever does not at all mean that they rejected his work or that he had failed as a teacher. Brooks *was* an excellent teacher in that he introduced his students to practice in the field, which they found exciting. He taught them the value of knowing the current literature relating to their subject, of producing very careful descriptive work that could prove definitive because of its thoroughness. And he provided a sense of the "larger problems." Although they did not generally follow his philosophical inclinations, they did share the conviction that careful tracing of embryological development and life histories of diverse organisms could illuminate broader important questions about the phyletic and embryonic development of complex forms. This led them to at least an implicit concern about those broader morphological questions which they revealed in their early papers and retained throughout their careers.

We see in Brooks's students, therefore, a concern with problems of tracing developmental stages of marine organisms. They considered relationships among organisms, often looking at homologies and dichotomies as evidence of evolutionary ancestral connections, just as Brooks did. Yet the students retained a focus on achieving positive knowledge about narrower questions rather than on the generation of larger theories. They accepted Brooks's research problems and his fundamental morphological program, that is, and employed traditional morphological research methods. They continued throughout their careers to refer to the books read and the problems discussed back in Hopkins days in Brooks's circle.

But they increasingly turned to an approach closer to Martin's than to Brooks's. Epistemologically, they largely agreed with Martin's emphasis on definitive, positive results from narrowly defined experimental cases. The students all accepted the validity of generating working hypotheses, but they moved more toward Martin's emphasis on hypotheses as tentative and ancillary to observation than to Brooks's desire for having theories, even when wrong. And they began, as biology itself changed rapidly in the course of the 1890s, to look more to what were later called *model systems* and to generalized patterns than to seek unities in comparing individual life histories or relationships.[72] They participated in and became part of the morphological tradition, even as they began transforming it in various ways in light of their Hopkins experience and subsequent work.

Wilson at Hopkins

All four of our protagonists began to publish during graduate school. Generally their early papers have been overlooked or deemphasized because they contributed relatively little that was strikingly new and exciting, or because they were only short notes; they may even seem rather dull reading today. Yet precisely because they reveal the continuities as well as discontinuities in traditions, they must be considered carefully in order to understand the significance of the research programs that emerged from them.

As the oldest of the four, Wilson amassed a considerable record of publications during the 1880s and best represents the earliest young American-trained biologists.[73] Typically, Wilson worked closely with Brooks and on marine organisms. The summer before moving to Hopkins, he had participated in a dredging project of the United States Fish Commission under Spencer Fullerton Baird, which resulted in a set of descriptive papers on the sea spiders. Then, while at Hopkins, he attended the summer laboratory sessions in 1879, 1880, 1881, and 1882 and returned for research in some of the locations later, serving as Brooks's assistant in the latter years. His early work reflects his close contact with Brooks and the morphological tradition.

In 1880, his first Hopkins product was a brief abstract of his paper presented at a university meeting. The paper reviewed embryological evidence of the genealogical connection between annelids and vertebrates, thereby gathering support for the "annelid theory" then endorsed by Anton Dohrn, Nicolaus Kleinenberg, and Carl Gottfried Semper.[74] Comparison of embryonic layers revealed homologies that suggested a common ancestor for annelids and vertebrates, Wilson concluded. Two similar papers appeared as notes in the *Johns Hopkins University Circulars* and revealed his emerging emphasis on annelids and early developmental stages in particular.[75] Interestingly, Wilson already devoted considerably more effort to the early, pregerm layer stages than Brooks did. He found the cytological changes that the fertilized egg underwent in preparation for cleavage and the pattern of early cleavage worthy of special note. Indeed, the formation of the germ layers rather than their eventual fate formed his primary concern even in these early papers. Not stimulated by Brooks, this fascination with early cellular changes originated elsewhere within the morphological tradition, probably in Edward Laurens Mark's two-hundred-page study of snail development up to the two-cell stage, and was reinforced as others moved in the same direction. In his work at the Chesapeake Zoological Laboratory, Wilson certainly exhibited a strong interest in the problems of cell development and heredity outlined in Mark's paper, within the general context of the morphological tradition.

In his dissertation work, nonetheless, Wilson sounds much more like Brooks.[76] He cited many of the same people as before, but he focused on a different problem. Instead of asking what happens to the egg during the early cleavage stages, as he had in his other work, he asked here about the origin and the significance of a later stage of development, namely metamorphosis. The work of 1881 showed that the worm *Actinotrocha* undergoes a major metamorphosis in which the structure changes so radically that the early and later forms had until very recently actually been classified as two separate species. Yet the radical changes result from relatively simple and minor alterations in the growth processes. Most of the changes follow from a few simple evolutionary adaptions, Wilson showed. He sought to address two different questions. First, he wished to outline the mechanism by which the radical metamorphosis occurs. Was it the migration of the anus or a folding of the animal that effected change? Second, he wished to explain why that mechanism might have come into play in the evolutionary past; what adaptive advantage such a rapid metamorphosis might have brought. The larval and adult forms experience such different conditions of life, he concluded, that interests of efficiency dictate a rapid change through metamorphosis. This work on *Actinotrocha* fell very much within the

Hopkins tradition, for much of it resembled Brooks's own work. Indeed, Brooks had suggested the project.[77] Though this dissertation study lacked the detailed cytological concerns of his more well-known, later papers, he began to express and develop those interests by the next year. He also continued to ask his two types of questions, about mechanism and about the historical past.

In 1882, during the year when he remained an "assistant" at Hopkins after receiving his degree, he recorded results of studies of earlier developmental stages in marine annelids. Citing the paucity of work on the problem, he specifically wished to correct the lack of satisfactory discussion of the early stages that give rise to an unequal segmentation. In fact, a close similarity exists between segmenting polychaete annelids and other animals, he said, pointing out that the similarity had previously been overlooked. With careful descriptions of the eggs from the time of deposition through several cleavages, Wilson compared five different genera and then drew general conclusions. Segmentation in polychaetes is peculiar, he concluded, but perhaps the early differences result from later adaptations to "modifying conditions." With such inconclusive findings, he moved on, temporarily, to the polyps.

By 1881, Wilson had turned to the polyp *Renilla*. In Brooks's report to Gilman on the Chesapeake Laboratory session of 1880, he recorded that another student, Mitsukuri, was studying the adult form of *Renilla*, a sea pansy, which they had found in abundance while dredging.[78] A side benefit, he reported, was that *Renilla* are beautiful at night. This colony of polyps proved abundant, and Wilson soon turned to *Renilla* as well. By the summer of 1881, Brooks was convinced that Wilson's work would prove important, for he wrote to Gilman, "If he succeeds in finishing his paper this year it will be a monograph which will be highly creditable not only to us but to American science and if a way can be found to publish it in good shape it will make as valuable and handsome a paper as those from Dohrn's laboratory."[79]

Instead of publishing that major work right away, however, Wilson continued to explore development of *Renilla* and presented only short research notes for publication. Despite their brevity, they reveal that he had made an unexpected observation that called into question some of the basic assumptions within the morphological tradition, though in a way that called for modification rather than rejection of that tradition itself. Eggs exhibit a considerable amount of variety in their patterns of segmentation (or cleavage), Wilson discovered.[80] Yet early changes ought not to be expected to provide a useful guide to genealogical relationships, since they represent only vegetative cell divisions, according to Ernst Haeckel and other morphological leaders. For the traditional morphologists such as Haeckel, these early changes were supposed to be reasonably stable among all eggs within a species, at any rate. Since even that was not true, Wilson suggested that the early variations might perhaps foretell later variations. His research had

> brought to light a number of new and interesting points, of which perhaps the most novel is the existence of great variations in the earliest stages of development. It has been generally assumed—possibly on account of the lack of sufficiently extended observation—that the segmentation of the egg is constant, or nearly so, in each species of animal. In *Renilla*, however, there are a number of widely different types of development, yet bringing about the same result. This is of great interest, as showing how readily modification in early development may be produced, and how little weight consequently can be attached to the early changes of the egg as a guide to the affinities of animals. Of more importance, perhaps, is its bearing on the laws of variation, since it is quite conceivable that variations appearing at a very early period of development may result in symmetrical or correlated variations at a later period.[81]

The significance of early cleavages clearly remained open to question, and hence to investigation. His larger papers on *Renilla* reflect Wilson's growing interest in that direction: how are hereditary stability and developmental adaptations all built into the egg cell and its cleavage processes?

In the summer of 1882, Wilson was working madly on *Renilla*. In fact, so many eggs were mating so fast that Brooks organized a team to help follow the developmental stages.[82] The results appeared in 1883 and 1884 in Royal Society of London publications. In that study, Wilson returned to his earlier concern with individual cells and their structure and development. Part I in particular recounted details of cell cleavages and formation of the germ layers, revealing considerable variation. This variation is bound to prove important for later development, Wilson concluded, and "this fact of extremely early variation is, I believe, one of great importance. It is evident that a structural variation in one of the segmentation spheres must make itself felt, to a greater or less extent, in the structure and development of the cells derived from it, and may therefore appear ultimately as symmetrical and correlated variations in the larva or adult organism."[83] That suggestion, growing out of the Hopkins morphological tradition and influenced by his exposure to additional factors, played an increasingly central role in his later work. He continued to ask that fundamental morphological question: just how does the individual lie within the inherited but developing germ?

Conklin at Hopkins

Of the four, Conklin was most sympathetic to Brooks's philosophizing and to his interest in evolutionary explanations of biological phenomena. In addition, Conklin had the least direct exposure to Brooks's research in action. Wilson worked with Brooks during a number of summers at Beaufort, North Carolina, and elsewhere; Morgan joined him in Jamaica and at the Fish Commission; and Harrison participated in the Jamaica trip and may have spent some time at Beaufort. Conklin was with Brooks at the Fish Commission for part of the summer of 1889 and may possibly have spent a short time at Beaufort, but he did not take part in the longer expeditions as the others did. For, unlike the others, Conklin was married: in 1889, he had married the music teacher at Rust University, Belle Adkinson, who herself studied biology and who took a course at the MBL in 1891; he therefore felt less free than the others to spend long periods in Jamaica or Bermuda. That Conklin was older than the others during their graduate school days and was left much more on his own to develop his approach perhaps helps to explain why he was more sympathetic to Brooks's philosophical writings and to his theorizing than were the others, who directly experienced and adopted much more of Brooks's practical everyday research side.

Conklin recalls that he became interested in the development of siphonophores, which he had begun to study when Alexander Agassiz sent some samples to Hopkins. When Conklin went off to the United States Fish Commission in 1889, Brooks urged that he continue the study by looking at northern siphonophores. Unfortunately, there were none in Woods Hole, or at least not in sufficient abundance. Since Brooks had remained in Baltimore that summer, Conklin was left on his own to discover a more suitable organism to study. After some consideration, he turned to the gastropods, choosing the abundant and largely unexamined slipper snail *Crepidula.* He began by examining the development in detail from the very earliest stages of cell division and became entranced with the early cleavages. The exacting work, carried out at the Fish Commission in 1890 and 1891, led to his long and meticulous dissertation, published finally in 1897.[84] This work, as well as the several longer publications that Conklin produced starting in 1892, reflect a changing morphological tradition. Charles Otis Whitman and

Wilson, both at the MBL, exerted considerable influence, and they introduced Conklin to the latest work in cytology from Europe. This took Conklin, as it had Wilson, away from Brooks's particular emphasis on later developmental stages, though it was not in conflict in any way with Brooks's own work.

Conklin soon came into other spheres of influence which shaped most of his published work, but his first paper was a descriptive piece coauthored with Brooks.[85] Throughout his life, even as others turned to other emphases and different approaches, Conklin retained a fundamental commitment to the mainstream of the morphological tradition. More than the others, he was concerned with the philosophical compatibility of science and religion, perhaps because only he was truly religious in a traditional sense. He kept his sights on the patterns of development of structures and on the broader relationships among patterns in different organisms. He, too, embraced epistemological norms that moved him to more narrowly focused problems. Yet he looked more for the larger explanatory theory and remained less concerned with small bits of positive knowledge than the others.

The research program that Conklin built remained most closely allied with the problems and approaches that had been in the mainstream of the morphological tradition during his graduate school days. Conklin wrote after he graduated in 1891: "I cannot begin to describe adequately the stimulus for scholarly work and research which I received at Johns Hopkins. It was as if I had entered a new world with new outlooks on nature, new respect for exact science, new determination to contribute to the best of my ability to 'the increase and diffusion of knowledge among men.'"[86] Conklin stayed in that new world throughout his career.

Morgan at Hopkins

Morgan's early work also retained unquestionable continuity with Brooks's sense of the morphological tradition. Though he entered the Johns Hopkins University only in 1886, seven years after Wilson, Morgan published several papers while under Brooks's influence and before moving in other directions. In particular, his experience around Baltimore and at the Chesapeake Zoological Laboratory pointed Morgan in traditional research directions. In 1888, he began study of the fate of the blastopore in *Amblystoma* (now *Ambystoma*), *Rana*, and *Bufo* amphibian forms. What exactly happens to the blastopore tissue, Morgan asked; for example, to what later parts does the blastopore give rise? Relying on traditional sectional studies, he showed that the anterior end becomes the neurenteric canal and the posterior end the anus in at least one species. In fact, he concluded, "I believe these propositions give good evidence to support the view that in Amblystoma we have found a form which retains during its larval development for a short period the ancestral conditions of the blastopore. . . . It seems to me that the following hypotheses or suggestions may at least tentatively be proposed . . ., that in Amblystoma we have in the behavior of the blastopore the changes which have in general taken place in the phylogeny of that organ."[87] This work closely paralleled some of Brooks's and Wilson's, except for Morgan's inclusion of vertebrate land as well as invertebrate marine species.

In 1888 and 1889, Morgan occupied the Hopkins table at the United States Fish Commission laboratory, an arrangement that allowed one or two Hopkins men to work there in exchange for financial support of the Fish Commission by Hopkins.[88] One advantage was independence for the men chosen; but lack of guidance also accompanied the privilege. Though Brooks also spent some time in Woods Hole in 1888, he had his own projects and largely left the students to define and follow through their own work. At Woods Hole, Morgan began his first marine research with the study of the origin of ascidian test cells, or ova, from their surrounding material.[89] He had intended to work with the tunicate *Salpa*, which Brooks had studied, but evidently decided that ascidians and development of their test cells (the cells that make up the hard outside covering) were more promising.[90] He gathered material over two summers, then developed a critique, based on his own comparative studies, of a current theory. This work reveals Morgan's emerging morphological interest in the significance of early developmental stages, before fertilization and embryo formation, for he asked about the structure and significance of the germ cells.

In 1889, Morgan preserved a series of the young soft-bodied *Balanoglossus* to illustrate metamorphosis, a study to which he returned the next year. He also turned to the sea spiders, the *Pycnogonids*, which Wilson had studied before. Using standard methods of preparation, he followed early cell divisions in several species. The formation of layers, or delamination, and appearance of ganglia, legs, and eyes were of particular interest in revealing developmental parallels and differences among species.

On phylogeny, Morgan took on Anton Dohrn's theory that the spiders are descended from annelids separately from other arthropods and proposed an alternative close connection with the arachnids.[91] Morgan's consideration of two alternative hypotheses of spider ancestry closely parallels similar consideration by Wilson of alternative working hypotheses and some of Brooks's parallel discussion of alternative theories. The particular structure of Morgan's paper falls distinctly within the boundaries of an expected morphological approach, with considerable emphasis consistently placed on related research and on assessing the various alternative views. This work on the development and phylogeny of sea spiders formed Morgan's doctoral dissertation, though he continued his study of *Balanoglossus* as well.

To this point, therefore, there is no compelling evidence that, as Garland Allen has suggested, "although trained as a morphologist, Morgan's thought processes demonstrated a critical quality that was often lacking in morphologists of the older generation." It is simply not the case that Morgan saw issues in a "clear-cut, rigorous way" while others did not.[92] Yes, perhaps Morgan was at times able to focus on central issues, but so were others, including Brooks and Wilson. And Morgan's paper on sea spiders covers the usual range of problems typical in morphology: phylogeny, metamorphosis, the particular interest of the eyes because of variations there (as Wilson had found variations in the eggs of *Renilla*). It does not in any significant sense break through to a new style of presenting research or of dealing with earlier work on related subjects.

In 1890, Morgan again spent July through September in Woods Hole, this time at the MBL.[93] He formally thanked director Charles Otis Whitman for the opportunity to work there, but his work remained strongly Hopkins-like, in questions asked, organisms chosen, and methods used. He continued to ask questions about the way in which the individual develops out of its inherited germ material. At the MBL in 1890, he returned to *Tornaria*, the larval form of *Balanoglossus.* Widely regarded as a key organism for unlocking relationships, *Balanoglossus* had gained popularity. "Few animals present so many possibilities to the morphologist," Morgan wrote.[94]

Back in Baltimore, Morgan continued his study in Brooks's morphological laboratory. Referring to much the same literature as Wilson or Brooks, he concentrated on later developmental stages, on changes after formation of the germ layers, as Brooks did and Wilson did not. After considering the earlier literature (and its errors), he concluded that *Tornaria* resemble echinoderm larvae in a profoundly significant way and that *Balanoglossus* also bears a genetic relationship to the chordates. So, an ancestral relationship must hold for the larvae. But do not go farther, Morgan warned, to try to place each on an exact phylogenetic tree. The separation occurred so long ago that the facts do not allow such general conclusions.

In his last sentence, Morgan finally revealed the theory against which he wished to argue: the annelid theory favored by Wilson and others. Modern-type annelids were not the ancestors of vertebrates. Rather, the adult *Balanoglossus* closely parallels chordates, but the larval *Tornaria* parallels the echinoderms. Thus, "if Balanoglossus be related through its larva with the Echinoderms, as I have attempted to show in the preceding page, we see how old a phylum that of the Vertebrates must be, and hence the futility of attempting to derive them from any such highly specialized animals as the Annelids of to-day."[95]

In 1891, Morgan and other students at Hopkins, including Conklin and Harrison, lobbied for a Chesapeake Laboratory session for that summer. They succeeded in convincing both Brooks and Gilman and so set off for Jamaica (though Conklin did not join them in the end). Brooks sent Morgan, as that year's Bruce Fellow, ahead to help make arrangements. From Port Henderson, Jamaica, Brooks reported later in the summer that the party was rather too large for his taste but that everyone was busy. He fretted a bit about Morgan's enthusiasm for everything and his lack of focus: "Morgan is in a constant state of excitement, and has mapped out work and gathered material for a course of research for ten years or more, and he has not yet decided what to undertake first, but I think he may be relied upon to complete something this season. He is working very hard indeed, but as he seems quite unable to pass by any good opportunity, he is overwhelmed with material."[96] As his former student A. H. Sturtevant said later, "He wasn't happy unless he had a lot of different irons in the fire at the same time."[97]

Morgan did, in fact, manage to complete a great many projects. Brooks need not have worried, except that the results at first appeared as a number of apparently independent bits and pieces rather than as the more sustained discussion of central problems that Brooks preferred for his own work. In 1892, for example, Morgan turned briefly to explore the origin and significance of metamerism (or segmentation), a problem that Wilson and Whitman both found interesting. Other notes on frogs' breeding habits and eggs, on lady crab dancing, on larval forms from Jamaica, and on sea bass appeared.[98] Then 1891–93 brought a brief flirtation with cell lineage studies (see chapter 4), since Morgan spent parts of the summers 1890 through 1892 at the MBL in Woods Hole, where cell lineage was becoming a hot research area. But he also worked, respectively in 1890, 1891, and 1892 at the Fish Commission, in Jamaica, and in the Bahamas. During that period, he exhibited an interest in the revived issues of preformation and epigenesis and in a shift to revised problems and methods as he came into contact with other forces within the evolving morphological and physiological traditions.

Throughout, Morgan always regarded the Hopkins influence as basic to his thinking about biology. In 1891 when he received his first teaching position at Bryn Mawr and had to give up his Bruce Fellowship at Hopkins, he wrote to Gilman, "It is with the greatest regret that I find myself adrift from the University & will always look back with pleasure to the four years spent with Dr. Brooks in his laboratory."[99] Like his colleagues, Morgan carried with him Brooks's morphological influence, even as he began his own search for driving problems that he could attack in a way that satisfied the physiological as well as the morphological side of his Hopkins training.

Harrison at Hopkins

As the protagonist most exposed to the physiological tradition prior to his work with Brooks, Harrison demonstrates the least direct influence of Brooks. Yet Harrison was definitely a morphologist. His first summer of graduate school, 1890, he spent at the Fish Commission table, alongside Conklin and without Brooks's direct leadership. There he undertook the careful observation and drawing of the external characters of the actinian form, the sea anemone. For later work, he proposed "a critical study of some of the philosophical problems of biology," through examination of the internal structures and of development of the actinians.[100] Yet that work never appeared in print, because other forces drew him elsewhere.

In 1891, Harrison joined the group in Jamaica, continued looking at sea anemones, and also took up the study of sex and sex cells in mangrove oysters. He later recalled that "main profit came from browsing on the coral reefs and in mangrove swamps. It was my first and only visit to the tropics and the memory of it is still vivid." It was also the only time he grew a beard.[101] With emphasis on careful description and drawing of his results, he proceeded along traditional morphological lines.

In 1892 and 1893, Harrison went to Germany and, after some exploration of possible research sites, ended up at the University of Bonn. There he encountered a host of different sorts of morphological work, as well as medical studies. In 1893, he sought to remain in Germany until the year's end as a Bruce Fellow, finishing his degree by examining the development of muscles and of symmetry in teleost fish. In requesting permission of Gilman to remain in Germany, Harrison stressed the advantages of learning new methods, particularly cytological ones. Brooks had no objections, he said, but was not convinced either of Harrison's real intentions or of the value of learning new methods. Why not apply those he had already acquired to achieve "the production of results"?[102]

In fact, Harrison was attracted to Germany partly by romantic interests—for he found a wife there and married in 1896. Yet he also had begun to move in directions that reflected new currents within the morphological tradition in Europe. In particular, he worried about the emerging question of when the differentiation processes of the developing germ bring determination. He had begun to ask, that is, additional sorts of questions, about additional sorts of organisms, and using additional methods. By the time he received his degree in 1894 and went back to Germany to add a medical degree to his credentials, he had moved beyond Brooks in a way that drew on his physiological as well as his morphological background at Hopkins. Yet, like the other three, Harrison remained very much a product of Hopkins even as he moved beyond the Hopkins sphere to Europe and on to forge his own research program.

PART II

Traditions

3

Morphology Abroad

WITH HIS DOCTORATE in hand and an additional year of experience as William Keith Brooks's assistant at Hopkins, Edmund Beecher Wilson followed the American tradition in scientific and medical education by continuing his study in Europe. Probably buoyed by the prospects of meeting the people whose works made up the required reading list for morphology students at the Johns Hopkins University, Wilson embarked on his scientific pilgrimage with the help of a loan from his oldest brother.[1]

For a fresh American graduate in biology, Wilson had considerable success abroad. Arriving in England in the late summer of 1882, he just missed working with the much mourned popular embryologist, Francis Balfour, who had been appointed to a professorship in June 1882, only one month before he died in an alpine hiking accident. Balfour left behind him an active group of young researchers, which Wilson joined for a short time.[2] Wilson also became acquainted with the leading British biologists Thomas Henry Huxley and Michael Foster, each of whom had a special relationship as advisor to Hopkins. Huxley looked over Wilson's *Renilla* study and invited him to read and discuss it at a meeting of the Royal Society of London. He then published the paper in the society's *Philosophical Transactions*—a tangible and rewarding sign of success for the young American. Wilson thus entered successfully, if only briefly, into the English scientific community. He reported that Huxley "received me very kindly and said a number of pleasant things about Baltimore in general and Johns Hopkins in particular." Foster was likewise friendly and kind. And of Cambridge generally: "My work there was extremely profitable, and I was greatly pleased with the laboratory and the *spirit of the men.* I think Balfour has founded a school of morphology at Cambridge which is bound to produce telling results in the future."[3]

Yet after his two months at Cambridge, it was time to move on to visit the continent. He decided to spend some time at Leipzig, working with Rudolf Leuckart, as Charles Otis Whitman and other Americans had done.[4] He had been led to expect great things in the way of modern techniques and equipment in the Leipzig laboratory; therefore he was "rather disappointed in it and in their methods here. . . . So far as laboratories go, Baltimore is ahead of Leipzig." In fact, Wilson himself taught the Leipzig group the latest in section-cutting techniques from Cambridge.[5] Yet he found Leuckart "a very interesting and brilliant man, and he has been very friendly to me." From Leipzig, Wilson intended to go on to spend a while at Jena, working with the much publicized Ernst Haeckel and Haeckel's former student Oskar Hertwig. "Haeckel is a most friendly and agreeable man, of perfect simplicity and kindness, but he somehow impressed me strongly with a sense of his *power*." Unfortunately, Haeckel's new laboratory was not yet completed. Wilson was invited to spend some time at Hertwig's smaller facility and eventually decided to do so, since "for my purpose he is one of the best men in Europe."[6] In addition to Jena, he also stopped in Vienna to visit Carl Claus and Berthold Hatschek, two more who had appeared on the morphology reading list at Hopkins.

During his moves through the European biological world, Wilson visited and worked in the laboratories of the leading morphologists. He learned what techniques they were adopting, what problems they addressed, what organisms they used, and what sorts of broader goals they had in sight. In short, Wilson encountered the morphological tradition in full swing, a rich tradition of which the Hopkins experience with its one professor of morphology could have touched only parts. It is to that tradition that this chapter turns.

Before I introduce the morphological tradition in detail, it is important to consider what sort of thing is being discussed. What, that is, do I mean by *tradition*? Since the term has been used in various ways, I will also explain why I consider the tradition a legitimate and useful unit of analysis for understanding the nature of the science this study is examining. The next section will deal specifically with morphology.

The second problem for this chapter is not, however, to argue the case that a morphological tradition existed. Rather, if one assumes that it did exist, the task is to display the character of that tradition, its core commitments, and the way that it changed up to the time when Wilson encountered it. (That such a discussion illuminates the historical events is, of course, itself strong evidence that such a tradition existed). Thereafter, Wilson and the other

Americans contributed to transforming the tradition. Thus, after examining the nature of the tradition in Europe, we will return to Wilson's experience there. Chapter 4 will explore early changes in the morphological tradition in the United States, and chapter 5 will turn to the interplay of American and German innovations as the pace of change increased and the tradition began to undergo significant transformation.

Traditions

It is crucial, before embarking on a more detailed examination of the changing traditions, to consider what such things are. Despite increasing interest in the subject, there has been surprisingly little of use in history and philosophy of science to enlighten us. Apparently some such groupings of research commitments do exist to earn the label *traditions.* There are some sets of shared commitments (or assumptions about things that are thought to "matter" for doing good scientific work) that most historians would be willing to say have exhibited historical continuity and thus can be regarded as traditions. Yet defining traditions precisely proves problematic, though this does not mean that we must or even should abandon the effort. Indeed, we should willingly embrace the complexity of scientific change, learn to be comfortable for the moment with temporary suggestions rather than definitive assertions about how that change occurs, then set to work to discover whether and how these suggestions do make sense and apply to the actual scientific cases.

Larry Laudan has provided one such introduction. He offers as a "preliminary, working definition" a very sensible combination of basic ontological and epistemological commitments: "a research tradition is a set of general assumptions about the entities and processes in a domain of study, and about the appropriate methods to be used for investigating the problems and constructing the theories in that domain."[7] This seems basically right, as does Laudan's discussion of the relation of traditions to other traditions and the role of theories within traditions. Yet, since mine is primarily an historical story, my emphasis here is different from Laudan's. I tend to see traditions in more practical and less theoretical terms. It is worth looking at why that analytical unit "tradition" is helpful.

Contrary to an older popular impression, today it is recognized that science is not carried out by lone individuals isolated from current intellectual and social forces. Instead, scientific work and workers fall into clusters affected by their contexts. Of all the possible basic assumptions about the way the world works and the way science ought to work, groups of researchers at any given time choose confined and shared sets of these assumptions.[8] Research efforts congeal as researchers share common concerns: interest in subject matter, shared problems, conviction about method, the search for particular types of solutions, an institutional setting, or a social bond, for example. At the most general level, a broadly shared set of concerns underpins a tradition. Thus, we have such examples as a Judeo-Christian tradition, a physicalist tradition, a morphological tradition. Yet, though such traditions exist, they have no clearly circumscribed essences that persist and serve to define them through all time. We must reject any such platonic or essentialist searches for clear definitions that would allow us to draw neat lines outlining traditions that would remain just the same for all times. Traditions change as the commitments of which they are made change. True, at any given time, there will be boundaries that set some efforts as inside and others as outside the tradition in question. But these boundaries will inevitably change.

Thus, a tradition can maintain historical continuity even if its components change. An early embryo looks very different from an adult organism in most cases, yet the historical (and material) continuity strongly supports the claim that all stages throughout development represent the same organism. Analogously, a research tradition may look different at different times, but historical (and intellectual rather than material) continuity insures that the tradition persists as long as it is viable and has participants. Individual organisms can appear, as offspring of parents with some similarities but also some dissimilarities, within a range of constraints. Similarly, new traditions can arise as offspring of earlier traditions. In addition, both individuals and traditions can lose their vitality or die.

Some traditions may more closely resemble the remarkable horseshoe crab, *Limulus*, which has changed very little throughout its long evolutionary history. Other traditions change relatively rapidly or are relatively short-lived. While there is no list of necessary and sufficient conditions that will define the elements of a tradition for all times and places, this only means that the traditions are more difficult to identify. It does not mean that they do not exist. Historical continuity is key. As in evolution, the idea that an entity (species or tradition) has reality even though it changes is difficult to grasp—but absolutely fundamental.

A suggestive elaboration of the character of these traditions will demonstrate how other groupings that historians and philosophers of science have identified might also fit within the picture of traditions. Within a tradition of general assumptions, at times research congeals around a problem or method or other key element; it then yields what many have labeled *disciplines* or *fields*. At any given time, several disciplines or fields might coexist within the same tradition (such as astronomy and geology within physicalism, or embryology and cytology within morphology). At times, an individual or subgroup may also operate within the shared framework of one field or discipline but may embrace elements of another, thus effecting what Lindley Darden and Nancy Maull have labeled *interfield connections*.[9]

In another sort of clustering, groups of individuals may form a school, which generally exhibits geographic localization and often has a strong leader as head. Such an individual may articulate a well-defined research program, for example, on which the work of the school will center.[10] A less directed effort by a group that shares only a very loose subject interest, such as bird watching, might more appropriately be identified as a club activity. Any of these subgroups can coexist within a larger tradition. At any given time, they fall within the general concerns of the tradition but experience narrower constraints on their basic assumptions about how to do good scientific work (or what others often call the *core commitments* or *leading principles*).

What are these basic assumptions, which ultimately lie at the base of any tradition or subgrouping? They fall into broadly epistemological and ontological categories. To the former belong assumptions about appropriate methodologies, appropriate problems, types of evidence thought legitimate, methodological commitments, and types of goals sought. To the latter belong convictions about the appropriateness of trying to understand the organism in terms of its pieces rather than as an indivisible whole, and convictions about vitalism, materialism, and mechanism. There are various ways to divide science into sets of working assumptions. At different times, some types of commitments will prove more important and more easily identifiable than others. Nonetheless, general groupings of types of commitments can help to elucidate the nature of traditions and of groups or fields within the traditions. For the purposes of this story, epistemological concerns remain central.

Within broad epistemological commitments are differing methodologies or approaches. We can, for example, determine the extent to which scientists have endorsed the one extreme of purely passive observation, in which the individual simply observes what happens naturally and says, essentially, "this is what I see," or the other extreme of active manipulation to test an experimentally derived hypothesis. Between these extremes comes the range of increasingly active observations and descriptions, for which the scientist may ask, "what happens in such-and-such a case," and then try to find such a case occurring naturally: or may ask, "what happens if I do such-and-such to this organism" and then perform the suggested manipulation. The scientist can then ask further, "how does the experimentally derived result affect such-and-such a theory or question?" In short, there is a range of increasingly more manipulative methods and an increasingly interventionist experimental approach using those methods to address carefully formulated questions. Keeping in mind that the range of possibilities captured by this continuum is useful for demonstrating how an individual shifted in methodological assumptions: whether by sliding gradually along the continuum of possibilities or by more suddenly employing, or at least rhetorically endorsing, a radically different methodological approach.

Similarly, there are continua of other epistemological factors, including expected research results, problems, and types of evidence. They also exhibit a range from more general to more specialized and narrowly defined. What are the types of problems addressed? Are they very general problems such as how to account for all the diversity of organisms in the world (Charles Darwin's problem), or how to deal with heredity, development, and evolution all with one unified "umbrella" theory (Ernst Haeckel's or August Weismann's desire)? Or are they problems that center on more narrowly focused questions, such as what happens to a developing egg if one alters the external pressure, say, by placing the egg between glass plates? The range from general to narrow or specialized problems is not as neatly defined or obvious a continuum as one might like, but it does offer a range of possible commitments along a roughly sliding scale. So, too, with the types of results sought: whether general explanatory (and suggestive) theories or narrower definitive additions to "positive," supposedly unassailable, knowledge. Types of evidence thought appropriate parallel convictions about legitimate methodologies. Is experimentally derived evidence that is acquired by interventionist manipulation legitimate and reliable, or is only passive observation appropriate? With embryology in particular, the question becomes whether microscopic work, serial sectioning, and more radical intervention best produce useful information, or whether only *in vivo* observation can be reliable.

Ontological commitments range from extreme vitalism to radical materialism, depending on the researcher's conviction about whether living phenomena can be explained in terms of inorganic materials and actions alone or require something uniquely organic. Separate, though often related, commitments range from holism to various versions of partism, depending on whether the researcher believes that studying less than a whole organism (or community, or species, or whatever the smallest unit is thought to be) can provide useful information for the problems thought appropriate and important.

Thus, a complex of commitments works at any given time. Most lie in the misty unarticulated background, accepted without much reflection as one learns from the existing community how to do "good science." Sometimes one or more undergo revision and may be brought under conscious scrutiny. Change may occur for a small splinter group within a tradition and may produce a new specialty or even a new field, thereby sometimes broadening the tradition. Or the change may even carry the individuals or group outside the tradition, perhaps toward a different tradition. Or sufficient change may so thoroughly affect the tradition that it is more appropriate to regard the old as having given rise to the new tradition(s). Awareness of the types of existing commitments and the range of possible assumptions within science facilitates understanding of both individual and group efforts in biology and of their changes. The character of the scientific grouping changes, but the groupings persist, experiencing historical continuity.

It is the historian's job to identify the core commitments. This requires going beyond presenting chronicles of the past, of course, but it begins with a careful examination of the details of the scientific research and discussion of the work. This can begin to reveal which commitments were the deep ones and which were only superficial. The deepest and most basic commitments are the important ones that drive scientific change. In order to provide a framework for understanding the nature of change in clusters of the deepest core commitments by a group of researchers, I look to traditions as the useful analytic unit.

A question does arise concerning the usefulness of traditions in this respect. If traditions change, how can we tell who is in and who is out, or what work falls inside or outside any particular tradition? For identifying the boundaries of traditions, we must turn pragmatically to more intuitive criteria rather than looking for only nice, a priori, essentialist definitions. As Ernst Mayr has pointed out, the "primitive Papuan of the mountains of New Guinea and the modern ornithologist each can set up a taxonomy of birds. Each recognizes as species exactly the same natural units that are called species."[11] They agree, despite their differences in training and their inability to articulate one clear set of criteria for differentiation. Similarly, I believe, we can distinguish traditions practically. By examining relevant characteristics such as the types of problems addressed, evidence accepted, methods used, and goals sought, we can recognize those clumping patterns of thinking that are usefully called *traditions*. We must begin by carefully looking in some detail at the scientific work done. Though the scientists' own rhetoric about what constitutes an appropriate style and appropriate commitments is undoubtedly useful information, it cannot be accepted as the only, or even primary, evidence for what assumptions actually were adopted. We must also step back and assess what is going on with our knowledge of the larger intellectual and social context.

To complicate the picture yet further, individual scientists may fall within more than one tradition because of the way we cut up traditions, because each tradition depends on a shared *set* of commitments and not the sharing of *all* commitments. At different times, for different people, the morphological or physiological tradition in biology may have predominated, for example, as individuals were preoccupied with form or with function. At other times, primary concern with materialism or vitalism, for example, may have dictated participation in different or in more than one dominant tradition, say in a materialist as well as in a morphological tradition. The details need to be worked out more fully to elucidate what it might mean to belong to more than one tradition. For the purposes of this study of four diverging research programs, however, that issue does not arise centrally since the focus remains on the morphological tradition and on changes within it.

The Morphological Tradition

In discussing the morphological tradition, I take it as given that such a tradition existed. I do not try to establish criteria that would "prove" its existence. Rather, I take as evidence the fact that researchers at the time considered themselves as morphologists within a larger ongoing community of morphologists with shared commitments. Further evidence comes from perceptions in the early decades of this century that there had been a morphological tradition (whether it still existed or not). In addition, understanding the study of form and change in form in the nineteenth century in terms of tradition works. It fits the facts in a way that other units of analysis do not by showing how a set of core commitments could remain stable and could guide the body of research called morphology even when many of the details of theory and method changed. The "morphological tradition" is useful as the unit of analysis. The task here is to examine the specific nature of that tradition.

In his classic historical treatment, *Form and Function,* in 1916, biologist E. S. Russell identifies a pure morphology and a less pure version. The "pure" morphologist regards structure as primary, with functions arising only secondarily and revealing nothing about the all-important homologies or historical relationships. This morphologist follows comparative descriptive methods to explore the unity of structure or "the principle of the unity of type" that underlies the apparent diversity of organic forms. This pure morphologist also "describes, classifies, generalizes; he does not seek for causes."[12] This implicit distinction between the search for generalizations from the data, perhaps in the form of laws, and an alternative search for causes, or explanations of the phenomena, became especially important for morphologists later in the century as they began to transform their tradition and to embrace both goals.

Prior to the 1880s, though, Russell's pure morphologist focused on structure alone as the appropriate subject and on generalization about establishing unities of structure and function, homologies, and relationships (and eventually evolutionary phylogenies after Darwin and Haeckel) as the desired goals. Concentrating on describing and classifying groups of organisms, they tended toward what Ernst Mayr has designated as ultimate rather than proximate concerns, asking about historical patterns rather than local mechanical causation.[13]

Yet at the other end of the spectrum, there also existed the not so pure morphologists. These might look at structure in terms of function and might draw conclusions about homologies or phylogenies on the basis partly of the functional role of structures. Or they might seek causal mechanical or proximate explanations of the structure and its development in order to explain individual form and its development rather than concentrating on finding unities of types or phylogenetic relationships.

While all of these pursuits remained within a morphological tradition, the difference between the pure and the not so pure lay in the way the limits of acceptable practice were defined along the range of possibilities. Epistemologically, pure morphologists sought underlying unities, expressible as laws. Relying on comparative study of structures, they sought patterns and similarities and largely looked beyond differences. They asked more general questions and also tended to look at larger structures such as whole organs, organisms, or even species instead of focusing more minutely either on smaller parts or on narrowly defined questions about those smaller parts. In fact, however, most morphologists diverged from those pure goals and considered some of those smaller and more proximate questions of interest as well.

Ontologically, morphologists could be strict materialists or could embrace various sorts of teleological thinking or vitalism. Yet along with most scientists and the public more generally, as the nineteenth century progressed, they became increasingly mechanistic and materialistic. Some were ontologically relatively holistic and others relatively reductionistic or willing to accept the organism as a sum of its parts, with the particular inclination somewhat dependent at any time on precisely what questions they asked. Embryological questions, for example, usually demanded more holistic discussions, while comparisons of cell or tissue structures allowed greater analysis into parts. Both ontologically and epistemologically, then, morphologists embraced a range of possibilities as appropriate in doing good biological work—not all possible sets, but more than one narrowly defined or "pure" set of practices. They remained morphologists because they were all focused on the discussion of form and in particular on establishing the underlying patterns of structures.

Doing morphological science by the end of the century also included a range of activities. Some might work with dead museum specimens, generally in the form of skeletal material or preserved whole organisms, tissues, or body parts. Careful observations to uncover details of structures as well as comparisons between structures in different "types" of organisms dominated this sort of work. At the same time, other morphologists worked with killed material of another sort: embryonic or cellular material. Histologists had, by century's end, developed a complex of sophisticated techniques for killing living soft material while embedding it in a fixative such as paraffin. They then could slice the material into thin pieces to facilitate seeing "inside" the organism with the microscope. Staining could bring different parts, such as the chromatin, into focus to provide more and more detailed new information about internal structures. Cytologists carried out a similar process with cytoplasmic material. In addition, other morphologists studied living materials. This might mean traipsing to the seashore and examining whole organisms of previously unstudied sorts. Comparing details of structure to identify the organisms or to establish similarities of parts or relationships within the whole formed a central part of such study. Others examined living material but, again, devised methods to look inside the organism, to gain information about the parts and pieces that cannot normally be seen. Study of regeneration fell into this category of work. All these different kinds of enterprises made up the broad morphological tradition.

That diverse group of morphologists had two major institutional homes during the nineteenth century: museums and universities.

The former housed those concerned more directly with classification, and often more oriented toward comparative studies. The latter included those in medical schools and, increasingly, in separate zoological or anatomical research institutes, both directed more toward delineating how organisms are structured. Early in the century, morphology and physiology had already emerged as separate areas of concentration within the life sciences and within medicine, but they only became institutionalized as such in Germany as the nineteenth century progressed and as new opportunities for specialization and diversification appeared. In many cases early in the century, for example, a university appointment might actually come in both anatomy and physiology, as for Johannes Müller at the University of Berlin. By mid century, however, German universities had begun to divide the chairs, and researchers had begun to specialize more. Theodor Bischoff taught physiology and anatomy at Giessen, then moved to Munich to teach only anatomy. Wilhelm His assumed the chair in anatomy at Leipzig in 1872, while Carl Ludwig held the position in physiology, two chairs that had previously been one. Lynn Nyhart's excellent study of German morphology demonstrates the pervasiveness of this pattern.[14]

Because of the willingness by some of the not so pure morphologists to consider the function as well as structure of parts and because of the imperfect separation of the institutional roles, it might seem that the two traditions blended together. Yet the morphological and physiological traditions nevertheless did remain essentially distinct in all interesting senses. Prior to the 1880s, few practitioners occupied a middle ground where they did not belong properly to either one or the other tradition, and few by that time could carry out research comfortably within both. Even though some taught both anatomy and physiology, by mid century they generally identified themselves with either one or the other tradition.

Carl Gegenbaur, whose *Elements of Comparative Anatomy* the Hopkins students studied carefully, demonstrates this point. As a morphologist inclined toward physiological concerns, he presented the dominant opinion that morphology and physiology remained distinct, with different subject matters, problems, and methods: "The task of physiology is the investigation of the functions of the animal body or of its parts, the referring back of these functions to elementary processes and their explanation by general laws. The investigation of the material substratum of these functions, of the form of the body and its parts, and the explanation of this form, constitute the task of Morphology." Classification, he wrote elsewhere, must be guided by morphological rather than physiological principles.[15] Morphology and physiology thus remained effectively distinct, though both were properly parts of biological research.

Naturphilosophie to Evolution

Earlier in the nineteenth century, a strong influence on morphology came from German philosophical ideas, through *Naturphilosophie.*[16] Most importantly, the Germans added to the general idea of unity of plan the idea of parallelism between individual development and development of the individual's "race" or animal series. This concern with parallelism came increasingly to introduce functional as well as structural evidence in order to illustrate or to establish the parallels. Advocating parallelism also implied an assumption that individual and species development arose by the same causes and therefore moved toward a concern with those causes.

Johann Friedrich Meckel, for example, saw both of what later emerged as morphology and physiology as dependent on the enlightenment brought by comparative anatomy and especially by comparative embryology:[17] "The development of the individual organism obeys the same laws as the development of the whole animal series; that is to say, the higher animal, in its gradual evolution, essentially passes through the permanent organic stages which lie below it; a circumstance which allows us to assume a close analogy between the differences which exist between the diverse stages of development, and between each of the animal classes."[18] Though the German *Naturphilosophen* did not all take up research programs in morphology, of course, their focus on parallelism and on comparative developmental patterns as a manifestation of that parallelism gained attention as the morphological tradition evolved. Their emphasis particularly attracted those interested in comparing groups of organisms.

Evolution brought further changes to morphology. In fact, the most familiar aspect of the morphological tradition in the late nineteenth century is surely that influenced by Darwin, especially Ernst Haeckel's morphology. Haeckel, some historians say, presented the program for morphological studies that others then pursued. His program was important and did gain considerable attention, if partly because, in Wilson's words, the man emitted considerable "power." He was also good at publicity, and his *Generelle Morphologie* and *Natürliche Schöpfungsgeschichte* brought public attention to the study of morphological problems.[19] His Gastraea Theory summarized there set the tone for many, including those who did not accept its conclusions, if only because it provided a strongly articulated and clear starting point for research.

The original ancestral organism was essentially a raw gastrula, Haeckel insisted, a Gastraea with germ layers defined. We know this because the developmental stages prior to the gastrula have no significance for later development, he claimed. Once we know this fact, there is no further reason to look at earlier developmental stages; morphologists need only begin with careful study of the gastrula and subsequent stages. The three germ layers of the gastrula themselves are homologous for all organisms. Later stages remain essentially homologous from one organism to another as well, with some special adaptations, and with extra stages added on at the end. Thus, more advanced organisms recapitulate their ancestral past as reflected in the way that the germ layers undergo change. Careful study of these later changes in the germ layers reveals ancestral relationships, therefore, and also reveals unities of form. Comparisons will show which organisms are older, or more primitive, and which more recent, or advanced, by demonstrating which have adapted away from the normal shared developmental progression. Eventually, this comparative research will generate full phylogenetic trees. With the trees in place, all embryonic development can be understood in terms of the doctrine of recapitulation, or what Haeckel termed the "biogenetic law." What Haeckel called for, therefore, was

1. concentration on the germ layers as the starting point of significant development, and thus as the key for assessing homologies, with the Gastraea Theory as the basis for research;
2. attention to recapitulation of stages that had been by definition (since they necessarily would have been eliminated otherwise by selection) adaptive in their ancestral, evolutionary past. The phylogenetic past, in effect, explains the ontogenetic present and thus focuses interest on the embryo, though only indirectly; and
3. the construction of genealogies, or phylogenetic trees, to reveal ancestral relationships.[20]

From the 1860s through the 1880s, some morphologists did embrace this program and expended considerable effort on the third step, constructing speculative phylogenies. Many, in the absence of contradictory evidence, adopted the assumption that the germ layers are, in fact, homologous and that differentiation really begins after the formation of these layers. The classical picture that has it that, until the 1880s, "the chief concern of the younger morphologists was the construction of genealogical trees" is thus at least partly correct.[21] Ernst Haeckel did serve as one important influence on the direction of this increased work in morphology.

Yet Nyhart's and Russell's works establish beyond question that the tradition remained much richer, with many alternative strands of research.[22] Some rejected the excesses of Haeckel's grand theorizing, saying that he erred in placing speculative theory as more important than careful establishment of concrete and reliable facts.[23] Others besides Haeckel also had considerable influence, and in other directions. Carl Gegenbaur, E. Ray Lankester, and Anton Dohrn, for example, put forth alternative theories within the context of evolutionary programs which explicitly provided alternative approaches for studying the past. While Haeckel and others directed considerable attention to recapitulation and to later stages of development, other morphologists introduced other methods and concentrated on other problems. In particular, embryological study of early developmental stages and even of the activities of single cells assumed importance for many. Cytological and embryological approaches came to play central roles in transforming the dominant morphological tradition through these alternative strands of morphology.

The Rise of Embryology

Earlier in the century, before Darwin and evolution, Karl Ernst von Baer had begun to attract attention to careful descriptive embryological study with his classic detailed study of chick development.[24] Von Baer concentrated on structure and sought to understand underlying unities as well as deviations from those general patterns, all with predominantly a descriptive, comparative approach that sought general laws rather than consideration of proximate explanatory causes. Von Baer, along with Christian Heinrich Pander and others, brought attention to developing embryonic types.

In particular, they showed the significance of the primary germ layers—ectoderm and entoderm (now endoderm)—which already exhibit organization and give rise in regular ways to the later organized parts of the embryo. A third and fourth layer arise later, according to von Baer:

> First of all, the germ separates out into heterogeneous layers, which with advancing development acquire ever greater individuality, but even on their first appearance show rudiments of the structures which will characterize them later. Thus in the germ of the bird, so soon as it acquires consistency at the beginning of incubation, we can distinguish an upper smooth continuous surface and a lower more granular surface. The blastoderm separates thereupon into two distinct layers, of which the lower develops into the plastic body-parts of the embryo, the upper into the animal parts; the lower shows clearly a further division

> into two closely connected subsidiary layers—the mucous layer and the vessel-layer; the original upper layer also shows a division into two, which form respectively the skin and the parts which I have called the true ventral and dorsal plates—parts which contain in an undifferentiated state the skeletal and muscular systems, the connective tissues, and the nerves belonging to these. In order to have a convenient term for future use, I have named this layer the muscle-layer.[25]

As these germ layers arise, the initially homogeneous egg becomes heterogeneous, an epigenetic fact often referred to as von Baer's law. His work, along with ideas of parallelism, coexisted with other parts of the developing morphological tradition, with embryology coming to occupy an increasingly important position.

Von Baer's version of epigenesis and his evidence for the gradual but regular emergence of form was convincing. It also met a receptive audience, as others began to question the older preformationist assertions that form preexists and is just, in effect, waiting to be let out. It made sense that features emerge gradually: that is what one sees through the microscope, after all. Yet form does not come randomly, from nowhere. For von Baer, each individual somehow gets its proper form from internal causes, because it belongs to some established type: elephants become elephants, sea urchins, sea urchins, and *Drosophila*, *Drosophila*.

By the century's end, some rejected his particular version of essentialism but still embraced epigenetic development. The key lay in the egg and therefore in internal structure rather than external forces on it. Since the egg is, after all, inherited, it is capable of conveying substance. It serves a conservative function in providing some sort of directions from one generation to the next. Embryonic development of individuals depends on heredity but brings differentiation in the course of time. Inheritance, that is, is not separate from development but is part of it. Heredity provides the initial material, which continues guiding all development thereafter, and epigenetic development takes place from the very first stages of egg production.

Von Baer provided a solid foundation for embryological study, but even he left open to question the significance for later determination of parts of the early developmental stages, prior to the germ layer stages. His work suggested, as Haeckel felt later, that early stages simply involve growth of essentially homogenous material, with only very gradual appearances of significant differentiation. Yet with improvements in microscopic and histological techniques, by mid century these earlier stages proved irresistibly interesting to some—namely to those morphologists concerned with form at the cellular level.

Indeed, even to one of phylogenizer Haeckel's own prize students, the early stages and alternative problems and approaches held primary appeal. In a letter to Richard Hertwig, Haeckel lamented that Richard's older brother Oskar had gone astray and moved toward cytology and away from evolution. In other words, he had become concerned with form for its own sake rather than for the sake of what Haeckel would have seen as larger and more appropriate evolutionary interpretation. On a trip to Switzerland, people had assaulted Haeckel with questions about Oskar's opposition to Darwinism. It seemed to them that Oskar must have rejected constructing phylogenies as the primary goal of morphology since he had presented a recent lecture without even referring to Darwin or to Haeckel's own biogenetic law. Nor had he concerned himself with investigating parallelisms, homologies, or underlying unities of type. Instead he had looked at cells and at their development. Haeckel lamented that poor Oskar had lost track of the proper realm of biology because he was preoccupied with cellular details of development, specifically of fertilization and early cellular actions. In Oskar's denial of the importance of systematics and paleontology, Haeckel saw an abdication from a true biological program toward such limited cellular anatomical studies as Rudolph Virchow's.[26]

Yet to some such as Oskar Hertwig, those early stages and cellular details obviously assumed legitimate central significance for explaining later development. They led their advocates in directions other than that of Haeckel's program, though not to a rejection of all its elements. British zoologist E. Ray Lankester, for example, saw that the early developmental stages might involve changes that have significance for later structural arrangements. With his Planula Theory, developed more or less simultaneously with Haeckel's Gastraea Theory, Lankester emphasized the value of constructing genealogies and thus of identifying homologies, as Haeckel did. He did not, however, downplay the significance of the early stages or the role of different types of cells, as Haeckel did. Indeed, he suggested that there had been an earlier, primitive, two-layered planula, modifications in which had given rise to later forms.[27] Though he developed no program to explore these early stages himself, he influenced others, such as Charles Otis Whitman and, indirectly, Wilson and Conklin, to move toward detailed cytological comparisons.

Cells

One search for unities had, in fact, led into the cell. In seeking the fundamental unit of organic beings during the late 1830s, Matthias

Schleiden and Theodor Schwann determined respectively that plants and animals share the cell as basic. Their resulting cell theory grew out of a tradition of microscopic work and the search for unities characteristic of *Naturphilosophie* and stretching back into the seventeenth century.[28] It also inaugurated a line of biological research concentrating on the cell, termed *cytology* by the 1870s.[29]

Concentration on cells fit with either primarily structural or primarily functional questions, so that study of cells found a home in both the morphological and physiological traditions in the latter half of the nineteenth century. It also fit into medicine, for Virchow, for example, recognized that cellular pathology involves physiological processes that cause the cells to behave abnormally.[30] He saw the cell as both the structural and developmental unit of organisms and, further, maintained that cells arise only from preexisting cells, in the form of eggs.

Since eggs were generally held to be cells, it seemed that they must behave as cells do. This suggested that all developmental processes and patterns must proceed along with, if not because of, cell division. A host of such researchers as Theodor Bischoff, Carl Vogt, Rudolf Albert von Kölliker, Johannes Müller, Franz von Leydig, Edouard van Beneden, Charles Otis Whitman, and Robert Remak all joined in to advance work at the intersection of cell theory and developmental study. In particular, Remak demonstrated in the 1850s that segmentation, or division into differentiated parts, definitely occurs as a process of cell division. That conviction that cell division is guided, in turn, by nuclear division provided a strong foundation for further cytological work in embryology.[31] Van Beneden's and Whitman's examinations of early localizations of later body parts in different cells and work on cell lineage both stimulated increased interest in the cellular details of the very earliest stages of development.[32]

Wilhelm His's microtome made it possible to section prepared organisms uniformly, and other advances in preparation techniques helped lead more and more researchers farther into the cell and on into the nucleus by the 1870s.[33] Otto Bütschli, Leopold Auerbach, and Oskar Hertwig, in particular, demonstrated that fertilization entails the union of male and female nuclei within the egg cell. Van Beneden, Eduard Strasburger, Hermann Fol, Walter Flemming, and many others pursued the details of cytoplasmic and nuclear changes during embryonic development during the 1870s and 1880s. Many errors, many advances, many new ideas emerged.

This cytological work, carried out especially in anatomical institutes, provided an important element in the background of Edmund

Beecher Wilson's and Edwin Grant Conklin's work, though not as much for that of William Keith Brooks, Thomas Hunt Morgan, or Ross Granville Harrison, who found other elements of the morphological tradition more appealing. Cytological study of embryology generally involved a willingness to discuss development and form in terms of cellular parts, though with the danger of losing sight of the whole interactive organism.[34] Cytological investigation into the causes and patterns of differentiation also lay behind the self-conscious attempts by His, Wilhelm Roux, Gustav Born, Eduard Pflüger, Moritz Nussbaum, Theodor Boveri, and others to transform the morphological tradition and to forge what they saw as new directions of research. With cytology came increasing divergence among morphologists.

The Middle Ground

While physiologists sought more proximate, or immediately causal, accounts of living phenomena, morphologists came increasingly at the century's end to concern themselves more with questions about why, historically, such phenomena occurred.[35] This began to leave something of a gap between the two traditions. In particular, interest in the causes of development fell into this gap since they concerned what processes brought the given pattern of morphological change, and those processes seemed to involve at least some physiological functioning. These intermediate questions were becoming more interesting to an increasing number of investigators, in part because of cytological successes and related interest in understanding the significance of nuclear and cell division for subsequent developmental stages. In part also, this greater interest came about because even traditional concern with problems of classification had begun to place emphasis on the developing embryo. To a growing group of morphologists, then, phylogenetic history and germ layers alone could not provide an adequate explanation of the various samenesses and differences of embryonic development, which they came to take as an important set of phenomena demanding explanation.

Nor did phylogenizing or comparative morphology address the proximate causes of these developmental phenomena. These researchers began to recognize that evolutionary morphology did not and could not explain how a particular feature came to exist materially in a given organism. Operating within the sorts of morphological programs inspired by Haeckel or Gegenbaur, for example, a researcher inclined to seek a proximate or material explanation of individual features and their origin would have trouble. Yet Haeckel pretended to explain ontogeny. Thus, he made the problem of causally explaining embryonic development a legitimate and visible one, but his program did not make it a viable problem for study. Those increasing numbers of researchers who took the problem seriously had to look elsewhere for an approach or for answers.

Questions about the material causes of form and development of form, though ostensibly about structure, also introduced a concern with process and change rather than with stasis. Traditional morphology as it had been practiced by most morphologists had few adequate tools or methods for assessing such changing developmental processes. Von Baer and others had concentrated on patterns of emerging structure but not on physiological processes of change as they affect (or effect) those patterns. As a result, during the 1870s and 1880s, a group of individuals began to work at generating effective approaches and techniques. The group included morphologists Wilhelm His, Wilhelm Roux, and Gustav Born, as well as Eduard Pflüger, who moved toward the middle ground from a physiological position. Beginning with morphological concerns, then, their shifting focus toward cellular and tissue changes in early developmental stages took them beyond what had traditionally fallen within morphology. Yet, even as they undertook to transform morphology, their programs all remained closely tied to the cytological work within the morphological tradition.

Otto Bütschli also exemplified the growing interest in the middle ground and reflected the commitment to material explanations of organic processes which had begun to underlie at least some such work. Professor of zoology and paleontology at Heidelberg, Bütschli had received training in mineralogy, zoology, and chemistry at Heidelberg, then had worked in Rudolf Leuckart's laboratory in Leipzig. His turn to cytology and cell theory in the 1870s brought him to conclude:

> Morphology composes only a part of the nature of organic forms. Each form must be capable of being explained in itself from given bases and influences. Only when it is shown that one organic form proceeds from another, and when the conditions of this appearance are known—as is hardly true in a single case today—is a material present in which it might be possible to seek a causal-mechanical explanation. . . . If we conceive of the elementary organisms [cells] as the building blocks of morphology, our understanding of the elementary organisms is altered, for the type of morphological observation of [multi]cellular organisms [previously] used loses its justification and physiological methods come to the fore. The phenomena on and in the elementary

> organisms can be more precisely stated only through a language of the physical-chemical conditions of their appearance and disappearance.[36]

There was, that is, a move by some within a morphological and some within a physiological tradition toward embryology and a move toward proximate, materialistic causal explanations. Embryos became the focus of attention in their own right, not primarily because they revealed phylogenetic relationships or because they might uncover some secrets about adult physiological processes.

This move toward embryos brought a shift toward new problems. What, exactly, does the egg do as it develops into an adult: what structural sequence of patterns does it run through, starting at the earliest unfertilized egg stage, and what processes act to direct those changes? The shift in questions in turn brought a move by morphologists toward a willingness to think in terms of parts of the organism, to study whole organisms as sets of parts in at least some respects. At the same time, physiologists began to look at the embryo more as a whole interacting organism and not just as a set of autonomous functions, as they had earlier. The outlooks of the two groups began to converge in some essential respects, as each embraced an approach between radical holism and extreme reductionism. Each also moved toward more interventionist, manipulative, and analytical or experimental approaches in order to achieve progress in answering the sorts of questions thought legitimate. Both traditions, in short, made epistemological shifts and came to embrace a larger range of acceptable problems, methods, and types of evidence considered appropriate.

As I suggested earlier, anatomist Wilhelm His made possible the move toward cells and embryos with his improved microtome. Only with effective and regular serial sectioning, which allowed a researcher to keep track of the entire embryo, and only with other significantly improved microscopic techniques could researchers observe and follow the minute changes in nucleus and cytoplasm which they increasingly recognized as important. At least this was true for most organisms. Of course, surface observations of living organisms also provided vital information for some species and supplemented the complex microscopical studies of prepared materials. This proliferation of types of evidence brought confusion as well as useful data, though, and appeals to different pieces of the assorted evidence brought a variety of alternative approaches to discovering the causes of development. His's was one of the first.

His's Program for Physiology of Development

In 1874, Wilhelm His, trained in cytology by Virchow, Remak, and others, set forth as *Unsere Körperform* a series of letters ostensibly to his nephew. In that published volume he attacked Haeckel's biogenetic law and his emphasis on the gastraea as key to phylogenizing efforts. In fact, His showed, even the earliest embryonic stages are already different in different organisms. It is absolutely not the case that all embryos experience more or less equivalent early development until the essentially universal gastrula stage, as Haeckel had claimed.[37] In other words, it is not a historical, evolutionary past that determines the development of all organisms. Instead, development is primarily a mechanical process dependent upon the physical condition of the current organism, His insisted. He thus clearly regarded his work as being in opposition to or in direct rebellion against Haeckel's. Instead of demanding a rejection of morphological work generally, however, his predominant concern lay with rejecting Haeckel's particular explanation of development.

Haeckel's appeal to evolution and his facile discussion of material causation His found laughable. In his tenth letter, he ridiculed Haeckel's understanding, or lack of understanding, of physics. Offering talk about material continuity, monism, and such mechanical formulas as "$mv^2/2$" by itself explains absolutely nothing about development and remains only a word game, His urged, as he continued, "All these words, which are capable of strengthening a heart thirsty for knowledge, come into use: parental material, molecular movements, life quantities, protein, form and protoplasm. 'Misce, fiat explicatio!' so runs the enlightening formula of our clever Doktor, and with this stroke he opens the eyes to all secrets of generation and life."[38]

Instead, His maintained that work in anatomy and cytology provides examples of how to pursue the proper sort of proximate, mechanical explanation. As a result, in 1874, he put forth his own "theory of transmitted movement," according to which various foldings and rollings of elastic plates and tubes bring about unequal growth and hence differentiation. These unequal foldings follow from the initial unequal differentiation of the egg. From the beginning, according to His, the egg has prelocalized areas, or "organ-forming germ regions." These important regions lie in the cytoplasm, not the nucleus, and therefore he insisted that study of the embryo must begin with the important principle of cytoplasmic prelocalization.

These convictions put forth in his early writings found expression in his later work as well. He insisted that "the principle, according to which the germinal disk contains the preformed germs of organs spread out over a flat surface and conversely that every point of the germinal disk is found again in a later organ, I call the Principle of Organ-forming Germ-regions."[39] In other words, organs already lie prelocalized in the embryo by the time the germinal disk is formed. Each prelocalized germ region correlates with a later structure, and, indeed, gives rise directly to those later body parts. Internal structure and action rather than environmental factors external to the embryo determine development. Developmental differentiation is the visible manifestation of complex, interacting chemical and physical processes internal to the organism beginning with its initial egg stage. His therefore called for a mechanical "physiology of development" in order to uncover the way in which initial differences translate into later differentiated parts. Look at structure, but also understand the mechanical processes that bring about the developmental change of structure, he urged.

In a later paper of 1888, His summarized the general conclusions from his earlier work and cited confirmation of his line of research from the parallel with geology:

> The principle of layer-folding is therefore a fundamental one in embryology, and the study of its consequences must be one of the most important tasks of this science. The germinal layers are elastic plates under the influence of certain pressures, and these pressures are to be derived from an unequal increase of dimensions in different directions. The laws of this increase must determine the fundamental occurrences which bring about the formation of the body of the higher animals.
>
> I cannot as yet find any fault in the short chain of these arguments, and I daresay that they are in full harmony with all our notions about other natural processes. Geology also has much to do with layer-foldings and with their consequences.[40]

Yet he acknowledged that, despite such evidence about the facts of mechanical layer-folding, not all morphologists agreed with his emphasis.

His lamented that too many still looked to the phylogenetic history to provide what they saw as an explanation of the development of an individual, as Haeckel and others had stressed they should. They thereby overemphasize the role of heredity in directing development, though he pointed out:

> I should be the last to discard the law of organic heredity, or to deny the immense progress that biological science has made, by introducing

> this grand conception on our horizon. Questions of phylogeny will be for long of the utmost importance, and of the greatest interest in biology; but the single word 'heredity' cannot dispense science from the duty of making every possible inquiry into the mechanism of organic growth and of organic formation. To think that heredity will build organic beings without mechanical means is a piece of unscientific mysticism.[41]

So, whoever wants to understand the nature of development must accept heredity but look to mechanical, or physiological, causal explanations. His elaborated what he saw as the proper goals for biological investigation of development:

> By comparison of different organisms, and by finding their similarities, we throw light upon their probable genealogical relations, but we give no direct explanation of their growth and formation. A direct explanation can only come from the immediate study of the different phases of individual development. Every stage of development must be looked at as the physiological consequence of some preceding stage, and ultimately as the consequence of the acts of impregnation and segmentation of the egg.
>
> Some of the modern publications may be considered as symptoms that embryological studies are about to take a more physiological direction. The important inquiries of O. Hertwig, Fol, Pflüger, Born, Roux, Gerlach, and others, regarding impregnation, the first axes of segmentation, and the artificial formation of deformities, are based upon physiological ideas and physiological methods, and they open new and large fields for biological investigation.
>
> Physiological considerations in morphology are far from interfering with phylogenetic inquiries; rather will the phylogenetic worker find in them a mighty help in his efforts. He has only to open his eyes to the actual processes of life and development.[42]

His's continued emphasis on embryology and his particular choice of the phrase *physiology of development* emphasizes that he saw the move toward mechanical explanation as a move toward accepting the mechanical approach of physiology. He did not reject morphological interests as unimportant, however, nor did he say that morphologists should all become physiologists. Rather, morphologists should incorporate elements of what had been physiology into their morphological investigations, particularly of embryonic development. They should, in effect, begin to transform both traditions.

His's view might seem preformationist. But he rejected such an interpretation. Instead, he believed that theories of development fall into four categories rather than the usual two of preformation and epigenesis. His own preference was for neither strict preformation nor epigenesis but for a somewhat mysterious "theory of transmitted movement."[43] Development of form is the external manifestation of a complex, interacting, lawlike physiological process, His insisted. As with a telegraph message, we cannot see the process but do witness the result. And the message must have had a concrete organized beginning, as development does. It is not the form itself nor the specific building material of the body which begins development, but "the stimulation of form-producing growth."[44] Production of form involves both the material medium, or the egg, and the processes of change which are stimulated by fertilization, according to His's strongly materialistic and mechanistic process. His called for a rearticulation of developmental morphological problems and, unlike Haeckel, suggested a way through mechanical analysis to attack them. Though he attracted few direct followers, he provided an attractive alternative research program and stimulated many to look at it and think about it in shaping their own work.

From the pure morphological classification programs and search for unities of type to Haeckel's and Gegenbaur's evolutionary search for unifying laws, to His's physiology of development: morphology reached across the century. From Huxley to Leuckart to Jena and Vienna: Wilson moved through the tradition. A rich and changing tradition. In the 1880s the best of that tradition could be found in Naples, at the Naples Zoological Station, where Wilson soon went also.

Naples as Biological Mecca

This premier research station began when German morphologist Anton Dohrn joined his teacher Haeckel and his fellow students on a trek to the seashore. Haeckel had gone with his own professor Johannes Müller to the Baltic Sea to do "pelagic sweepings" in order to collect simple marine organisms. Only in the early nineteenth century had people begun to realize the great richness of marine life, and only occasionally did anyone make the demanding trip to study it. As the century progressed and as recreational areas developed along the coasts, it became easier to find sufficiently hospitable surroundings to entice researchers to explore. And Haeckel's evolutionary views suggested that the direct descendants of the simplest, most primitive, and oldest ancestral forms would be found in the sea. So Haeckel took his students to the seashore in the tourist town of Messina.[45]

There Anton Dohrn first experienced marine research. But he found the temporary nature of the facilities less than ideal. Why should researchers have to haul their increasingly cumbersome collections of glass and odds and ends back and forth for each research session? At first he persuaded the Swedish consul in Messina to supervise the equipment for his own students. Yet Dohrn envisioned something bigger and better, even the best. He resolved to build a permanent marine research laboratory to offer the most advanced possible facilities and to attract the very best researchers. Despite the intervention of the Franco-Prussian War, his Stazione Zoologica opened in Naples in 1874.

Designed for serious research, the impressively solid and classical building housed researchers upstairs. Downstairs, the public was allowed and even encouraged to visit the aquarium, specially conceived to attract attention and therefore funding through public support. Upstairs, only researchers were allowed. The Naples station would not teach classes, Dohrn decided, but would remain a facility exclusively for advanced scientists carrying out their own research. Zoology for Dohrn meant much the same, familiar sort of thing that his inspiring dissertation director, Ernst Haeckel, advocated. Therefore, the first department established at the *stazione* was morphology.

Yet botany, physiology, and bacteriology soon had their own departments as well. Dohrn wished to allow the morphological and physiological traditions to exist side by side, with work in a variety of established and emerging disciplines. An evolutionary perspective and a commitment to the best-quality advanced research were the basic requirements for researchers at Naples. They were to supply their own microscopes, microtomes, and microtome knives, though an American manufacturer had supplied two microscopes for American visitors to use. But aquaria with salt and fresh water, desks, paraffin ovens, gas equipment, mercury pumps, kymographs, and a whole range of other morphological and physiological accouterments were provided for researchers to carry out their wide range of researches.[46] Both resident staff and visiting researchers were to use the *stazione* as sort of a zoologists' greenhouse, with access to a range of materials and the best possible equipment.[47] It also afforded access to the fine library, augmented by the Naples Zoological Station's own publications. Gathering good workers and providing them with the best facilities for either year-round or seasonal visits would yield results, Dohrn said, and he left considerable flexibility in the selection of what work went on. He clearly felt that zoology should move beyond the static and unenlightening descriptive classification work that had dominated it earlier in the century, though he also encouraged detailed monographic studies of the flora and fauna of the region.

Dohrn was a committed Darwinian and Haeckelian. As a result, he felt that leading research should focus on study of natural selection and of phylogenetic recapitulation during individual development.[48] Fortunately his convictions matched those of leading zoologists throughout Europe closely enough, so that those in powerful positions began to subscribe to research tables and to visit themselves or send their best advanced students to Naples. The Stazione Zoologica became a training ground and a meeting place for the growing group of preprofessional biologists. Established professionals and others with independent resources gathered to pursue their research programs in morphology and more specialized work in cytology, embryology, physiology, and other areas.

In addition to its research opportunities and its scholarly excellence, the Naples station offered a slice of European culture. For the Americans, coming from middle-class and midwestern backgrounds, as many did, this was a special attraction. As the Stazione's archivist Christiane Groeben has put it, "Dohrn was also good company, taking excursions, enjoying jokes and even childish games." He loved music and art, was of refined taste, and had had a traditional education in his wealthy family, where his father's passion had been entomology. Dohrn welcomed his visitors with concerts, literary discussions, a billiards table, and a bowling alley to supplement the daily diet of scientific exchange, collecting trips, and even diving adventures. Life at the *Stazione* was unique and irresistible for the curious, eager, budding American biologist.[49]

Wilson at Naples

By March 1883, Edmund Beecher Wilson had resolved that he must move on to Naples. Everywhere he heard about the place, about its advanced techniques, its rich research materials, and its unsurpassed facilities. The best people obviously gathered there, and he longed to join them.

Charles Otis Whitman had been the first American to work at Naples, in late 1882 as he was returning to the United States from a two year visit at the Imperial University of Tokyo. Inspired by two summers at Louis Agassiz's Penikese Island school to seek his doctorate in zoology, Whitman had gone to Leipzig to work with Leuckart. With his degree in hand, he had considered taking a fellowship at the Johns Hopkins University but went to Tokyo instead after Huxley turned down the position and when Whitman's predecessor, Edward S. Morse, recommended him strongly. Whitman enjoyed his time in Tokyo and produced four professional biologists out of the

Fig. 8. Wilson (*seated, left*) with friends at Naples, 1883. From the collection of Linda Timmons.

four graduate students he taught there. He also continued his own research on the life history and development of leeches, including the study of early developmental stages and cell lineages. Yet by 1882, he had decided to return to the United States and to stop at Naples on his way.[50]

Fig. 9. The Naples Stazione Zoologica as seen by Wilson, 1883. From the collection of Linda Timmons.

At Naples, Dohrn accepted Whitman as his personal guest and invited him to carry out his work there despite the lack of any official institutional affiliation. Whitman savored the experience and always sought, in his own later positions as head of the Marine Biological Laboratory and of Clark University and the University of Chicago Biology Department, to send students to the seashore and advanced researchers to Naples.

Naples was a wonderful place, he wrote in *Science,* so that its "mild and equable climate," as well as "the unsurpassed richness of the fauna and flora of its bay, and the best equipped laboratory in the world, conspire to give the Naples station pre-eminence among institutions of its kind, and to render it probable that it will remain what it is now acknowledged to be,—the world's great biological station." It was a "Mecca of biologists, and a seat of unprecedented prolific activity."[51] To a friend he wrote that he was "having a delightful time" at the station with its "greater advantages than are to be found anywhere else in Europe."[52] American biologists should join their European fellows and go to Naples in order to become biologists, Whitman urged.

Emily Nunn went. Later Whitman's wife, Nunn at the time was a frustrated would-be biologist. She had tried to enroll in the graduate program but had been politely rejected on the grounds that the Johns Hopkins University could not admit women to the biology program, where they would, after all, discuss sex and other unladylike subjects.[53] Emily Nunn had thereafter visited England and worked at Naples at the English university table. She, too, recorded her impressions and helped to advertise the Naples station with her own detailed description of the offerings. In particular, she praised the library and the facilities for collecting and keeping diverse marine organisms. She saw study of the earliest egg cell stages of development and study of the "process of the formation of a highly organized animal from the almost invisible germ" as the high point of Naples research.[54] She stressed, as Whitman had, the importance of plentiful financial resources to make the laboratory possible and also called for American support to help keep it going. Unlike Whitman, she was not judged sufficiently advanced to stay long and carried out no published research during her stay.

Whitman and Nunn had gone to Naples; so would Wilson.

Wilson was probably especially attracted by the musical opportunities at Naples. After initiation to the musical world through singing lessons and playing the flute, Wilson expanded his horizons in Baltimore. He attended Baltimore Conservatory of Music concerts and thereafter switched allegiance from the flute to the cello, in time becoming an accomplished quartet player. Though he believed that he "was too old to take up so difficult an instrument with any hope of mastering it," his love of playing chamber music gave him great joy and led his daughter to play the cello professionally. The visit to Naples, following his Baltimore conversion to cello and chamber music, would yield opportunities for a "musical debauch" amidst the research work.[55]

Unfortunately, he encountered difficulties in his aspiration to work at Naples because of Dohrn's system of subscribed tables for research. An institution would subscribe $500 and then have the privilege of sending whomever it selected to carry out research there. But no American institution had ever subscribed. While Dohrn had been willing to allow Whitman, who then had no American affiliation since he was returning from Japan, and Emily Nunn, as a visitor to the English table, to work at the laboratory, he decided to stop that practice. If Wilson wanted to work at Naples, Dohrn felt it only fair that he should have to do so at a properly subscribed table.[56] Quite possibly Dohrn, the experienced politician and fund raiser, felt confident that the well-connected Hopkins graduate would be able to secure support from America.

Wilson wrote to Daniel Coit Gilman that he needed help. Dohrn insisted on at least a reasonable prospect of a subscription, and Wilson appealed to Gilman to make Hopkins the first from the United States; Naples was "practically the headquarters from which most of the leading European laboratories derive their best meth-

CARL A. RUDISILL LIBRARY
LENOIR-RHYNE COLLEGE

ods, and where, indeed much of their most telling work is done."[57] It was embarrassing that no American institution had officially joined the group of leading institutions represented there. Gilman responded that the Johns Hopkins University would take a table. But in the meantime, Wilson had already worked out an arrangement with his cousin Samuel Clarke, who then taught at Williams College. Williams would pay for a table, which Wilson would use. Then the next year, Clarke would work at the table in Naples while Wilson filled in for him at Williams. Wilson had his stay in Naples even without Gilman's help.[58]

Wilson wrote much later: "That first year at Naples—it was not quite a year—was the most wonderful year of my life. I despair of conveying any notion of what it meant to me, and still means, as I look back upon it through the haze of fifty years. It was a rich combination of serious effort, new friendships, incomparable beauty of scenery, a strange and piquant civilization, a new and charming language, new vistas of scientific work opening before me; in short, a realization of my wildest, most unreal dreams."[59] At the time he was a little less ebullient. It is worth recording his words from Naples at some length since they reflect so clearly what the exposure to European biology at Naples meant to a young American Ph.D.:

> It is in every respect the best laboratory I have seen and my high expectations have been fully met.
>
> Two things especially strike me as characteristic of this laboratory. The first is the perfection of the technical methods of research. It is now almost proverbial for zoologists to say: 'For methods go to Naples' and in the same breath is usually added: 'A good method is half the battle.' Certain it is that many of the best modes of work now used at Leipsic, Cambridge and elsewhere have originated here. The secret of this is simply that fifteen or twenty zoologists are usually at work, who come from laboratories in all parts of the world and bring this experience to a focus here. They are all experimenting and comparing results and new methods can thus be very thoroughly tested.
>
> The second thing which particularly impresses me is the thoroughness and solidity of the work which is being done and the importance to general zoology of the papers which go out from the station.
>
> We have not far to seek for the cause of this superiority in the work—a superiority which must strike every unprejudiced man who compares it with the zoological work produced by our American laboratories taken as a whole. It lies simply in the fact that *money* has not been wanting; so that the management has been able to offer good facilities for work and has thus attracted the best workers. The laboratory is most fully equipped, the work-tables are convenient, chemicals, aquaria and other needful things are supplied without stint, an endless amount

> of material is constantly being brought in by the two steam-boats and other boats belonging to the station, and a large corps (between 30 and 40 in all) of assistants is always on hand to render any kind of assistance. This, as it seems to me, is the real secret of the success of the Naples laboratory—that it has a good money basis. The day has gone by when a man can go to the seashore with a carpet-bag and a microscope, and do good zoological work—or be sure of doing it. The difficulties of research are constantly increasing, since to meet them the zoologist must have more and more perfect appliances.[60]

At Naples, Dohrn encouraged Wilson to extend his *Renilla* studies to examine development and evolutionary relationships among the Mediterranean version of this sea polyp. Wilson's one publication from his Naples stay did suggest some relationships among different groups of polyps, but the thrust of the paper remained on the contrast between sexual and asexual reproduction in the class of polyps *Alcyonia*.[61] A dorsal pair of filaments appears near the very last in the embryo, much later than they do in the asexual bud, where they are among the first such organs to become visible. Why does the particular pair have such greater variability than the other filamental pairs? Wilson asked. After examining a variety of species and genera of polyps, he determined that this one pair has an ectodermal origin, while the others do not.

This variation accounted for the different timing and other structural details. It also helped to explain the different functions. For, with further close examination, Wilson found that this particular filamental pair does not take part in digestion, as the others do. Instead, they seem to be involved in circulation. Direct observation as well as their ectodermal origin confirm this suggestion. Thus, physiological and morphological evidence worked together in his study of embryology, as they worked together in Wilson's education at Hopkins.

This time at Naples afforded Wilson the opportunity to examine closely a wide variety of related organisms and especially to look for crucial similarities and differences among them. In turn, this careful comparison pointed to causes of differentiation of developmental patterns. The advanced histological techniques, as well as the rich and abundant range of organisms, made possible the observations and descriptions Wilson needed to ground his more theoretical and speculative conclusions about the functions of the filaments in each organism and the phylogenetic relationships among organisms. It also provided him with the tools, the experience, the materials, and the questions to direct a host of further inquiries. In addition to the wonderful opportunities for research, Wilson loved the cultured environment with classical music and serious conversation, and he became great friends with Dohrn.[62] At the end of his official visit, Dohrn invited Wilson to stay on for three years to prepare a monograph based on the comparative work on evolutionary relationships of polyps. But Wilson felt he must decline. He noted that the necessary material for the project was "not abundant and the most important questions depend upon the embryology which is clearly not an easy matter to study here," presumably because of the difficulties in making prepared sections in the absence of a fully equipped laboratory for such work.[63] In addition, Wilson recognized that the monograph would take too long to complete. He felt obligated to return to the United States to spend his year at Williams so that cousin Sam could follow him at the Stazione Zoologica. And he wanted to enter professional American biology. He enjoyed teaching at Williams for a year, then moved on to other positions. With his European jaunt, Wilson had enthusiastically joined the morphological tradition.

He knew the best techniques, the best men, and the best work in the morphological study of zoology. He understood the reasons for Dohrn's evolutionary emphasis as well as the cytologists' interest in cells. Even as he absorbed the best of the European morphological tradition, however, that tradition had begun to undergo transformation. Naples was one center for that transformation, but there were others as well. And Wilson, Morgan, Conklin, and Harrison each found his respective way to different parts of the changing tradition.

4

Transforming Traditions at Home

EDMUND BEECHER WILSON'S year at Williams replacing cousin Samuel Clarke provided useful teaching experience, but "on the whole," he wrote later, it was "scientifically a dead loss; I had no time nor appliances for research, no scientific stimulus, no incentive to research."[1] When Clarke returned, Wilson was happy enough to move on. He resolved, with his old Yale and Hopkins friend William Sedgwick, to write a textbook for general biology. Since Sedgwick was at the Massachusetts Institute of Technology, Wilson went there too. With a lectureship to provide income, he set to work and spent the next full year constructing what was, in effect, a revision of the approach to biology introduced by Henry Newell Martin at Hopkins.

Martin had used the model developed by Thomas Henry Huxley and himself in England. Huxley explained in his introduction to *Practical Biology* that the work was intended as a laboratory guide to introduce people to biology.[2] The approach relied on the simplest, commonest, and most readily obtainable plants and animals. The students progressed from yeast through other organisms to bacteria, through ferns and on to polyps, mussels, crayfish, lobsters, and frogs; dissecting, observing, and describing the morphological structure and physiological functioning of the different parts as they went.

Wilson and Sedgwick had been impressed with this general approach. Yet they also had new ideas about how to modify the textbook to make it even more effective. They originally set out to produce two volumes to show the beginner how to enter biology. Instead of surveying a range of plants and animals from the simplest on upward toward the more complex as Huxley and Martin had, however, they decided to begin with a basic introduction to the principles of biology. "Biology," they said, is the "science which treats of matter in the living state." There is not some completely new and different subject matter of life, they believed, nor is there some radically different science of biology. Rather, biological study involves examination of the physics and chemistry of matter that happens to be in its living state. Therefore, they felt, they should begin with an introduction to the fundamental properties of matter and energy. Following that would come a more exhaustive investigation of one representative plant and one animal, rather than a necessarily superficial study of a larger number of organisms. They would address respectively problems of morphology, or the "science of form, structure, etc.," and physiology, or "the science of action or function." For this purpose they selected the earthworm and the fern.[3] They intended a second volume to provide the laboratory guide but finally decided to publish a single volume in 1885 without a laboratory companion.

In the second, revised, edition, they explained that they had given up the idea of a second volume. In fact, they had eliminated the detailed laboratory work altogether since it had proven too difficult to satisfy all the wide range of constraints operating upon the many instructors interested in the textbook. In the second edition, they also added some discussion of unicellular organisms, and they reorganized the text to place animals before plants. Plant physiology had proved simply too difficult for beginning students to understand without more introduction, they discovered.[4]

The book was written by both Sedgwick and Wilson, with Sedgwick appearing as first author, and it represented their shared Hopkins perspective as modified by their own practical experiences with teaching. It is difficult to assess just which parts reflect more of Wilson's particular views and which Sedgwick's. Perhaps they agreed on everything. Yet the beginning chapters, addressing the fundamentals of biology, sound very like Wilson in his later writings. The reader is informed that the organism consists of organs, which are heterogeneous and differentiated to produce an effective "division of labor." The cells that make up the organs and tissues act much as counties or townships on a map, dividing up the material into units. Yet these are not mere material divisions, for the cells are the "ultimate units" of life and hence of biology. Indeed, cells must clearly be basic since each living being begins as a single cell. Consisting of protoplasmic and nuclear components, the cells become differentiated for their respective physiological functions. Within the cell, the protoplasm is the "physical basis of life" and of the cell functions. It seems that oxidation or combustion provides the source of protoplasmic energy, and chemical activities clearly take place in the cells, but the details of how these processes work remain as yet unknown.[5] This view of biological fundamentals, focusing on the central role of the cell and its protoplasm, must have characterized Wilson's views as of 1885 and is certainly compatible with everything else he had written by then and soon after. Only later, after a second visit to Europe in 1891–92 did his view of the cell begin to undergo greater transformation and modification.

The presentation of the earthworm's development probably also reflects Wilson's work at least as much as Sedgwick's. The descriptions of the formation of germ layers follows along the traditional morphological lines of the time. But the discussion of the way in which particular cells go into making up the respective layers and the detailed illustrations went beyond current standards for an introductory text. The cells, from their beginning in the unfertilized egg, are already different in size. Furthermore, while the primary germ layers (the outer or ectoblastic layer and the inner or entoblastic) look very much alike and hence are morphologically similar, they seem already differentiated functionally.[6] Yet the text does not positively commit itself to a belief in preorganization of parts or any sort of predifferentiation in the unfertilized egg. Indeed, the suggestion is that up to the blastula stage relatively little differentiation has yet occurred. The questions of exactly when and how differentiation occurs remained unresolved. Wilson and Sedgwick had to be satisfied to set aside such lingering questions and to concentrate on presenting the range of material coherently for beginning students.

Wilson at Bryn Mawr

With the book completed, Wilson went the next year to Bryn Mawr, to head the Biology Department. Here was a special opportunity to combine teaching with research in a way that had proven impossible at Williams. Bryn Mawr was purposefully organized to provide a full-scale undergraduate liberal arts education. From its inception, it was also intended to offer high-quality graduate programs as well. This meant that for biology, as for other fields, the new school would seek the best possible research-oriented instructors it could find. Wilson, with his new, coauthored textbook for biology and his growing list of publications, fit the bill nicely.

In addition, Wilson came from the Johns Hopkins University, with which Bryn Mawr had close connections. Joseph Taylor, whose philanthropic bequest had established the new women's school in

Pennsylvania, was a close friend of one of the first Hopkins trustees. A number of Hopkins trustees also served Bryn Mawr, and that college's dean and later president Martha Carey Thomas was the daughter of one of the Hopkins trustees who had urged President Gilman to allow women to attend Hopkins on an equal footing with men. One historian suggested that Bryn Mawr's first personnel so overlapped with Hopkins's "that it was aptly labelled 'the Miss Johns Hopkins.'"[7]

From 1885 to 1891, Wilson remained at Bryn Mawr. This was the time in his career when he solidified what he had learned, established and extended his research program examining early embryonic development, and matured as a scientist and as a teacher. It is also the time when he came into contact with Charles Otis Whitman, whose own work on early cell divisions in leeches raised questions closely related to Wilson's own. Wilson's work throughout the period concentrated especially on development of those relatively simple organisms, the earthworm and other annelids.

The earthworm had attracted Wilson's close attention in his collaboration with Sedgwick on their textbook. Finding himself at Bryn Mawr, teaching introductory and graduate students as well as establishing his own research program, he continued work with this readily available organism. In particular, he grew increasingly interested in the origin of the germ layers. Given that the germ bands exist and contribute the crucial cells that then differentiate into embryonic and adult parts, how do those germ bands and, more generally, the germ layers arise? Which cells give rise to which germ layers? And, further, what do homologous patterns of development in different species tell us about the relationships among various organisms; or what can such embryological detail reveal about phylogenetic relationships?

These are questions firmly placed within the morphological tradition. As such, they have seemed uninteresting and old-fashioned to some historians. As Alice Baxter has put it, "In most ways Wilson's work—up to this point—was neither extraordinary nor pioneering." It was traditional, falling within the standards and style of the best of the European morphological tradition. Yet, as Baxter goes on to point out, there were elements in his work at Bryn Mawr "which marked a new emphasis in his research."[8] He gradually dismissed Haeckel's simplistic recapitulation doctrine from his work, then repudiated it directly. He rejected the importance of germ layers as the starting point for understanding development of phylogenetic relations. He also delved deeper into cellular details of development. It is the latter of these changes that came to dominate his work and deserves more careful attention.[9] For this emphasis, shared as it was by others, such as Whitman, helped to effect a transformation of the morphological tradition in the United States.

Wilson, Whitman, and Changing Epistemologies

In his first research after completing the textbook, Wilson was clearly inspired by having read a paper of Whitman's. Appearing in a German publication, Whitman's study of embryology in the leech *Clepsine* followed up on his dissertation work of a decade earlier. Whitman asked about the significance of the germ layers, but he also concentrated on the origins of those layers. In particular, did they arise from pre-existing "organ-forming germ-regions," to use Wilhelm His's term, or from particular cells? Can differentiation be traced farther back to some early existing state in the egg, that is, or does it arise later. If it arises later, does that development occur gradually because of either internal historical (phylogenetic recapitulation) or external (epigenetic) factors? This was, fundamentally, a version of that age-old traditional embryological question of whether development is essentially preformationistic or epigenetic. Yet Whitman approached it differently, by appealing to a different kind of evidence. He gave detailed descriptions of which cells do what in the course of the earliest developmental stages.[10]

As Whitman had said in his dissertation a decade earlier, the cell is like a quarry from which complicated structures will emerge later, with proper excavation. "In the fecundated egg slumbers potentially the future embryo. While we cannot say that the embryo is predelineated, we can say that it is predetermined. The 'Histogenetic sundering' of embryonic elements begins with the cleavage, and every step in the process bears a definite and invariable relation to antecedent and subsequent steps." In fact, the later differentiation is "anticipated in the early stages of cleavage, and foreshadowed even before the cleavage begins."[11]

Whitman had found His's idea of organ-forming germ regions suggestive, though he disagreed with His's conviction that differentiation comes about essentially through a mechanical unfolding of preexisting parts. Some sort of predeterminism must take place, he felt, but not strictly preformation or even prelocalization in His's sense. E. Ray Lankester's idea of "precocious segregation" seemed closer to the mark. "Though the substance of a cell may appear homogeneous under the most powerful microscope, excepting for the fine granular matter suspended in it," Lankester had written, "it is quite possible, indeed certain, that it may contain, *already formed and individualized,* various kinds of physiological molecules. The visible process of segregation is only the sequel of a differentiation already established, and not visible."[12]

Whitman agreed with Lankester and concluded that in the early developmental stages, one finds a forecast of the future organism. Thus,

> although there is scarcely anything in the external appearance of the eight-cell stage to indicate the relation of its parts to the future embryo, yet we know by what follows that an immense work has already been accomplished. All those fundamental conditions and relations implied in the terms anterior and posterior, right and left, dorsal and ventral, are now definitely established. The ground-plan of the future structure is there, and the segregation and distribution of the building material have advanced far toward completion.[13]

For example, the annelid teloblasts, those particular cleavage products that undergo rapid and frequent division and that Whitman called the "specialized centres of proliferation," always give rise to the germ bands—a phenomenon in which Wilson was particularly interested.[14]

Comparative studies, examining the origins of later body parts in different species, show that germ layers cannot always be effectively identified. Therefore, contrary to Haeckel's assumption, it remains unclear just which layers are homologous. Further, germ layers may not always give rise neatly to the same adult parts, and thus phylogenetic relations cannot easily be discovered there, as Haeckel and others had assumed they could. "When, as in the case under consideration, we find an organ arising sometimes from the ectoderm, and at other times from the mesoderm, we have to admit that there is no fixed and impassible boundary-line between these two layers; and that its association with this or that germ-layer is not an infallible guide to its morphological identity."[15]

Instead, Whitman continued, one must look to the "precise genealogy of the cells," or cell lineages. Look, that is, to the way in which every single cell moves through its divisions and gives rise to particular later germ layers and body parts. Traditional concern with evolution and homologies of germ layers continued in Whitman's work, but the focus on germ layers had given way to an emphasis on earlier, cellular changes. This is just the direction that Wilson's own work was taking when he encountered Whitman's.

There is no evidence that Wilson was exposed to Whitman's point of view prior to 1886. Whitman had been working as an assistant in zoology at the Museum of Comparative Zoology at Harvard from 1882 to the summer of 1886, and he had been productive. Yet nothing would obviously have commanded Wilson's attention until the leech work of 1886, which raised questions similar to Wilson's own. Since Wilson could well have met Whitman the year before, when both were in Boston, he may have been further inspired to look at that paper on leech development when it appeared. Even though 1886 found Wilson off in Pennsylvania and Whitman moving to the Allis Lake Laboratory in Milwaukee, Wisconsin, the two had begun a long and significant working relationship and friendship by then.[16]

Wilson found Whitman's work suggestive. It focused attention on the particular cells that give rise to each germ layer in leeches and showed remarkable similarities to the parallel patterns in Wilson's annelids. Also, the cells that give rise to the embryologically active germ band of Wilson's organism are parallel in Whitman's *Clepsine.* This suggested a close historical relationship between the two groups of organisms and, in Wilson's view, called for further careful study to establish exactly which cells are homologous with which others. Not a fundamentally new question driving the research here, but a reflection of the new focus on cells, cell fates, and cell homologies which a few researchers such as Whitman shared.

In his major opus on the earthworm, published in Whitman's new American *Journal of Morphology,* Wilson lamented the limitations imposed by the available morphological and, in particular, embryological techniques. In pursuing exact details of egg fertilization and the first cell cleavages, he found problems. In order to examine any particular ovum at any one stage, it was necessary to kill it, fix it, preserve it, slice it up, and put it under a microscope. There was no other way to look inside the otherwise opaque individual. Yet treating it in this way obviously makes it unusable for studying further developmental stages. It no longer exists as a developing organism. And each other ovum has slight differences, Wilson found, for, try as he would, some pathological variation or other would inevitably intervene. He finally decided that the only way to proceed was to compare what he found in the various different individuals which represented different developmental stages and which appeared together in a single egg capsule. Yet this gave a limited sample since not all stages were represented in any one capsule. Though necessarily incomplete, such results were nonetheless the best he could achieve, and he believed them "to be trustworthy as far as they go."[17]

The next year brought help in that crucial respect. It also brought a visit to the Marine Biological Laboratory (MBL) in Woods

Hole, Massachusetts, and important direct contact with Whitman and the community of researchers he was beginning to collect there.

Wilson reports that during the summer of 1889 he obtained abundant quantities of *Nereis* worms for the first time. Through the efforts of Hopkins graduate student Ethan Allen Andrews, who evidently spent the summer at the United States Fish Commission, Wilson was introduced to this wonderful new research material. The eggs of two species of this marine polychaetous worm that Andrews had located (*N. limbata*, Ehlers and *N. megalops*, Verrill) exhibited an "extraordinarily favorable character" for observation. "They are transparent, of comparatively large size, and they may be procured in abundance."[18] In addition, because of the particular patterns of cleavage, the embryo's axis remained visible throughout the complete cleavage and gastrulation process. For an embryologist asking the questions Wilson wanted to ask, *Nereis* were nearly perfect. The rate of progress in answering specific questions about precise origins of different germ layers and about homologies would increase radically with the improved research subject.

Other important changes occurred as well. As Wilson realized that he could follow the progress of each cell, division by division, from its earliest form through all its sequence of cleavages and on to its role in the origins of each of the later parts, he grew more confident of his results. The tone of this paper seems different from that of earlier works. Instead of sustained theoretical discussion, accompanied by much consideration of the various European disagreements and by suggestions for further work, here Wilson writes more positively of his results. He asserts confidently, for example, that "*each of the two divisions of the ventral plate may be traced back to a single cell (pro-teloblast), which is obviously homologous to a corresponding cell in the early embryo of Clepsine.*"[19]

After following the cleavage process many times in *Nereis*, he found it "extremely constant." Each division produces two cells, which follow a regular pattern as to size, shape, orientation, and future significance. This constancy, together with the ability to watch one developmental process all the way through, changed the face of embryological study of cells. Wilson could observe each developing sample, draw it, watch a few more individuals if necessary to fill in details, then polish the drawings. He, and others working in parallel on other organisms, could begin to produce the beautiful standard illustrations that represent cell development for others to see and assess. The drawings and the schematic diagrams they also began to generate provided a standardized way of capturing, representing, and labeling information on the basis of complex observations. The researchers could then use these representations to compare the results of their cell lineage studies, as these pursuits came to be called. With this solid and standardized data in hand, they could begin to achieve their goals of establishing definitive patterns of homologies and presumed ancestral relationships among organisms. Furthermore, they could begin to uncover the patterns and possible causes of embryonic development in each individual.

What sort of change was this? First, this was plainly no change in theory in any traditional philosophical sense. With time, Wilson realized that he was "going to destroy the germ-layer theory of development."[20] But that came later. Rather, what was primary was a new set of organisms and a new way of doing scientific work. This approach to embryological problems brought facts that were much more definitive and standardized across different organisms. And those definitive facts made it possible to address questions about the proximate, mechanical causes of embryonic development. One no longer had to appeal to some recapitulation of historical past to explain just how each developmental stage in the current organism gives rise to the next in a regular, constant way. Instead, embryological study of the detailed morphological changes during any individual's development could uncover the causes of its own development, here and now. Or so it seemed. The regularities that Wilson found showed that the causes of development and differentiation lie largely inside the egg in some way. Or so it seemed. For the moment, new methods and new organisms brought more solidly established facts and greater confidence about the regularity of phenomena and the validity of observations.

These recognitions in turn set new standards of evidence for embryological work. No longer was it acceptable to Wilson and his growing American cohort to appeal to germ layer homologies as an adequate basis of phylogenies. Instead, one had to look to the more specific units, the cell, and to establish with empirical observation the developmental similarities and differences. Further, the researcher should ask what caused the individual developmental changes—what sorts of internal mechanical as well as historical, inherited causes. This was not a change in theory. Nor was it simply a psychological change, wherein the participants became more confident in their work. Neither was it primarily a change in instruments or techniques. Rather, there was a change in the organisms used, and at root there was a subtle but profound change in the epistemic setting within which any embryological theory was to be judged. Now every such theory had to face higher standards of evidence based on definitive, demonstrable, reproducible empirically observed facts, because it had become possible to achieve such results. Haeckel's guesswork and reliance on incomplete data or conjecture about past causes and relationships was no longer sufficient in the face of a more productive approach. The epistemological change reflected here was to have far-reaching consequences.

At first, Wilson probably did not realize the full potential of his new work. His tone changed, and he expressed more confidence in his results, but he certainly did not announce a new biology or any radical change. In fact, he continued with his traditional morphological questions, asking not "simply *what is?*" but also "*how came it to be?*" These are the dual questions of morphology, Wilson pointed out, pursued as morphologists strive to uncover fundamental genealogical relationships.[21] They ask about homologies in form and in pattern. As they do so, they seek to understand the significance of such recurrent phenomena as concrescence (the building along the median line of the embryonic body) and metamerism (the segmented structure of body parts). They simply approach their questions differently and expect something different at the end.

Various theories had arisen to answer each of the morphologist's concerns, Wilson acknowledged. So far, what existed was a set of working hypotheses, some better than the others, but none sufficiently well founded as yet. "To those whose interest in science lies in the consideration of its positive results only, the outcome of this discussion will doubtless seem rather unsatisfactory." Yet it is precisely the unsolved problems and the unanswered questions that most lure the curious scientific researcher on. Wilson then suggested that embryology could help to choose among alternative interpretations. For example, similarities in the earliest developmental stages could reveal ancestral relationships that are otherwise concealed by later adaptations. That the basic problems of morphology would "ultimately be solved, there can be little doubt; but the present need is for new facts, not for new theories. When the facts are forthcoming, the theories will take care of themselves."[22] New facts, definite facts, those facts that would allow the theories to emerge naturally out of them and to "take care of themselves"—that is what Wilson and his contemporaries sought. And now they had a new way to succeed, with cell lineage study of living embryos.

In many important ways, despite appearances that have led historians to conclude the contrary, this work was both extraordinary and pioneering. It fell squarely within the morphological tradition but also helped to transform that tradition. Cell lineage studies became the standard at the MBL and for other American embryologists as well.

The Marine Biological Laboratory

Wilson attended the second summer session of the MBL, in 1889, and thereafter spent at least part of most summers there. In 1890, he was made a trustee of the laboratory, and he supported the facility in a number of ways that helped make it a stronger and better place. Many Bryn Mawr students also went along with Wilson through the years, since the MBL from its inception encouraged participation by both female and male students. Each of our other Hopkins protagonists joined Wilson at this Massachusetts marine laboratory as well, so that it played an important role in the development of their careers as well as of American biology more generally.

The idea of a seaside laboratory clearly did not originate with the MBL. Aside from Naples, Louis Agassiz's school on Penikese Island and Spencer Fullerton Baird's laboratory in Woods Hole for the United States Fish Commission both inspired its creation. Alpheus Hyatt then led the way in translating the ideals of both those efforts and his own Annisquam school into action in the form of the MBL.[23]

After a trip to South America and the Galapagos Islands, Louis Agassiz had carried out his ambition to open a summer school in natural history. Financial backing came in 1873 from a wealthy New York merchant, John Anderson, who read of Agassiz's appeal for funds in the newspapers and was sufficiently impressed to offer a site on his island off Cape Cod, with a dwelling house and barn, as well as $50,000 for equipment and support.[24] Agassiz accepted immediately. He called the undertaking the Anderson School of Natural History and immediately began constructing dormitories and cleaning the barn, which was destined to become the classroom and lecture hall. Barely ahead of the announced opening, all arrangements were finally completed, and the experiment in summer practical education for American teachers began. Agassiz acted as center of the show but also brought a steady stream of leading scientists to supplement with experienced teaching the practical exposure to nature on which the students' days centered.

While focusing on natural history, Agassiz's school incorporated what he saw as new directions in research. Experimentation as well as descriptive work, physiology as well as morphology, found places in the summer program. Agassiz wrote with enthusiasm in a highly instructive letter of 1873 to the science editor of the *New York Tribune* that "Natural History is no longer a mere descriptive science today. It aims at improving knowledge by experiment as well as by observation. And from the first day I have known the intentions of

Mr. Anderson, I have wished to combine physical and chemical experiments with the instruction and the work of research to be carried on at Penikese. That physiological experiments lay at the very foundation of an exhaustive study of Zoology is as plain as the simplest truth."[25] His hopes for physiology and for experimentation did not, in fact, reach fruition the first year. But the emphasis that his students, especially William Keith Brooks and Whitman, continued to place on both attests to the more general success of Agassiz's ideas.

One of Agassiz's problems the first year came from oversubscription to his school. After considerable weeding of applicants and writing to some whom he had initially accepted to suggest that they perhaps should not attend after all, Agassiz settled on a group of forty-four students, twenty-eight men and sixteen women. According to one contemporary reporter, the average age of the students was roughly twenty-five to thirty. The women were very "schoolma'amy" in appearance and "the gentlemen are not a whit behind."[26] The enthusiastic, if unglamorous, group gathered at the steamer landing and cruised together on the prescribed day to Penikese. After a stirring introductory address by Agassiz, the group ate dinner and settled in for a summer of work.

Each student received a table assignment, an aquarium, collecting jars, and instruction to "study nature, not books."[27] Morning lectures and an hour or so of dissection gave way to afternoon freedom to roam about the island, explore, collect, and study—anything but to loaf. Asked later what he thought of Agassiz's method, Whitman responded that each student "did not think so much ot if [sic] at first, but as time goes on he thinks more and more of it. We do the work for the student. What we should do is to set him a problem and let him work it out."[28] This is what Agassiz tried to do, and both Brooks and Whitman took his lessons to heart. One excited reporter described the students' general attitude toward this regimen as a "spirit of indomitable work—humble, conscientious, methodical—which, in so far as it is possible is bound to overcome difficulties."[29] There were minor disagreements. One Sunday, Agassiz wanted to organize a steamboat excursion, but only Brooks, Whitman, and two others were willing to undertake even such a tempting adventure on the Sabbath.[30]

What Agassiz offered was an opportunity for group learning, with each student pursuing individual investigation but guided by a structured set of lectures and suggestions for study. Elementary instruction rather than independent research dominated. The second year, a number of the students, including Whitman, returned, partly to pursue their independent studies.[31] Had the school survived, it might well have gone on to combine more advanced research and instruction, as the MBL eventually did. Unfortunately, Agassiz died in December 1873, and his wife, Elizabeth, and son Alexander managed to keep the school going for only the next summer, which had already been advertised.[32] Afterward, Alexander kept a collection and invited some students to his Newport laboratory to work in the summers, and he also sustained a table at the Fish Commission and subscribed to a table at Naples for American students, but the Anderson School itself did not continue.[33] The inspiration did persist, especially through Alpheus Hyatt who, Edwin Grant Conklin suggested, "was the real father of the MBL."[34]

As curator of the Boston Society of Natural History, Hyatt had believed in the value of establishing a seaside laboratory in connection with the society. He had opened part of his house in Annisquam, Massachusetts, to several students, thereby founding the marine biology school at Annisquam. With the support of the Woman's Education Association, Hyatt had then opened the Annisquam Laboratory in a separate building in 1881. Informal instruction and individual investigation went alongside each other in the laboratory. From 1881 through 1886, groups varying from ten to twenty-six men and women attended, including Thomas Hunt Morgan for one summer. The low-budget school had proven its value sufficiently that members of the society and the Woman's Education Association determined to found a more permanent and independent marine biological laboratory. They appointed a board of trustees and underwent incorporation in 1888, the beginning of the MBL.

They chose Woods Hole for the same reasons that Spencer Fullerton Baird had: the relatively pure water and the accompanying exceptionally varied sea life available. The absence of any large freshwater streams in the area maintains a nearly constant salt concentration despite tidal and seasonal changes. Baird recognized this unusual condition as a particular advantage for "biological research of every kind and description."[35] In addition, the tidal and current conditions support a great diversity of organisms.

After identifying the advantageous setting, Baird, the commissioner of fisheries, had established the first scientific research laboratory in Woods Hole in 1875, in the form of a permanent station of the United States Fish Commission.[36] Oriented toward improving commercial fishing, the laboratory naturally encouraged study of behavior, breeding patterns, and embryology of marine organisms. Baird worked hard to make the commission a research center, and he tried to attract investigators to use the subscribed research tables. Brooks worked there several years, for example, and took along or sent his students from Hopkins to carry out research. Yet, for various reasons, including President Grover Cleveland's opposition to the sort of independent research facility Baird had envisioned, the Fish Commission never succeeded as a vital biological center. It offered no instruction or organized research for the inexperienced and so did not appeal, as the schools did, to teachers seeking further summer training.[37] Yet Baird had shown that Woods Hole was a fine location for a marine station. And the Fish Commission was a great boon to the MBL in its first years, providing specimens and equipment as well as conversation with a larger group of researchers. In their early meetings, the MBL trustees expressed their desire to locate in Woods Hole and to cooperate with the Fish Commission.[38]

With Hyatt unanimously elected as first president, the MBL board also included others who had been at Penikese or Annisquam and wished to pass on the ideals embodied there.[39] When the question of directorship arose, the group at first considered Samuel Clarke, Wilson's cousin from Williams College. He declined, politely but resolutely. "Let us consider some of the points," Clarke wrote:

> The offer is, for me to pay my expenses to Wood's Hole, to continue paying my expenses through the summer and my return expenses home; also to take all the burden of organizing the Laboratory, adjusting it to the conditions there and in Boston; the receiving of and answering of applications, the reception and settling of those admitted; the labor of giving lectures, arranging courses and providing material; and the establishment and preservation of a cordial feeling between the U.S. Fish Company and that of the Laboratory—as well as between all those in the Laboratory.[40]

With no obvious remuneration. No thank you.

So one member recommended Brooks because of his "special qualifications," presumably including his experience with the Chesapeake Zoological Laboratory. Also, Brooks might possibly come without pay. Further, the Johns Hopkins University might, after the Naples model, "take a table which would help us financially." After some discussion, the trustees decided to ask Brooks to serve as director, with an assistant director under him. Meeting in April 1888, the trustees hoped to secure his services for that summer. Since Brooks did sometimes spend his summer in Woods Hole, this was not an unreasonable expectation. By 28 April, Brooks had been interviewed and had "looked favorably" on becoming director. Only by 12 May did the trustees learn from Brooks that he did not intend to accept after all, largely because he had decided that the one Fish Commission laboratory should be enough for Woods Hole. A letter went immediately to the only other American who had directed a similar laboratory, albeit an inland lake version, Charles Otis Whitman.[41] He accepted, also immediately, and the trustees quickly completed plans for the imminent summer session.

Botanical and zoological studies should coexist, they said, thereby making the laboratory a biological rather than a zoological station, as Naples had initially been. As Whitman wrote: "It is not a zoölogical, nor a botanical, nor a physiological station that will best meet our need, but a biological station large enough to include all three sciences of life. In my opinion—and I believe I express the conviction of the leading naturalists of the country—such a station is to-day the greatest need of American biology."[42]

Independent investigation and instruction should also coexist, bringing together the best of the Fish Commission, Penikese, and Annisquam.[43] Not all agreed with the latter goal, but, as Whitman reported later, "on the basis of ten years' experience, and a previous intimate acquaintance with both types [laboratories based on either instruction or research], I do not hesitate to say that I am fully converted to the type which links instruction with investigation."[44] Unlike either Penikese or Annisquam, the MBL should also attract a strong national support system, eventually gaining allegiance beyond that of the local Boston group, the trustees felt. To achieve this, the model of offering research tables for subscription to universities and other groups served well.

In 1888, the trustees distributed a circular announcing the opening of the MBL in Woods Hole, with Professor Whitman as director. Fortunately, as the second MBL director, Frank Lillie, said later, "In Professor Whitman the trustees had found a man not only fitted to carry out their purposes but possessing imagination adequate to transform their shadowy ideas, the zeal and determination required to give them form and substance, and the courage to face whatever difficulties might arise."[45]

The laboratory opened as scheduled in 1888—almost. A professor from Mount Holyoke who became a longtime supporter of the MBL, Cornelia Clapp, attended that first session. When she arrived on 10 July, carpenters still labored on the new building, and Clapp discovered "that Dr. Whitman had not arrived; that he was delayed by illness in his family; that the equipment for the building was still on the road, probably side-tracked somewhere; that it might be some time before the laboratory opened; that no arrangements had been made for boarding, and that I must look out for myself."[46]

She found a place and settled in, as did the other fourteen attendees. Eventually a "mess" opened to provide food for everyone, and the laboratory building became available for use. Although difficulties remained, that first MBL group coped.

Whitman determined that Clapp had had sufficient background to join the other six investigators, so she said that her "next question was, what subject should I investigate?" Whitman suggested the patterns of development of the lateral line system of toadfish and urged that the investigator's business "was truly to *see* and to *get the results.*" Of that first somewhat disorganized and low-key year, Clapp reported that "the atmosphere of that laboratory was an inspiration; the days were peaceful and quiet; there were no lectures nor anything else to distract the attention from the work in hand."[47] Subsequent years brought expansion and increased activity, including lectures, but the goals and flavor remained the same.

Whitman's Program for Research

A paper of Whitman's from 1887 underlines one focus of interest which he encouraged at the MBL from the first. Examining maturation (or preparation of the egg to receive the "spermatic element") and fecundation (or "those attending changes in the protoplasm which form the concluding steps in the premorphological organization of the egg"), Whitman suggested that the majority had by then come to regard the nucleus as the "primum mobile" of development.[48] The movement of the two pronuclei was widely regarded as evidence that the nucleus controls developmental changes. But no, said Whitman. The pronuclei move through the cytoplasm. Indeed, cytoplasmic forces direct the nuclear movements, as seen with the formation of astral lines through the cytoplasm. The egg cytoplasm exerts just as much influence as the nucleus, even at these earliest stages. Cytokinesis (movement of the cytoplasm) and kinokinesis (movement of the nucleus) necessarily work together, cooperatively.

With this study, Whitman had embarked on an effort to establish definitely not only that the embryo experiences early organization, hence that it is preorganized in some way, but also that both the cytoplasm and the nucleus exert cooperative influences, hence that neither a simple cytoplasmic preformation nor nuclear predestination occurs. This work brought him directly into the revived debates over preformation and epigenesis as well as disputes over the significance of the cell theory.[49]

In shaping his answer, Whitman made it clear that he disliked what was later labeled *reductionism,* which he felt could delude the researcher. "The resolution of organs into tissues, tissues into cells, and cells into smaller units, does not disclose the secret of life, but it does extend our knowledge of organic mechanism. It is strange that experienced and acute biologists should so far misunderstand the spirit and language of cytological research as to imagine that anyone expects to explain life and get rid of its mysteries 'by imagining a living creature indefinitely divided into minute living parts.'"[50] Biologists, Whitman lamented further, have become too enraptured with that living part, the cell. They draw a line around it. They demand, for example, that structures not be considered homologous if they do not have the identical cellular arrangements. This is wrong, and "so far as homology is concerned, the existence of cells may be ignored."[51] In fact, cell divisions may hold only secondary importance, since the whole organism is the continuous individual of interest. The individual has formative forces acting within the organism, acting on all the cells. Differentiations in regions within the embryo occur independently of cell formation, Whitman believed.

Thus, cleavages do not determine differentiation or direct development. Rather, the cell divisions reflect a deeper predetermined pattern of development. Cell divisions and cell lineages are both the manifestations of this underlying pattern but do not themselves *cause* differentiation. Instead, Whitman suggested, biologists "will find the secret of organization, growth, development, not in cell-formation, but in those ultimate elements of living matter, for which *idiosomes* seems to me an appropriate name." What those idiosomes are and how they act remains the "problem of problems." As a starting point, they are bearers of heredity, the real builders of the organism, and not limited by cell boundaries. Whitman did not simply identify them with chromosomes but felt they transcended cell boundaries—somehow. For Whitman there exists "a definite organization in the egg, prior to any cell-formation."[52] This opened questions about the significance of cell divisions for others to address.

Whitman's suggestions and his deepening concern with traditional problems of preformation and epigenesis caught the MBL's attention, though not everyone agreed with his conclusions. Each summer starting in 1890, the growing group gathered for evening lectures to address the great major unsolved problems of the day. Some of these lectures then appeared in the *Biological Lectures,* intended as a more accessible publication to supplement the professional *Journal of Morphology* for lengthy major articles and the *Zoological Bulletin* (after two years renamed *Biological Bulletin*) for shorter research items. The lectures guaranteed interaction among the diverse members of the MBL community and sparked lively discussions, that, in turn, helped to forge a vital community with shared interests. Questions about exactly how and why development occurs dominated the discussion through the 1890s.[53]

Foremost among the reigning questions was whether the unfertilized egg is isotropic, as some physiologists felt, or preorganized or predetermined in some way, as Whitman believed. If it is isotropic, then epigenetic development must occur, presumably directed by factors external to the egg itself. Advocates of this view sought data by controlling external factors, for example. Alternatively, others held to some sort of internal directive factors, whether "idiosomes" or cytoplasmic factors. They generally argued for some version of preformationism. Whitman purposefully encouraged the coexistence of competing epigenetic and predeterminist views.

Whitman also sought to establish physiology and morphology on equal terms. The two must work hand in hand, ideally with specialists doing research in each, cooperatively, together comparing results and ideas. This happy community is what Whitman sought for the MBL. In 1892, the year he introduced physiology to the laboratory, he wrote, "As Morphology deals with the ground-structure on which physiology operates, it naturally takes the place of pioneer and guide; but if permitted to wander too far in advance it soon finds itself entangled with physiological problems with which it is not prepared to cope, and its efforts to release itself often end in sterile speculation." As morphology moves to embrace a wider range of embryological problems, it seeks to consider physiological factors that cause development to proceed as it does. Morphology seeks to achieve such explanations. But without sufficient facts or appropriate methods, it must appeal to speculative theorizing. Physiology provides the methods and can produce solid facts. In addition, Whitman suggested on various occasions, physiology has solid standards of evidence closely tied to empirical observations and is unwilling to speculate excessively. Therefore, Whitman agreed with the founders of the Johns Hopkins University, the two should ideally work together:

> The workers on both sides should therefore advance abreast in hand to hand contact. It is only in such reciprocally helpful relations that specialists can attain the highest possible individual development. . . .
>
> The association of morphological and physiological research enlarges the field of vision on both sides, reduces the chances of useless labor, corrects false notions, stimulates inquiry, converts half views into whole views, and withal secures mutual respect.[54]

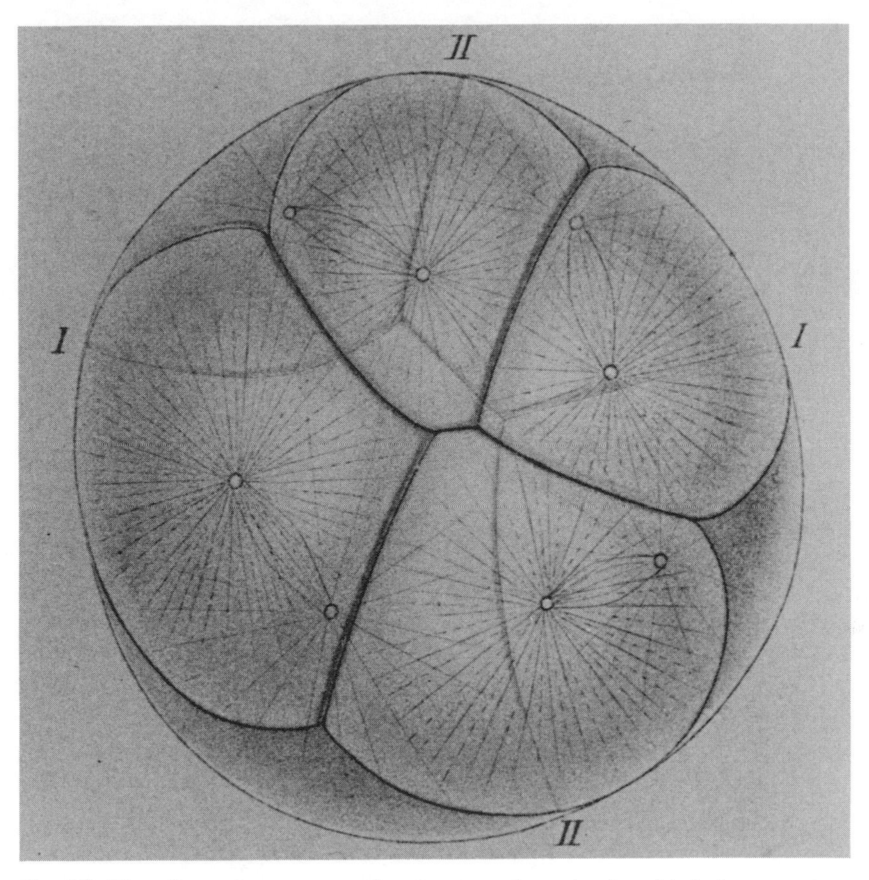

Fig. 10. View from the upper pole of an egg later in the third cleavage, to show the position of the spindles. Wilson, "The Cell-Lineage of *Nereis*" (1892), pl. XIII, fig. 8.

Whitman was not fully successful in putting this ideal into action, but such a dual emphasis would have seemed perfectly normal to the many Hopkins graduates who joined the MBL community. Their own experience in graduate school, which combined physiological and morphological studies, gave them a similar perspective. The two traditions remained closely allied rather more than clearly distinct, as they had in Germany.

The discussions that took place in the laboratory, at the beach, at the communal "Mess," and after the evening lectures often became quite heated. They forced the individuals involved to formulate their own views, which often changed through the 1890s in response to the continual influx of new research results, approaches,

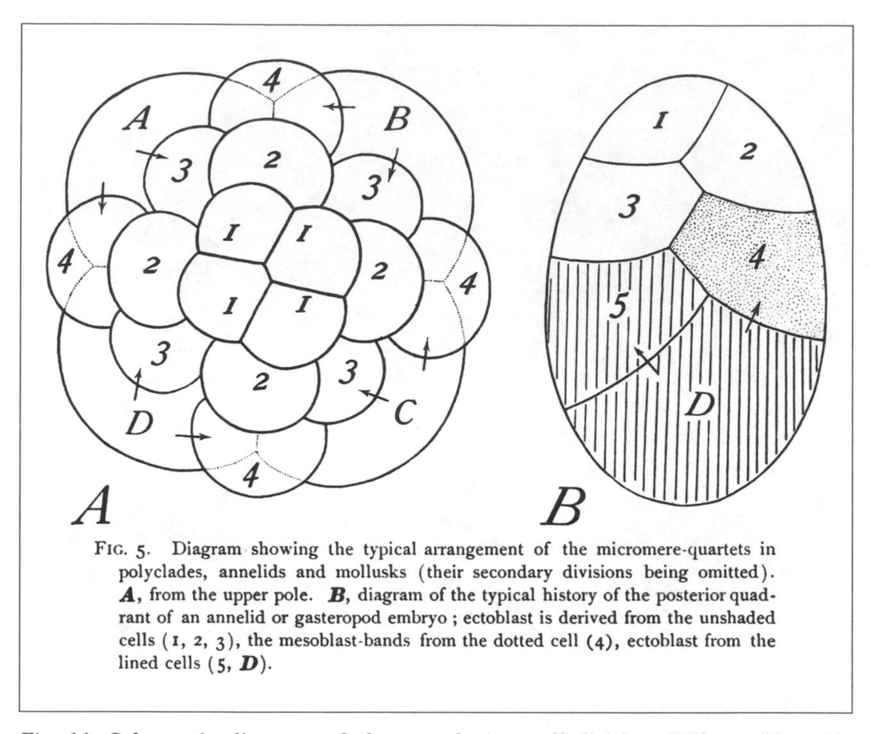

FIG. 5. Diagram showing the typical arrangement of the micromere-quartets in polyclades, annelids and mollusks (their secondary divisions being omitted). ***A***, from the upper pole. ***B***, diagram of the typical history of the posterior quadrant of an annelid or gasteropod embryo; ectoblast is derived from the unshaded cells (1, 2, 3), the mesoblast-bands from the dotted cell (4), ectoblast from the lined cells (5, ***D***).

Fig. 11. Schematic diagram of changes during cell division. Wilson, "Considerations on Cell-Lineage and Ancestral Reminiscence, based on a Reexamination of Some Points in the Early Development of Annelids and Polyclades," *Annals of the New York Academy of Sciences* 11 (1898): 1–27, fig. 5.

and theories. Reports of researches from Europe excited special interest, but at first the MBL group remained unconvinced by the new European ideas (see chapter 5). They found them too speculative and ungrounded in reliable empirical fact. Instead, the physiological explorations of external influences on development led by Jacques Loeb and others at the MBL itself attracted their attention in one direction. And the cell lineage work inspired by Whitman, Wilson, and others also held sway.[55]

Cell Lineage at the MBL

When Wilson first arrived at the MBL in 1889, he was worrying through just the sorts of problems Whitman had been addressing, and comparisons of details in their respective organisms revealed many similarities. It seemed that further careful study of the patterns of cell lineage was worthwhile, including recording precise actions

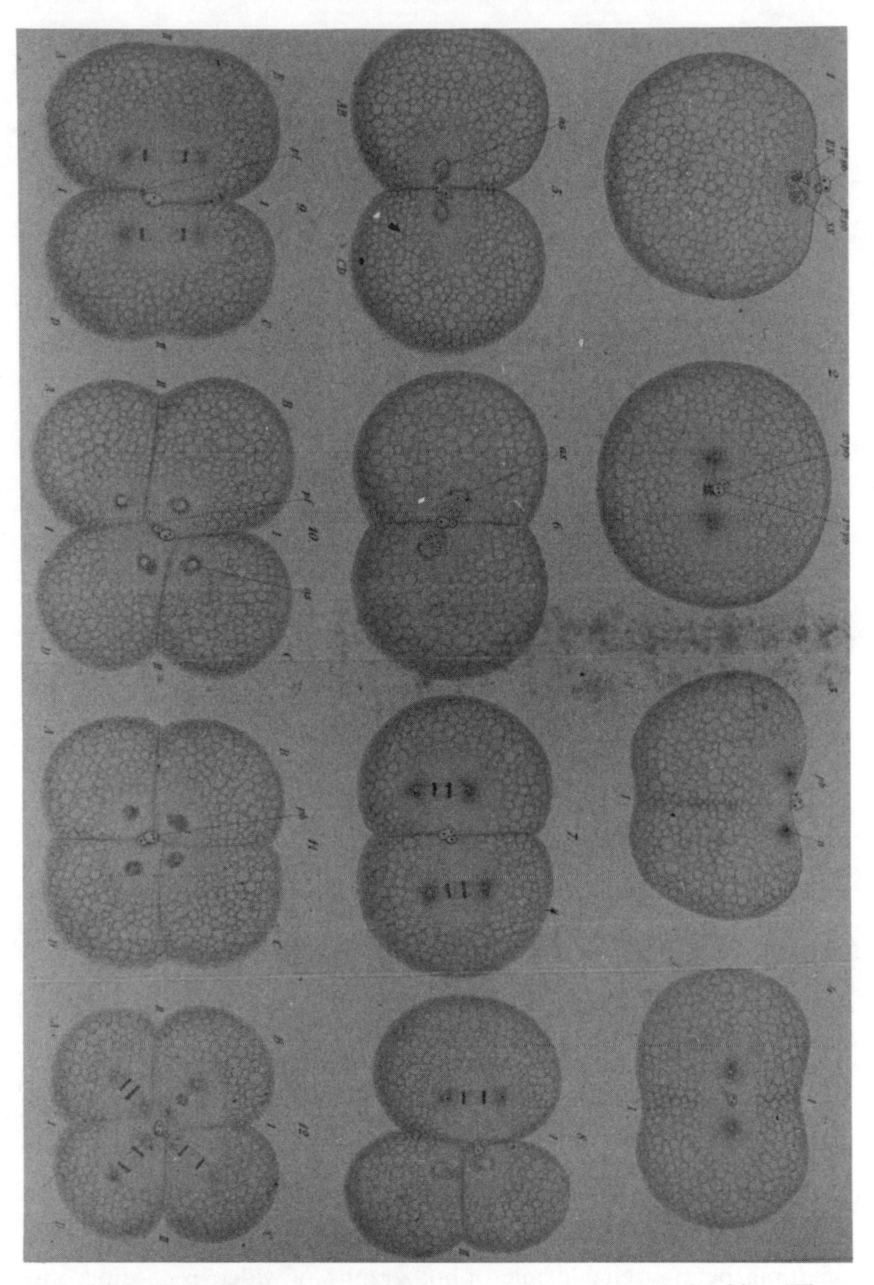

Fig. 12. The egg of *Crepidula fornicata,* to the third cleavage. Conklin, "The Embryology of Crepidula" (1897), pl. I.

of each nucleus and cytoplasm. Wilson examined the polychaete worm *Nereis*, publishing his results in a classic paper of 1892. In addition to recognizing its great advantages as a living research material, Wilson enjoyed collecting the animals. The beautiful red worms move gracefully through the water. They also carry out marvelous mating dances in the early summer moonlight, and Wilson could simulate that light and stimulate egg release by lying on the collecting dock of the MBL's Eel Pond and extending his lantern over the water. A simple net to grab the active worms, then a rush back to the laboratory to start looking.[56] Years later, Wilson's wife said to a young student as he began his own work on *Nereis*, "Oh, my husband loved that egg." But, she reported, that meant that, even the first year they were married, he was spending his nights out collecting and in the laboratory watching his worms. "That," she admitted, "was very distressing."[57] Yet it had to be done like that since Wilson knew of no way to induce the worms to release eggs and sperm during the more convenient daylight hours.

What Wilson did with his eggs was remarkable. He made meticulous observations of the living, dividing cells. In each case, he recorded exactly the shape and size of the cells and the lineages of each cell, as one gave way to two, and so on. Any irregularities of cellular material, such as areas of yolk material, were recorded. He also noted and recorded in detail other changes, such as the nuclear changes. Exactly what were the chromosomes doing, and the spindle fibers, and other parts of the nuclear apparatus? At what time did each change occur, and what else happened at the same time or shortly thereafter that might be causally linked?

Wilson observed the process of cell division in meticulous detail to the twelfth division and as well as he could thereafter, with little time for rest, through many individual specimens. Over and over, he watched, recording the sequence as accurately and in as much detail as possible. Aware of the limitations of human powers of observation, he repeated the sequence again and also compared his results with preserved material. The latter essentially each "froze" a single moment into permanence and could be cut open and stained to extend the information available within living material. Such double checking proved particularly valuable for those hardest-to-see innermost cell divisions, of course. The meticulous hand drawings, in this time before reliable microphotography or video recording, give testimony to Wilson's care and success.

Wilson's first paper on *Nereis* reflects concerns similar to those of his earlier studies, but with a reorientation. As he had pointed out, Darwin's theories had focused on two basic questions for the biologist, who must ask both "what is?" and "how came it to be?"[58] This second question necessarily depended on the historical, or evolutionary, past of the organism. Yet he had begun to realize that the only immediate way to investigate that question was by following patterns and processes of change in current organisms. Comparative studies would then presumably illuminate evolutionary relationships. Knowing the relationships would not *explain* the development or how it came to be. For by now he had rejected Haeckel's biogenetic law as having any explanatory power.

Rather than a primary concern with answering Darwin's second question, then, Wilson moved to seek an understanding of what steps the particular organism at hand had undergone to develop as it did, and how development of various apparently related organisms compared. Wilson did not yet really ask "how" in the causal, deterministic sense of Ernst Mayr's proximate causation. His answers did not yet give mechanical accounts of development, that is. With his cell lineage work, he instead focused on describing exactly what patterns of cell division and cell fates occur and what those particular patterns mean later rather than on the processes effecting change. He thereby attacked one of the fundamental problems of the day which commanded central attention at the MBL.[59] The work was more morphological than physiological in its focus and methods. It also showed a primary concern with development of the individual in response to its conditions rather than with heredity.

Wilson continued to work on *Nereis* for the next year, publishing his masterpiece in 1892. The manuscript was submitted from Munich in December 1891, but a number of references throughout the paper indicate that he had essentially completed the research in Woods Hole before leaving in the fall of that year. In addition, his observations and comparisons with Conklin's results clearly occurred at Woods Hole. Probably he did no more than complete the final drawings, add a brief reference to Theodor Boveri's work, add a note in the appendix, and finalize the paper while in Europe. Certainly, the paper gives some indication that he had heard of German embryologist Wilhelm Roux's ideas about mosaic development, according to which each cell division produces a qualitatively different piece of the larger mosaiclike organism (see chapter 5). Yet the reference is only minor and does not suggest that Wilson had really considered Roux's ideas of methods seriously at this point.

In this classic paper, Wilson definitively moved beyond the germ layer doctrine, which suggested that the germ layers are the real starting point for development.[60] Thus, he showed that the embryologist cannot take the germ layers as the homologous starting point for investigation. More significantly, he also pursued the themes of his earlier *Nereis* work in advocating the discovery of facts and in exploring the mechanical causes of development and differentiation. He had compelling successes on both counts.

He had undertaken his work, he said, to resolve "certain perplexing problems" "relating to the formation of the mesoblast" layer in polychaete worms. *Nereis* proved an ideal material, and the fundamental questions that had produced a proliferation of alternative theories had been resolved. Further than that, he had also established that the developmental process entailed a "strictly ordered and predetermined series of events." The second cleavage plane lies along the median plane of the adult body, which suggests such a predetermination. But there is no simple preformation since the micromeres go through a very complex series of spiral and direct cleavages before arriving at that adult form. The strict epigenesists' suggestion fails: environmental conditions external to the egg itself do not direct the cleavage patterns and therefore the formation of the adult. Rather, with the rotation of the micromeres during spiral cleavage, for example, the internal nuclear spindles already exhibit a spiral pattern before the cells divide. External factors could not have caused such a nuclear spiral, Wilson concluded.[61] Instead, the primary direct cause of rotation and of other divisions must lie internally, within the cells.

Whatever the primary cause, the cells divide according to a strict and regular pattern. That series of cell divisions, in turn, causes the embryo and adult to assume the form they have because of the way the material is divided into the cellular units. The underlying cause of the particular pattern of cleavage is therefore determined by internal, mechanical conditions and is not predetermined in the usual strict sense. Cells have to divide in certain ways once they have undergone the particular patterns of past cleavages. Because of the unequal cleavages, some areas of the egg become heavier and some lighter. This causes particular cell divisions to be accelerated in order to produce balance in the embryo. Ultimately, because each division mechanically directs the next, the cause of the entire pattern lies in the egg and, before that, in the parent cells.[62] Nonetheless, the pattern of development does not result from morphological prearrangements alone. Explaining development depends on understanding the process of cell division and therefore must call on internal physiological as well as morphological factors.

Here lay an exciting research program for the group of young Americans gathering at the MBL. The individuals in the community could each tackle a different organism while asking the same questions and using the same techniques and general approach. They could therefore make progress in their research. They could gather definitive and reliable facts about embryonic development and cell differentiation. They could also assess cell homologies and begin thereby to attack fundamental morphological problems of phylogenetic relationships as well as problems of ontogenetic development. This new embryological approach held out great promise. That appeal must have helped to attract followers.

It also brought with it a shift in approach. In seeking solid facts about development, the embryologist would appeal to physiological or process considerations as well as to details of structural patterns. An epistemological shift occurred here since the researcher must adopt an expanded set of methods and approaches in order to gain knowledge in answering the basic questions of development. The basic questions shifted as well so that what counted as "knowledge" came to involve proximate explanations of causes of development. Once again, this was not a radical new approach but a shift of emphasis. The morphological embryologist such as Wilson followed the implications of his Hopkins training and moved toward what had been the middle ground between the morphological and physiological traditions. The two overlapped, as they had for the not so pure morphologists in earlier decades. Each tradition also began to undergo transformation.

We see this move in Wilson's paper on heliotropisms as well. Undoubtedly stimulated by Whitman's increasing interest in animal behavior and intelligence, Wilson took on the question of why hydra undergo movements toward the light. Submitted as his last publication from Bryn Mawr, in the spring of 1891, the paper concluded that these organisms give the impression of consciousness or design in their movements toward light. They might seem to be maximizing their warmth instead, but experiments demonstrated otherwise. In fact, they are only moving into a position so as to maximize their food supply, which, in turn, increases reproduction.[63] Mechanical factors and physiological needs direct behavior, Wilson determined in this paper presented to the recently formed American Morphological Society. He agreed with the views about behavior that Whitman came increasingly to articulate in the course of the 1890s.[64] Physiology, Whitman said, must work closely with morphology, and together they can explain apparently purposive or conscious action. Whitman, as the society's organizer, may well have asked Wilson to write such a general-interest paper to respond to provocative claims by German-turned-American physiologist Jacques Loeb. Yet this side trip into behavior studies remained just that for Wilson. He continued to concentrate on *Nereis*, cell lineage, and embryology.

As Wilson spent his summers working through the *Nereis* opus, he undoubtedly discussed his work with Whitman. Whitman played something of a father role, encouraging and directing but remaining largely in the background—as one student fondly put it, Whitman "brooded" over the cell lineage work.[65] Wilson then acted as the older brother in 1891 when Conklin, a fellow Hopkins graduate student, arrived and began his own cell lineage work.

Other Cell Lineagists

Conklin chose a less beautiful but also intrinsically interesting subject with the slipper snail *Crepidula*. In *Crepidula fornicata*, for example, a number of snails pile up, sticking together, the small males on top and the large females on the bottom. As they grow, males eventually mature into females. Since the molluscs had received relatively little attention, Conklin felt that this particular gastropod would provide a valuable subject for his dissertation work. He began study in 1890 while a graduate student at the United States Fish Commission. Even the next summer, while Morgan, Ross Granville Harrison, and thirteen others went off to Jamaica for the Chesapeake Zoological Laboratory session, Conklin returned to Woods Hole. Unlike the others, he had a new wife, who had herself been a student at the MBL in 1889, and Woods Hole was a better place for them both to spend the summers and carry out graduate research.

Conklin reported that he found his dissertation subject on his own, without any real help from his advisor, Brooks. In one of his favorite stories, Conklin recorded that Brooks had sent him to the Fish Commission table to study siphonophores. Unfortunately, there were none in Woods Hole. Then Brooks had said to him, "Conklin, you will think that I have neglected you, but I have done it on purpose for I believe so fully in the law of natural selection that I think it is a kindness to all my students to have them find out whether they have it in themselves to go ahead with their own research work without outside assistance."[66] So Conklin decided that he should examine the very early stages of *Crepidula* development, including cell divisions. Brooks, who believed that the germ layers are the real starting point for development, then responded that he did not see the point of studying this "mere duplication of parts." The Hopkins morphologist further doubted that Conklin could ever get his dissertation published, as was then required for the degree, because Morgan's recent lengthy paper on *Pycnogonids* had virtually bankrupted the *Johns Hopkins University Studies.* Conklin nonetheless determined to continue his work and had his revenge when Brooks later tried to carry out his own cell lineage studies on the very uncooperative oyster.[67]

The next summer, 1891, Conklin, returned to the Fish Commission. One Sunday morning, Wilson came across the street from the MBL and said that he had heard of Conklin's work and hoped that they could compare results. Conklin latter recalled: "Then began one of the most interesting and exciting experiences of my life. We compared our drawings, his of Nereis and mine of Crepidula—an annelid and a gasteropod. The fundamental resemblances were amazing. Neither of us had ever dreamed that we would find homologies in the early development of two such distinct forms as gasteropods and annelids."[68] They did find many clear homologies between the two organisms in the way that cells divided and what parts of the body the cells later became. This convinced them both beyond doubt that early organization and cell division *must* have at least some morphological and developmental significance in directing later developmental patterns. Even if Whitman was right that cell division was a mere "histological sundering" of material directed by some behind-the-scenes idiosomic instructions, establishing the regularities of pattern would nevertheless uncover many facts about development.

Conklin reported that, when they discovered the recurring similarities in their two organisms, "Wilson was as excited by those results as I was and he reported this to Whitman. Whitman at once sent for me to come over to see him in the office." Of course, Conklin went right away. Whitman asked if he could publish Conklin's paper in his *Journal of Morphology,* thereby offering Conklin salvation for his dissertation. When it became clear that Whitman would demand quality color illustrations and that the cost of producing the lengthy article would run up to $2,000, Conklin feared that Whitman would withdraw the offer. When he expressed his concerns, Whitman characteristically responded that he saw no problem. "After all, what is money for?"[69] Eventually, after years of meticulous work, the masterpiece appeared in 1897 as a 226-page volume, with nine beautiful plates.

The decision to publish also augmented Whitman's influence over the growing cell lineage field, for Conklin revised his paper in light of his exchange with Wilson and subsequent active discussion with the growing number of cell lineagists at the MBL. It was then also that Conklin promised Whitman that he would join the MBL staff in 1892, as instructor. Conklin returned to the MBL nearly every year thereafter, joining the board of trustees in 1896 and remaining intensely loyal to the institution and to its director. He wrote of his experiences, "I feel that I owe more to the men I have met here and to the inspiration and facilities which I have received here than to any other factors in my professional life."[70]

Conklin's first short reports of his work appeared in 1891 and 1892.[71] He declared that he had developed a new method for determining the relation of the first cleavage furrow to the axis of the future embryo. This subject had attracted some considerable interest in Germany, but Conklin did not refer to that discussion. Instead, he emphasized his new method and results, which made up the subject of his second paper. Details about the morphological changes in cleavage furrows and about actions of the nuclear material both received attention in his diagrams, though the narrative stressed changes in cells. Clearly, in this preliminary work, Conklin was most enthusiastic about his new method. With careful observations of the exact points where each cleavage furrow appears and the way the cleavage line forms, he could determine the relation of the cleavages to the later axis of the embryo. As he went on through the next several years, he continued to use the basic techniques of cell lineage work: the use of both surface observations and sectioning to gain visual access to all details of cell divisions, accompanied by meticulous descriptions and drawings. While Wilson moved on to adopt other methods and approaches to the developing embryo, Conklin continued for most of his life to pursue cell lineage work and its implications.

Conklin summarized the significance of the great rush to cell lineage work at the MBL, explaining that:

> such work has shown the close genetic relationship of the groups named; it has also set a new pace in embryology. Now that we know the exact cell origin of these layers and organs, it will never again be possible in describing the development of these animals to refer the origin of certain organs to 'germ layers' merely, nor to refer the origin of these layers to certain general regions of the embryo. The importance of this line of work, not only in the study of the groups named, but also to the science of embryology as a whole, is fully recognized both in this country and abroad, and the credit for this service belongs in large part to the Woods Holl [*sic*] Laboratory.[72]

Conklin, like Wilson, realized that morphological study could now join understanding of the earliest cell divisions with later differentiations in the embryo. The results possible with cell lineage work brought a positive science of causal embryology that, as Wilson and

Whitman also recognized, drew on what had traditionally been both morphological and physiological problems and approaches.

In addition to Wilson's and Conklin's work on cell lineage, Morgan also briefly pursued such studies. For a while, he had studied the early cleavages and cell lineage of the flatworm Polycoerus. When he decided not to pursue that further, possibly because he became distracted by a move to Bryn Mawr and other research possibilities, he gave Harrison the slides and sketches he had made. Though Harrison did not actually work on the Polycoerus, he evidently dabbled with early cell divisions of that and several other marine organisms. Like Morgan, he never published any of the results.[73] In addition, a number of others in Whitman's sphere of influence at the University of Chicago, where he went as first department head in 1892, and at the MBL carried out comparative embryological and cell lineage studies. Morgan himself reported details of early cleavages and their significance in his study of frog development.[74]

Further, the 1898 summer session of the MBL brought a series of lectures that summarized a decade's pursuit of a whole range of problems related to cell lineage. The lectures addressed the significance for phylogenetic questions, the significance of cleavage, the role of cytoplasm and nucleus, and other questions and provided the last public outlet for cell lineage work as such, at least until recently.[75] By the late 1890s, the cell lineage approach to embryology had yielded pride of place to more productive alternative approaches.

As Wilson pointed out, cell lineage had established the strongly determined nature of cleavage and it had "thus given us what is practically a new method of embryological research." Yet he acknowledged criticisms about the significance of the work: "The value and limitations of this method are, however, still under discussion, and among special workers in this field opinion as to the morphological value is still so widely divided that most of its results should be taken as suggestive rather than demonstrative. Like other embryological methods, it has already encountered contradictions and difficulties so serious as to show that it is no *open sesame*."[76] The value lay more with the emphasis on embryology and on development generally and on the facts it provided than with the particular establishment of lineages. Different organisms showed such variations in their development that the significance of the early cleavage stages remained unclear, Wilson acknowledged. "These and many other facts, less striking but no less puzzling, can be built into a strong case against the cell-lineage program, and I wish to acknowledge its full force." In addition, it had proven difficult to trace the lineages all the way to adult organisms. Furthermore, it remained impossible to determine the physiological processes involved in directing the changes in cell patterns for most divisions. Despite the difficulties, Wilson nevertheless remained "on the side of those who as morphologists believe that the study of cell-lineage has demonstrated its value, and that it promises to yield more valuable results in the future."[77]

Traditions Transformed?

The Americans did not do anything radically new and different. They did not reject earlier morphological or physiological work as invalid. Nor did they endorse any new theory in preference to an older one. The dominant recent historical interpretations of this period, which suggest that they did have serious problems. A simplistic Kuhnian analysis, for example, might well pinpoint this as an exciting time of important scientific change and might try to find a revolution.[78] Yet most historians realize that such major shifts in paradigm that produce incommensurabilities and discontinuities are very rare. We should not expect to find such a revolution.

Yet Garland Allen has argued that, in fact, this was a time of rapid discontinuous change, which he has characterized as a "revolt from morphology." According to his interpretation, a group of biologists, which included the four friends from Hopkins at the center, undertook a generational revolt against the purported speculative excesses of their morphological training at Hopkins.[79] They moved, Allen says, from a naturalist to an experimentalist tradition. He supports this thesis by looking at part of the work of representative biologists and by focusing on several pragmatic papers. It is easy to see how he could have been led to his interpretation. It is also easy to see how Jan Sapp could have been led in his study of cytoplasmic genetics to emphasize a "struggle for authority" and for resources as basic to scientific development.[80] Yet such struggle does not seem to have played a significant role for our four since each rapidly emerged as an established leader who commanded considerable resources for his work, which was regarded as being in the mainstream of biological scientific research. A more detailed analysis of the whole body of work and its context places a different focus and results in quite a different picture. This refined picture portrays greater continuity in commitments with long-enduring traditions but reveals subtle changes in underlying epistemology that have quite profound effects. The result is compatible with Larry Laudan's interpretations of scientific traditions, and the embryological story fits very nicely with Fred Churchill's and John Farley's careful studies of heredity and development during this period.[81]

Rather than radically rejecting past errors, then, the young developmental morphologists turned to a new set of organisms that, it seemed, could give them reliable and reproducible data relevant to phenomena they already thought important. This new sense that reliable and reproducible results could be achieved had two significant consequences. First, it drew researchers into new problems and studies. These researchers were not repelled by the older work; instead, they were lured by the promise of positive progress. Along the way, they redefined what fell within morphology and physiology, transforming both and generating a new set of pursuits falling between the older traditions. Second, the expanded observational base brought by these developments profoundly and permanently changed the standards of evidence against which all scientific research in the area, both new and old, was to be measured. Thus, the change was deeper than the birth of a new area of inquiry together with minor redrawing of the boundaries between specialties. The new epistemological standards, developed as part of the new work centered on cell lineages, also applied to more traditional work, altering its epistemological credentials as well. True, a new way of working emerged, but more importantly, the existing traditions were gradually transformed.

Certainly when Wilson learned of the *Nereis* worm, the data he got seemed fundamentally different from and more telling than earlier data. The pattern of cell division that he and Conklin found seemed more regular and more predictable. Of course, even the new cell lineages had their limitations. Time and expanded studies revealed that some organisms and some cleavages were less regular than others. There were even indeterminate cleavages.[82] So the new way of working would not be a cure-all for solving all the old traditional morphological problems.

It could, however, provide information about an approach to some interesting new embryological problems: just exactly how do the fully differentiated embryo and adult come into being out of the fertilized egg? What causes the particular patterns of development and differentiation? Broadly, their answer was that these patterns of change are caused by mechanical pressures within the egg, responding to outside pressures. Thus, development is primarily epigenetic but also predirected by inheritance in some way. Development, it seemed, is a result of changes in structure, effected by a physiological process that is put into action at fertilization. Thus, with regard to the problem it addressed, embryology fell between the traditional programs in pure morphology and in physiology. In its questions and methods and especially in its careful detailed observation of structures and homologies, embryology was closer to traditional morphology. But it was more than this. It began to look, as well, at changing pressures and mechanical processes within individuals. Though all lineagists had not embraced a full study of the processes involved in causing development, they went much farther than traditional morphology would have allowed.

This was certainly no radical break. A variety of workers at the MBL and also in Europe had already begun to move toward this middle ground. In the United States, in programs like that at the Johns Hopkins University and in Whitman's spheres of influence at Clark, the MBL, and Chicago, morphology and physiology had always been more closely tied than in most of Europe. The cell lineage work just encouraged moves that were already under way. Moreover, the catalyst for these changes was no rejection of the past but the attraction of the future. Sometimes discovering what people thought they were fighting against can tell a great deal about their particular convictions and concerns. This is *not* such a case, because they were not primarily fighting against anything. The cell lineage researchers were much more taken with the positive promise of their new method and their well-chosen organisms than with any opponent, either real or perceived. They were excited about their discoveries and determined to push them into new areas, determined above all to make progress.

The emerging sense that the new data were harder, more durable, and more probing had another important consequence besides changing the area and character of the research. It changed the standards of appraisal as well. A good rule of thumb for scientific work is that evidence and arguments for any given result ought to be as strong as can reasonably be hoped for in that domain. What happened in the 1890s was that the sense of what could be hoped for changed. Standards of evidence changed in ways that the embryologists felt made them tighter. The new emphasis was on seeking more definitive and more reliable results that would hold for different researchers even if they were working on different organisms under different conditions. In the face of previous uncertainty and inconsistency, they called for regularity, predictability, and reproducibility of results. Willingness to rely on conclusions derived from intervention as well as from normal, living organisms resulted from the juxtaposition of and equal appeal to cytological and cell lineage study. A move toward increasing intervention also came in the course of the 1890s.

In addition, there was a search for proximate, immediate causes and a move away from appeal to historical, evolutionary explanations. One would pursue embryological study of existing organisms to get at evolutionary relationships, not phylogenetic pasts to get at current development. Such phylogenizing and similar appeals to inheritance from the ancestral past as determining all development seemed increasingly like an appeal to a distant, shadowy past rather than a real explanation. The American workers wanted explanations rather than merely plausible and unsupported stories. Besides, current development was accessible quite directly, and the phylogenetic past was not.

In retrospect, these changes in the standards of epistemological appraisal may have come to seem like a rejection of older scientific work. That work may no longer measure up. Putting the matter that way, however, ignores the fundamentally positive character of the shift. The older traditions are not ended but merged; not rejected but transformed. The character of that transformation is equally important, for the change was brought about not by a new theory or even by a new argument, and not by new organisms or new empirical results alone, but by a changed sense of what can be done.

5

Responding to Innovations Abroad

EDMUND BEECHER WILSON became a trustee of the Marine Biological Laboratory (MBL) in 1890. So did Henry Fairfield Osborn, a paleontologist from New York. As a Princeton graduate and nephew of J. Pierpont Morgan, Osborn brought with him valuable connections with a wealthy and influential segment of New York society. In addition, he represented the American Museum of Natural History and also Columbia University, where he had been hired in 1891 as dean of the new Faculty of Pure Science.[1] Osborn worried about the relations of problems of individual heredity to problems of evolution, a concern that fit in well with MBL interests in 1890. Charles Otis Whitman accordingly invited Osborn to join Wilson and himself, among others, in presenting an evening lecture in 1890.

Not surprisingly, when Osborn began to hire faculty for Columbia, he consulted Whitman, who was then (1889–92) solidly established at research-oriented Clark University. In response to a query about Wilson's qualifications, Whitman wrote: "I am sure that he is the best man we now have in invertebrate and general embryology. I have been doing my best to get him here, but I am beginning to think my efforts will be fruitless. He is an admirable lecturer, one of our best writers, a very quick and keen investigator. . . . Personally, I think Wilson is one of the immaculate."[2] After that recommendation, and given Osborn's own impression of Wilson from the MBL, how could he fail to offer him a position? In fact, he asked Wilson to join the Department of Biology, designed to develop graduate work and research and to supplement departments of botany and physiology.[3]

Undoubtedly, the opportunity to help put together a new department, oriented toward research, had its attractions. New York's musical offerings also would have held appeal for Wilson. In addition, the new position brought with it a year's study in Europe at the outset. Wilson accepted. After some time at the MBL for the 1891 session, he left for his tour abroad.

Wilson's Return to Europe: Boveri in Munich

First stop was a semester in Munich, at the Zoological Institute, directed by Richard Hertwig. On his first European venture, Wilson had spent some time at Richard's brother Oskar Hertwig's institute in Jena, learning about the latest in cytological techniques and discoveries. This time he selected Munich, but instead of working with either Hertwig directly, he chose Theodor Boveri. Boveri had received his doctorate in 1885 at Munich, then had become Lamont Fellow for independent research at the Zoological Institute. In 1891, the year when Wilson arrived, he began two years as assistant there. After a series of family tragedies and an illness and depression that kept him from doing any work through the summer of 1891, Boveri needed the change of pace brought by his promotion and by Wilson's visit. He wrote with pride to his sister-in-law to tell her that the American professor was coming to work with him.[4] By 1891 when Wilson arrived, Boveri had completed his series of three "Zellenstudien."[5] Marvelous works with beautiful detailed plates and meticulous descriptions of cellular and nuclear changes in a variety of phenomena, these *Studien* did just the sort of thing that Wilson had found particularly intriguing since his undergraduate admiration of Edward Laurens Mark's work. Where better to spend his first semester abroad, then, than with Boveri in Munich? The Hertwigs and the others they attracted to share their interests in cytology provided a further incentive.

Wilson loved his stay in Munich, and he and Boveri became close friends. Boveri was "far more than a brilliant scientific discoverer and teacher. He was a many-sided man, gifted in many directions, an excellent musician, a good amateur painter," and the two "found many points of contact far outside the realm of science." "The best that he gave me was at the Café Heck where we used to dine together, drinking wonderful Bavarian beer, playing billiards, and talking endlessly about all manner of things."[6] That talk and music and deep comradeship meant a great deal to Wilson at this pivotal time in his career. The keen intelligence and dedication that Boveri brought to his work also made a major impact. In commemoration, Wilson wrote that Boveri's work "enriched biological science with some of the most interesting discoveries and fruitful new conceptions of our time. But beyond all this it is distinguished by a fine quality of constructive imagination, but a sureness of grasp and an elegance of demonstration, that make it almost as much a work of art as of science. In this respect, as I think, Boveri stood without a rival among the biologists of his generation, and his writings will long endure as classical models of conception, execution and exposition."[7] In fact, Boveri's high standards so inspired Wilson that he dedicated his major opus, *The Cell in Development and Inheritance*, to his friend.[8]

While with Boveri, Wilson first began consciously to separate questions of heredity from those of development. Previously, he had treated the two phenomena as fundamentally interconnected.[9] For Wilson, as for Whitman and most of the other Americans at the MBL, the egg cell represents the product of heredity, of course. Each individual developing organism reflects inherited adaptations from its ancestral evolutionary past. It is the product of parental development and is therefore organized or predetermined in some way that continues to influence the entire process of the individual embryo's differentiation. Therefore, heredity and individual development are fundamentally intertwined. It is not the case according to the Americans, therefore, that heredity operates up to a point (by providing genes, for example) and then gives up so that development can take over. Rather, heredity and development each operate throughout the entire life of the organism, with heredity conservatively ensuring continuity with the past and development responding to the changing conditions of the current organism and consequently insuring that change and adaptation could occur. As a result, it did not really make sense to focus on the nucleus alone and to leave out the entire cytoplasm.

Yet that is precisely what a few of the German researchers had begun to do. Oskar Hertwig had demonstrated by then that fertilization brings the combination of two nuclei, but no more than a nucleus from the male. Perhaps, such results suggested, the nucleus is the bearer of heredity. After all, the male parent is represented in the offspring and the nucleus seems to be the only link between parent and that next generation.[10] Further studies throughout the 1880s had established that both male and female contribute to the fertilization process; that the gametes each go through a complex process of changes in the chromosomes which reduces the number of chromosomes in each cell; and that chromosomes maintain their individual integrity throughout the entire division and development process.[11] The upshot was that chromosomes seemed to carry hereditary material from parents to offspring and to be intact from the moment of fertilization. By the late 1880s, then, it made sense to this group of German researchers to focus attentively on the apparently inherited chromosomes and on their changes and significance during the subsequent developmental process.

Boveri did just that. In his "Zellenstudien," he made use of a research material particularly advantageous for such work, namely a threadworm of the horse, *Ascaris* (*megalocephala* and *lumbricoides*). Since this parasite normally has only four chromosomes (rarely two), it is easier to keep track of what each is doing as it undergoes its complex steps of division. Concentrating on exactly what happens during the very earliest stages of development, even before those first cell cleavages on which Whitman and Wilson had concentrated, had become feasible and appeared to be a productive line of research to follow. In addition, such work could begin to provide facts useful to assess the plethora of theories which had emerged to explain the significance of heredity and development. The question of precisely what role the chromosomes play held particular interest and inspired a number of suggestive hypotheses.

August Weismann, for example, had suggested that the process of chromosomal division and reduction which occurs in producing gametes serves a useful purpose. It allows a new combination of chromosomes when fertilization occurs. Since the chromosomes are the bearers of heredity, fertilization brings new hereditary combinations into being. Thus, heredity is a creative process and provides a source of variations on which natural selection can work. At the same time, fertilization involves a reduction of the number of chromosomes in each germ cell, Weismann agreed with Wilhelm Roux.[12] That reduction actually reduces variety within each individual because the chromosomes undergo qualitative as well as quantitative division as they prepare for fertilization. So the loss of individual variation accompanies the creative new recombinations from both parents, which in turn increases overall diversity among individuals. Or so Weismann's logic ran. Others offered alternative theories, some based on an assumption of qualitative and others on one of quantitative chromosomal division.

The more information gathered about exactly what the chromosomes do and about the relative contributions of male and female parents to the gametes, the further researchers would have gone toward determining which parts of which theories most conformed with the facts. Accordingly, Boveri set out in 1889 to acquire information from experimental intervention into the normal developmental process. In the spring, he visited the Naples Stazione Zoologica. There he decided to study the development of sea urchins and to pursue the method devised by the Hertwig brothers in 1887.[13] Shaking unfertilized sea urchin eggs (several species) causes pieces to break apart and to survive independently. Boveri assumed that some of these must contain the egg nucleus while others do not. Yet even some of those that do not can be successfully fertilized and caused to develop thereafter, apparently normally.[14] This shows that paternal chromosomes plus maternal cytoplasm are sufficient for development to occur. Apparently each set of chromosomes contains in itself whatever hereditary information or material is needed to make its proper contribution. Two sets of chromosomes are not necessary.

Boveri also thought that the whole chromosome remains intact in each cell; Weismann's assumption of qualitative divisions of chromosomes seemed wrong. Instead, Boveri's evidence suggested that each chromosome maintains its individuality throughout the various divisions of its developmental process. For Boveri, the cytoplasmic contributions also seemed important for development. Thus, differentiation of parts in the developing embryo requires something other than chopping up and doling out inherited bits of chromosomes. Some complex interactions between chromosomes and cytoplasm must occur.

Boveri's plan was to assess the relative actions of nuclear and cytoplasmic material as well as the relative contributions of male and female parents. He was trying to get at the same sorts of root questions that Whitman and Wilson had been asking at the MBL. While they had concentrated on the earliest cell divisions, after fertilization, Boveri and some of his German colleagues looked also to the fertilization process itself and even to the steps contributing to fertilization.[15]

To try to gain information about that otherwise invisible and inaccessible process, Boveri devised a set of experiments. He took the nonnucleated bits of sea urchin eggs after they were shaken apart and fertilized them with sperm from another species. If they developed along the pattern of the mother's species, then cytoplasm must play at least the major role in directing development. If they instead followed the paternal pattern, then the sperm must be more important. Further, since the sperm consisted primarily of nucleus, this suggested a nuclear determination of development. Unfortunately his results remained inconclusive. Yet he nonetheless believed that "practically every doubt has been eliminated and the proposition that the nucleus is the sole bearer of heredity has been confirmed."[16] Within a few years, other experiments and reproductions of his own, as well as exchanges with Morgan and others, pointed out some defects of his research and required him to change his views somewhat, but at the time Wilson arrived in Munich, Boveri was inclined toward the opinion that the inherited nucleus largely determines development. The question was, how?

Following on the sea urchin experiments at Naples, Boveri completed his third "Zellenstudien," which looked more closely at the functional or physiological role as well as the morphological changes in the chromosomal material in the fertilization process. Given his growing conviction about the importance of nuclei as the bearers of heredity and his interest in showing that they persist throughout development, he concentrated on showing the individuality of maternal and paternal chromosomes during and following fertilization. This concentration placed attention on heredity and therefore began to suggest that heredity might be separable from development rather than inextricably connected, as had been generally assumed.

During a trip to Naples, he then began work on the primitive chordate *Amphioxus*. This small lancet fish, later called *Brachiostoma*, which burrows into the sand, had been variously classified as a worm, as a fish, and finally as a primitive chordate. It attracted attention in the 1880s because of its unusual structure and development, both of which suggested that it represented one of the simplest and most primitive of vertebrate forms. Boveri examined *Amphioxus* and discovered nephridial tubes, which raised questions about the primitive nature of the excretory system. He also focused on head development as a key to the phylogenetic relation of this organism to others and decided that it did, indeed, suggest a primitive protovertebrate form.[17] These were pressing questions for morphologists which related closely to Wilson's own study of annelid development. With such a convergence of interests, Wilson's decision to visit Munich and to work with Boveri would almost seem to have been overdetermined. In addition, Boveri's careful way of working stressed the importance of establishing solid and reliable facts over putting forth a speculative theory. Such a theory explains the facts but has to extend too far beyond them. In his effort to establish the relative roles of each parent and of nucleus and cytoplasm in heredity and development, his approach was similar to Wilson's.

In addition to Boveri, Munich boasted Richard Hertwig, who was also interested in the cell as the fundamental unit of life. Hertwig's own study of the role of cells throughout development also looked internally, at the process of fertilization and thereafter and at changes within the egg and within cells. With this combination of interests and approaches related to Wilson's own, it is hard to imagine that Wilson could have found a better place to spend half of his year in Europe.

Undoubtedly, Wilson learned about a variety of new methods, new organisms, new questions, and new ideas. Yet he was not a naïve, uninformed American taking up research for the first time. He also carried with him a perspective informed by his own cell lineage studies, which informed and influenced the Munich work as well. The MBL community had been addressing much the same questions as the Munich group, only with different emphases. The Americans had devised their own approach and had discovered their own peculiarly useful organisms. What Wilson learned in Munich began to reinforce his own emerging sense of how one should do biology and what path research should take. It also redirected his specific research program in some ways.

Wilson's own interest in cells, manifest in his cell lineage papers and even in those early chapters of his textbook with William Sedgwick, reached a peak in 1892–93. That year, his first in his new position at Columbia, he gave a series of lectures that made up the fourth contribution to the Columbia University Biological Series. Wilson then continued to refine the work, so that it finally appeared in published form in 1896 as *The Cell*, a masterpiece of collected best knowledge about the cell.[18] Comparing the content and style of that volume with the earlier chapters of the Sedgwick and Wilson text suggests that Wilson had thought long and hard about the project during his Munich stay. He acquired information, ideas, and techniques during those warm congenial months with his new friend.

Next Stop: Naples

From Munich, Wilson wrote to Anton Dohrn to confirm that he wished to return to the Naples Stazione for the second half of his year abroad. He wanted to extend his studies of *Nereis* to compare with several other marine annelids. In particular, he wanted to look at a local species of *Nereis* (*Damerilli*), about whose development he was sure that other writers had been mistaken. If Dohrn could send him any specimens ahead of time, he would start his observations in December, he wrote, and he looked forward to arriving in Naples in January.[19]

He arrived on 11 January, traveling from Munich with another zoologist from Norway. In Naples, he immediately met the young and lively German zoologists Hans Driesch and Curt Herbst. After attending lectures by Weismann for two semesters and then studying under Ernst Haeckel, Driesch had received his Ph.D. degree in 1889 from the University of Freiburg. Trained solidly in the morphological tradition, he was concerned with problems of heredity and development. He was also directly exposed to Weismann's new theories about qualitative division of chromosomes and to Roux's similar views about nuclear determination of developmental differentiation. After graduating, he made his way to the Naples Zoological Station to carry out morphological research and explore the validity of these ideas.

Because of innovations within morphology itself, and because of transformations and reorientations of the respective domains of physiology and morphology, experimental embryology had become a hot topic. In the 1880s, a young professor of physiology in Breslau had turned from his work on respiration and heat production to begin asking what causes sex, specifically in frogs.[20] Eduard Pflüger's research largely depended on manipulating and controlling conditions external to eggs, such as semen concentration, and on determining the effect of these changes. This work, in turn, led him to further questions about frogs, including an interest in hybridization of frog species.

With such major differences as he saw in the appearance of different frog species, Pflüger believed that an observer should easily recognize hybrids. Yet he had never seen any. Perhaps this occurred simply because the breeding times were different. He could overcome that factor by obtaining from friends both frogs and a few toad species that had different breeding seasons. Aware that zoologists often found physiologists suspect, and believing that physiological experimentation was all too often based on a misunderstanding of morphology, Pflüger self-consciously and cautiously sought to work on that developmental middle ground between morphology and physiology.[21]

He could not get eggs fertilized with foreign semen to develop consistently, but the controls inseminated with semen of the same species always did so. Sometimes the artificially fertilized eggs divided briefly, then developed no further; some crosses developed further than others, some not at all. Because of the difficulty of raising frogs' eggs, Pflüger urged that the results remained incomplete and open to question.

Following his work and other studies of unripe eggs, Pflüger turned to more actively manipulative experimental examination of the fundamental problems of development. In 1883, he therefore inaugurated his much cited gravity experiments. He rearranged the embryo within its gravitational field by pressing it firmly between glass plates. A new cleavage plane developed, in a different place from the normal one.[22] He could detect the difference because of the strongly differentiated dark and light pigmentation in the egg of the frog he used. After ascertaining that gravity did, in fact, determine the direction of the first cleavage plane, he then demonstrated that the experimentally produced plane would persist throughout all later developmental stages. The organism would continue to develop in its altered new direction.

Since the first cleavage plane apparently gave rise to the bilateral axis of the organism, and since the plane could change with a changing gravitational field, then the embryo had not been lying preformed in the egg. It must instead be strongly directed by the environment, except that Pflüger held that the structure of the egg, with its differentially pigmented halves, also constrained development in some way. The egg experiences a "relative isotropy," he concluded. After considering the way in which polarized molecular arrangements might give rise to the embryo's proper orientation within the egg, he bypassed questions about precisely how such molecular actions might work. Instead, he concluded that the embryo actually arises anew "through later, purely accidental external influences on the organism."[23] There is no problem with explaining the hereditary transmission of characters, he insisted, because inherited characteristics do not even exist. Any hereditary effects arise later, through molecular actions—a clear call for a mechanical causal account of ontogeny, but a "simple hypothesis" rather than a full account. With their suggestive conclusions, Pflüger's gravity experiments stimulated a flurry of controversy and activity.

Pflüger's work closely paralleled some of his Breslau colleague Gustav Born's morphological work on frogs and stimulated further interest more generally. By 1883, Pflüger, Born, Roux, and others had begun exchanging materials, corresponding, and otherwise creating an active community of interest in frog development.[24] Perhaps Pflüger's precise physiological style of articulating problems and presenting results influenced the others. Perhaps their particular morphological and cytological concerns influenced him. At any rate, a new type of work appeared which lay between what had traditionally been morphology and physiology. Over the next few years, a rush of research papers appeared in this area.

This new work led the move to more actively manipulative experimental methods in morphology, to more direct concern with the material and proximate causes of developmental phenomena, and to disagreements of interpretation which helped to crystallize research efforts in experimental embryology and which transformed the morphological tradition. Epistemological convictions, especially about what sorts of questions one should reasonably ask and what methods and approaches one should employ, underpinned the transformation, as it did for the American cell lineagists.

Born's research followed lines similar to Pflüger's. Though trained as a morphological embryologist, and beginning with studies of salamander and amphibian development, he had also turned to experimental studies of the causes of sex differentiation.[25] Then he began to focus on hybridization and production of sex differences, also employing more directly manipulative experimental work such as Pflüger's. Born's continued experimentation on sex differentiation led him to study other questions of differentiation and to try tissue transplantation techniques as well, the latter ultimately stimulating the major approach to experimental embryology developed by Ross Granville Harrison and German embryologist Hans Spemann.[26] In 1883, Born turned to the influence of gravity on frogs, directly in response to Pflüger's paper. He disagreed with Pflüger's conclusions and did not accept that gravity really directly affected the orientation of the future embryo. Instead, he saw evidence only of a very interesting but indirect effect of gravity "which is caused by the eccentric position of the nucleus and the presumed least specific gravity in the special case of the fertilized frog's egg."[27] Pflüger had overlooked the vital importance of nuclear divisions, Born held, so his conclusions about the external effect of gravity on the egg's direction of division must be questioned.

Roux agreed with Born's skepticism about the efficacy of external conditions. He concluded that all the major orientations of the embryo are established by the time of the second cleavage, so that development thereafter follows a rigid pattern.[28] In later discussions over the next two years, he further specified that the first cleavage definitely establishes the two halves of the embryo and later cleavages bring equally irreversible differentiations. Though Born had expressed his hesitations about Pflüger's external determinism more tentatively, as Pflüger had expressed his own conclusions, Roux's tone was much more positive. Cell divisions *cause* differentiation, he confidently asserted.

The issues and approaches that Born, Roux, and Pflüger adopted set a new standard for experimental work in embryology. Bold assertions joined with new approaches in a move toward that middle ground previously lying between morphology and physiology, even while some questions remained the same. Roux, for example, continued to pursue questions about early egg orientations, generating speculative theories as well as data. Again, he decided that the orientation did not affect—or at least did not effect the details of—development. As he rotated eggs about an axis, he found no change in development, which proceeded normally in both speed and detail. From this evidence, he concluded that eggs are self-differentiating, directed by internal mechanical causes, and do not require external formative influences to guide their way to normal differentiated development.[29]

By 1885, Roux had begun to develop his initial speculations about self-differentiation of embryos. Though little was understood about development of individuals, he said, researchers should nonetheless seek a causal analytical account of the phenomenon. Individual development and phylogenetic histories operated as a double-barreled phenomenon, but he wished only to consider the first part. First, individuals pass through a stage of organ formation. Then follows a period of functional development. The first is a period of independent development of other than internal nuclear divisions, while the latter is dependent on a complex of factors. Roux charged that Pflüger had misassessed the significance of his results. Analytical work reveals the strong possibility, he insisted, that nuclear divisions rather than external environmental changes actually bring about differentiation. In fact, the nucleus holds the qualities of individual formation.[30] With considerable rhetorical style and little experimental or descriptive detail as yet, Roux presented elements of his theories of qualitative cell division and strong predirection of development by the earliest divisions. The nucleus held the crucial hereditary information, he felt. He did not fully address the suggestion, which Pflüger saw as an important possibility, that the egg's cytoplasm might have an initial and important organization already.[31]

August Weismann became the outspoken advocate of a similar nuclear predeterminist position. The cytoplasm remains isotropic, or unorganized, but the nucleus decidedly acts as the determiner of subsequent development. Nuclear changes during mitosis distribute the determinants that are parts of the qualitatively dividing chromosomes, differentially to separate cells, which then act effectively as individuals and undergo their programmed differentiation into various types of cells. Each cell assumes a particular cell type simply because of the action of determinants distributed to it. Thus the cells are self-differentiating, as Weismann later put it, "that is to say, the fate of the cells is determined by forces situated within them, and not by external influences." Internal factors alone determine development. Ontogeny is not a "'new formation' of multiplicity" or epigenesis, then, but the unfolding of multiplicity, or the evolution of previously invisible multiplicity. Determinants hold the key to that "invisible multiplicity," and the nuclear divisions that distribute determinants to different cells in effect explain development, according to Weismann. Amplified and developed further by Roux, the Weismannian nuclear determinism and resulting self-differentiation of cells became the standard preformationist position to adopt or attack.[32]

Weismann admitted that the inevitable attack might prove fatal. Putting forth a bold and definite theory was nonetheless justified, for

> the ceaseless activity of research brings to light new facts every day, and I am far from maintaining that my theory may not be disproved by some of these. But even if it should have to be abandoned at a later period, it seems to me that, at the present time, it is a necessary stage in the advancement of our knowledge, and one which must be brought forward and passed through, whether it prove right or wrong, in the future. In this spirit I offer the following considerations, and it is in this spirit that I should wish them to be received.[33]

Science, Weismann and Roux felt, should proceed with bold and productive theories such as theirs. Continued research would serve as the necessary corrective.

Roux, Driesch, Wilson, and Partial Embryos

In 1888 came Roux's experiments on frogs' eggs which strongly supported his earlier theoretical interpretation of the egg as a mechanically determined mosaic. In that classic and much discussed paper of 1888 on half-embryos, Roux had set out to test what he saw as the two leading alternative theories of development. As he reported, "The following investigation represents an effort to solve the problems of self-differentiation—to determine whether, and if so how far, the fertilized egg is able to develop independently as a whole and in its individual parts. Or whether, on the contrary, normal development can take place only through direct formative influences of the environment on the fertilized egg or through the differentiating interactions of the parts of the egg separated from one another by cleavage."[34]

Working with frogs' eggs, Roux cited assorted evidence for the self-differentiation of cells. True to his rhetorical pronouncements for "causal analytical experimental embryology," he maintained that only direct experimentation could establish that development acts as a sum of separate mosaic developments.[35] Roux's experiments in this case consisted of puncturing with a needle one of the two blastomeres after the first cleavage, and a few examples of puncturing one or two of the four blastomeres after the second cleavage. He did not remove the punctured cells, and the early results remained unsatisfying. After modifying his technique, Roux succeeded in obtaining, in about 20 percent of the operated eggs, the ability of the unpunctured cell to survive and continue developing. These cases formed his small experimental sample, which he examined carefully through traditional histological staining and sectioning techniques typical of current cytological research.

Roux found a regularity in these abnormal cases. The experimental embryos produced only partial blastula or gastrula stages. The remaining blastomeres did not compensate for the injured blastomeres; rather, they apparently developed as they normally would. Roux concluded that "in general we can infer from these results that each of the two first blastomeres is able to develop independently of the other and therefore does develop independently upon normal circumstances." Thus, "all this provides a new confirmation of the insight we had already achieved earlier that developmental processes may not be considered a result of the interaction of all parts, or indeed even of all the nuclear parts of the egg. We have, instead of such differentiating interactions, the self-differentiation of the first blastomeres and of the complex of their derivatives into a definite part of the embryo."[36] In short, the early developing embryo acts as a mosaic of independent parts, a mosaic probably caused by "qualitative separation of materials," though Roux suggested that proof would require further research. From the existence of a stable developmental pattern followed the suggestion that further study could reveal the exact role played by each blastomere, a suggestion the Americans pursued in quite different ways and for other reasons through their cell lineage work.

Roux did not go on in his paper of 1888 to conclude that every step of development produces independently developing cells and hence a perfect mosaic. Though he certainly did not reject such a possibility, he cautioned, "How far this mosaic formation of at least four pieces is now reworked in the course of further development by unilaterally directed rearrangements of material and by differentiation correlations, and how far the independence of its parts is restricted, must still be determined."[37] In this work, therefore, he clearly supported the cause of mosaic development, as a Weismannian sort of predeterminism. Yet he did not argue adamantly for an extreme form of predeterminism in which no regulative epigenetic action could occur. And he left open the possibility that later stages, as the embryo becomes more complex, might exhibit increased dependent differentiation, where cells respond to other cells and to external conditions.

Roux's continued research on half-embryos produced some cases where a whole embryo did result. By that time, though, Roux was so committed to his mosaic interpretation that he did not seriously question his theory. Instead, he generated an auxiliary hypothesis that there exists a reserve idioplasm that is called into action when regeneration or postgeneration (following injury) occurs.[38] Despite these modifications, he maintained a predeterministic mosaic position. Others followed Roux's lead and pursued similar experiments with other organisms.

Driesch, for example, took on similar work in experimental embryology. Following the Hertwigs and Boveri, he shook apart sea urchin eggs. Other such "shakers" had concentrated on eggs before fertilization and had gotten a large number of pieces, some with nuclei and some without. Driesch worked instead with the eggs immediately after fertilization. He separated the two blastomeres after the first cell division. Contrary to expectation, each developed into a gastrula and then even a pluteus larva that was perfectly normal in all but size. Each was smaller than normal but otherwise just the same. This result evidently surprised Driesch as much as it surprised others.

He had undertaken this work, Driesch reported in his published paper, to extend Roux's experiments to this particularly suitable organism. Sea urchins are easy to obtain and to observe, and their eggs are relatively resistant to damage. During March and April 1891, Driesch therefore shook his urchins at the Stazione Zoologica in Naples and watched them develop, expecting the sort of mosaic result that Roux had obtained when he got half-embryos from frogs' eggs. Driesch waited expectantly, he said, hoping for the odd sight of half-gastrulas swimming about his glass dishes but predicting the half-embryos would die instead. When each of the two blastomeres unexpectedly developed normally, he concluded that some kind of totipotency existed for the early cells. Each cell had somehow maintained its ability to become the whole embryo or at least to "regenerate" the missing parts from the very first moments.[39]

The results refuted Wilhelm His's hypothesis of organ-forming germ regions, Driesch concluded. For, if the embryo were truly preorganized already in the unfertilized egg, then cell division could only chop up already determined parts. The idea of cytoplasmic localization was refuted for Driesch. He was less certain about the hypothesis of qualitative chromosomal division. Clearly, the single blastomeres with only half the embryonic material recovered to form a whole embryo, but that could happen without quantitative chromosomal division. It seemed that each cell had a full complement of nuclear material and somehow managed to "regenerate" the part of the embryo that would normally have been produced by the other blastomere. This "regeneration" occurred so early with sea urchins that the appearance of normal development was preserved. Driesch recognized that his results were inconclusive. At this point, he certainly did not see his results as being in opposition to Roux's. He acknowledged that his method had been different from Roux's, in which the first two blastomeres were never fully separated, and that this might account for the different results. Only in separate, more theoretical papers did he begin to explore the various alternative explanations of the differences. Only there did he begin to develop his ideas about what sorts of approaches and theories properly belong as part of biology.[40]

Further, he saw the potential value of another phenomenon that had arisen as a side benefit: artificial production of twins which sometimes accompanied the shaking apart of blastomeres. Sometimes each blastomere developed into a whole embryo, so that two side-by-side gastrulas would appear together. This might help reveal the extent to which and the way in which the blastomeres accommodate for missing parts. It suggested an epigenetic interpretation, certainly, but he also realized that the work was more suggestive than conclusive and that the primary need at the time was for further research, for "it were futile to indulge in idle speculation without actual facts."[41]

Wilson arrived in Naples amidst eager discussion of this work and its implications. Almost inevitably, given the energy and enthusiasm that Driesch exuded in Naples, and given the similarity to questions Wilson already found interesting, Wilson became curious about how his own species would fare under similar treatment. Would they develop as Roux's frogs did or along the lines of Driesch's urchins? Perhaps he tried *Nereis* and other annelids without great success, but he did work with *Amphioxus* and found results that seemed similar to Driesch's own. He wrote to Driesch in June, after Driesch and Herbst had returned from Naples to Germany, "It is very easy to shake the blastomeres apart and I have got numerous half- and quarter-embryos exactly like the usual ones but ½ or ¼ as large." He further obtained one-eighth embryos, it seemed, and various intermediate forms as well. In addition, Wilson found some cases where the two- or four-cell blastomeres each gave rise to a little colony of cells. The result was a colony of gastrulas, with up to thirteen stuck together. Thus, it looked "as though *any* cell of the early cleavage-stages may, if slightly disturbed, give rise to an embryo." It would, he thought, "prove an interesting and important case."[42]

Further, Wilson had decided to experiment on the eight-cell stage, following the four cells that normally give rise only to ectoderm. Could these also generate a whole embryo, he asked in a letter to Driesch. "If it does so the case will be an important one, for in the case of the 2's or 4's, it might be maintained that each cell still contains material from all three germ-layers, since it contributes to all. If however a pure ectoderm cell can regenerate the whole, we shall have a demonstration of your views and a fatal blow at the theory of 'Keimplasm' and qualitative nuclear division."[43]

After postponing his return to the United States in order to be able to continue his work at Naples, Wilson wrote to Driesch again in September. His earlier interpretations had been in error, he said. In fact, only blastomeres up to the four-cell stage remain capable of generating whole embryos. Those after the next division cannot. They live a very long time, it is true, and may give rise to blastulas but never to gastrulas. Further, in the one-half or one-fourth embryo stage, the cleavage that takes place occurs exactly like that of the normal egg cleavage. There is no accommodation at any point for the fact that only half of the material is there. Thus, Wilson concluded, no regeneration takes place in the sense that "nothing is regenerated."[44] Driesch's interpretations and Wilson's own initial conclusions demanded revision. Some of this revision, Wilson wrote to Driesch, would appear in a paper to be published soon.

In that paper, he concluded that the one-eighth embryo could not develop into a gastrula because of qualitative deficiencies. It was not simply a lack of a sufficient quantity of material, since other, smaller pieces could develop properly. Rather, something about the cell divisions caused the one-eighth pieces not to have everything required for full development. Yet it is not the case that a mosaic type of qualitative nuclear division occurs, as Roux had seen in his frogs. "In this respect Amphioxus differs from the frog (Roux) and from the sea-urchin (Driesch), in both of which the development at first agrees with that of a normal embryo-half and only later gives rise to a perfect embryo by a process which may provisionally be called 'regeneration.'" This regeneration occurs much earlier in *Echinus* than in *Rana*, and seems even earlier in *Amphioxus*, but it is not regeneration in the ordinary sense.[45]

Finally, the time came to return to his new position at Columbia. On the way, Wilson sailed for Genoa, where he saw Paganini's Guarnarius violin: "The thrill that it gave me was only equalled by my ascent of Etna."[46] After returning to the United States and carrying out his series of lectures on the cell at Columbia, Wilson wrote once again to his friend Driesch. He had just received a copy of

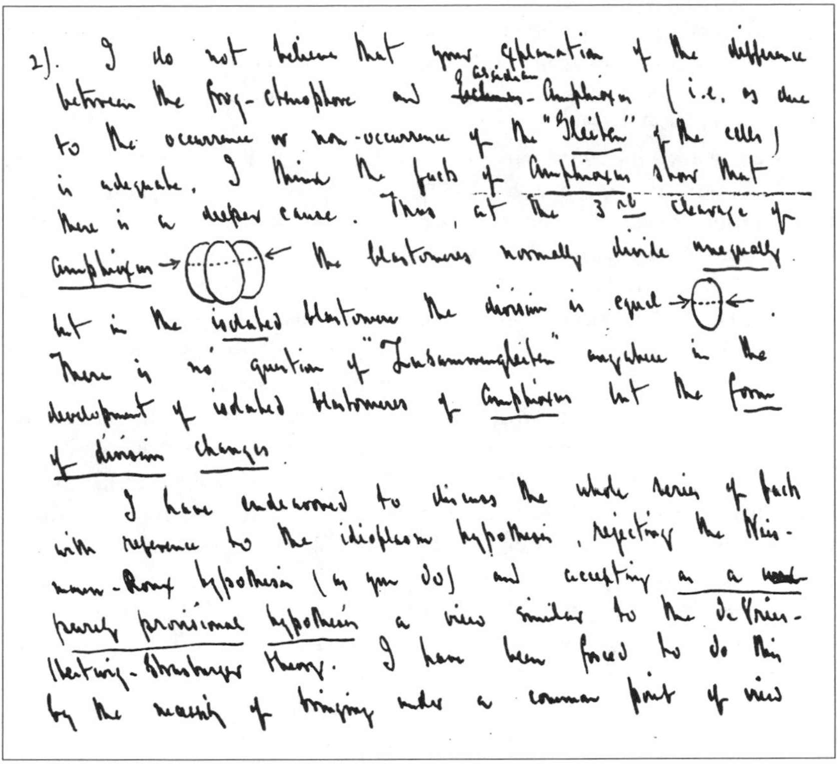

2). I do not believe that your explanation of the difference
between the frog-Ctenophore and ascidian-Amphioxus (i.e. as due
to the occurrence or non-occurrence of the "Gleiten" of the cells)
is adequate. I think the facts of Amphioxus show that
there is a deeper cause. Thus, at the 3rd cleavage of
Amphioxus → ◯◯ ← the blastomeres normally divide unequally.
but in the isolated blastomere the division is equal → ◯ ←.
There is no question of "Zusammengleiten" anywhere in the
development of isolated blastomeres of Amphioxus but the form
of division changes.

I have endeavored to discuss the whole series of facts
with reference to the idioplasm hypothesis, rejecting the Weismann-Roux hypothesis (as you do) and accepting as a
purely provisional hypothesis a view similar to the De Vries-
Hertwig-Strasburger theory. I have been forced to do this
by the necessity of bringing under a common point of view

Fig. 13. From a letter from Wilson to Driesch, 5 June 1893. From the Anton Dohrn Archives, Stazione Zoologica, Naples.

Driesch's last paper and found that, even though he largely agreed with the conclusions, he also had some disagreements. Since he had recently completed a lengthy paper that considered the same issues and directly addressed Roux's and Weismann's theories, he was prepared to explain his point of view.[47]

Driesch had hypothesized some kind of sliding (*gleiten*) of the cells to explain why isolated blastomeres in frogs behave differently from those of ascidians and *Amphioxus.* Wilson rejected this idea and tried an alternative explanation. He demonstrated that in the experimental case with *Amphioxus,* at least, the division with isolated blastomeres occurred differently from the normal: division at the third cleavage divided the cell equally, rather than unequally, as it did in the normal case. This altered division could account for the blastomere's ability to accommodate and generate the whole organism. His conclusion, Wilson wrote to Driesch, was that the Weismann-Roux hypothesis should be rejected, as Driesch had also said. Instead, he adopted something like what he called the "de Vries-Hertwig-Strasburger hypothesis," as Driesch did also. Further, he accepted this only as a "*purely provisional hypothesis*," again much as Driesch did.

"Taking *all the facts* into consideration," Wilson said, he could not "escape the conclusion that the idioplasm must become modified as development goes forward—but *not by qualitative division* as Roux-Weismann suppose." The results from the ability of sea urchins and *Amphioxus* to accommodate contradicts that hypothesis. Perhaps some sort of "physiological specialization" occurs instead. At root, he saw "the ontogeny as a series of actions and interactions between the blastomeres wh. results in a steadily increasing specialization of the idioplasm, and in which each form conditions the following form." Somehow there is some "hereditary element in cleavage-forms" that differs from Driesch's "purely mechanical conditions," but Wilson recognized that he had to work at discovering what those were.[48] He felt, as he said in his published paper, that the best hypothesis was some compromise version: it should begin with a mosaic theory, though without the qualitative nuclear division that lay at the heart of Roux's version, and with Driesch's emphasis on a mechanical regulation of development.[49]

That summer, 1893, Wilson addressed the problem again in a popular lecture at the MBL. There he outlined the current state of embryological study and explained his (and the more generalized American) reaction. Three factors had recently brought about a change in morphological work, he explained. First, the germ layer theory had been discarded, largely through the efforts of cell lineage studies that had demonstrated the importance of the earlier developmental stages, before the germ layers formed. Second, correlations of early cleavages and later body relations had been discovered, largely through the research of Pflüger, Roux, and others. Third, the application of experimental methods to embryology and to cleavage in particular had brought a move to concentrate on the earliest stages and the subsequent significance of cleavage for later development.

The latter emphasis had begun with Pflüger's experiments on the effects of gravity and pressure on development. As Wilson explained:

> These pioneer studies formed the starting-point for a series of remarkable researches by Roux, Driesch, Born and others, that have absorbed a large share of interest on the part of morphologists and physiologists alike; and it is perhaps not too much to say that at the present day the questions raised by these experimental researchers on cleavage stand

> foremost in the arena of biological discussion, and have for the time being thrown into the background many problems which were but yesterday generally regarded as the burning questions of the time. It is the purpose of this lecture to consider, briefly, the most central and fundamental subject of the current controversy.[50]

That controversy focused directly on new versions of the old question of preformation or epigenesis, Wilson felt. Given a focus on proximate causation, then, what causes embryonic cellular differentiation? The disagreement rested on interpreting the significance of cell division: does it cause differentiation or does differentiation only follow cell division secondarily and chronologically?

The complex of theories by Roux and Weismann explaining differentiation, reproduction, variation, heredity, and regeneration awaited full trial, Wilson explained, but the evidence so far was negative. "Brilliantly elaborated and persuasively presented as they are, they do not at present, I believe, carry conviction to the minds of most naturalists, but arouse a feeling of scepticism and uncertainty; for the fine-spun thread of theory leads us little by little into an unknown region, so remote from the *terra firma* of observed fact that verification and disproof are alike impossible."[51] In particular, Roux's mosaic theory had received an early and ironic blow from Driesch's pursuit of Roux's own experimental approach, Wilson held, since Driesch's 1891 results proved incompatible with the strictest form of Roux's theory.

Wilson realized in this 1893 lecture that Roux's introduction of his auxiliary hypothesis of "post-generation" (or regeneration) to explain how the sea urchin blastomeres might produce (or regenerate) whole gastrulas logically saved the possibility of some sort of predeterminist position, but it failed to qualify as good science. The ad hoc nature of the addition

> really abandons the entire mosaic position, by rendering the assumption of qualitative division superfluous and, aside from this, its forced and artificial character, places a strain upon the mosaic theory under which it breaks down. . . . The "explanation" is, therefore, unreal; it carries no conviction, and no real explanation will be possible until we possess more certain knowledge regarding the seat of the idioplasm (which is an entirely open question), and its internal composition and mode of action (which is wholly unknown). In the meantime we certainly are not bound to accept an artificial explanation like that of Roux, however logical and complete, unless it can be shown that the phenomena are not conceivable in any other way.[52]

Excluding one of two theories did not render the other a winner, however, for radical externalist epigenesis had its own difficulties. How could such a position explain the regularities of cell division, for example? Pflüger had suggested "that the fertilized egg has no more essential relation to the later organization of the animal than the snowflake to the size and form of the avalanche which, under appropriate conditions, may develop out of it."[53] Wilson sharply disagreed with such an extreme view. Instead of either an extreme preformationist or an extreme epigenetic view, an intermediate position would most likely prove appropriate. Drawing on his own careful comparisons of the details of several cell lineage patterns, he concluded that cell formation must play some central role in differentiation since the lines of differentiation so often follow along cell boundaries. Yet the whole organism also exerts an important corrective influence on differentiation, so cell division does not alone directly cause the differentiation. Further experiments on *Amphioxus* supported Wilson's convictions. From Europe, Wilson had gained a particular experimental method and an interest in alternative approaches to studying the causes of differentiation. From his training and his detailed comparative work in the United States, he retained a healthy skepticism about such bold general theories and a desire for centrist rather than extremist positions—as did the rest of the MBL community.

Jacques Loeb at Bryn Mawr

When Wilson accepted the Columbia position in 1891, Bryn Mawr invited Thomas Hunt Morgan to join the faculty to replace him. Morgan had already accepted a second Bruce Fellowship at the Johns Hopkins University to continue his postgraduate work and had gone off to spend the first part of the summer in Jamaica with the Chesapeake Zoological Laboratory. He then visited the MBL. There he received a telegram offering him an associate professorship in biology at Bryn Mawr. If he chose to accept, he would also replace Wilson as head of the Biology Department. Morgan agreed.

The other member of his new department was the recent German import Jacques Loeb. He would replace Wilson's companion, fellow Hopkins graduate and physiologist Frederic Schiller Lee, who had moved on to Columbia with Wilson. Loeb had become unhappy with his situation in Germany and with the lack of opportunities available to him as a Jewish scientist. When he married an American woman, Anne Leonard, he let it be widely known that he would like to move to the United States to pursue his biological career.[54] The problem was religion. Bryn Mawr already felt that it had compromised enough in hiring Morgan, for "the college trustees were complaining that Bryn Mawr, as a Quaker school, ought to build a faculty made up of Friends. In proposing Morgan, who was not a churchgoer, [President] Rhoads had reassured the board that Morgan had at least been raised an Episcopalian and was 'thoroughly respectful to Christianity.'"[55] But what, they worried, could they do about this German Jew? Fortunately for Loeb, his new wife was a respectable graduate of Smith College, and Bryn Mawr finally relented and offered this outstanding researcher their position as physiologist.

Loeb managed to avoid the request that he teach taxonomic botany, but he did agree to teach elementary embryology even though he had little familiarity with the field. His wife reported that a friend tried to teach him basic sectioning and staining methods, but that "he said it bored him almost to death to make those stupid sections."[56] He never did really learn, so that in various publications later, he thanked Edwin Grant Conklin and others for their assistance in making sections and preparing his materials for him.

Despite their different backgrounds and interests, and despite disagreements about how to divide up tasks in the two-man department that Morgan officially directed, the two became friends.[57] Loeb had a powerful influence on Morgan's work, and Morgan undoubtedly influenced Loeb as well. Their basic concerns in biology really did not lie so far apart, considering that one was officially a morphologist, the other officially a physiologist; one an American embryologist, the other a German trained in a medical environment. Despite his confessed lack of interest in traditional embryological and cytological techniques, Loeb concerned himself with the sort of approach that Whitman, Wilson, Conklin, Morgan, and the MBL community regarded as central to biology.

Loeb's most influential early work began with his studies of heteromorphosis. These considered many of the same problems and same organisms that had become the focus of the new embryological work, but Loeb had a different interpretation of the results and an alternative research emphasis. As Philip Pauly has convincingly demonstrated, Loeb wanted to achieve a causal explanation of development, just as His, the Hertwigs, Roux, Weismann, Driesch, and the others did. Yet Loeb had a further synthetic goal; he wanted to control life as well as to understand it.[58] To do that, he believed, would require a chemical rather than a strictly mechanical explanation.[59]

Since Pflüger had turned attention to the importance of external stimuli in directing development, Loeb decided to pursue that suggestion. He also resolved to explore experimental cases in which he destroyed parts of simple organisms that would normally regenerate the missing or injured parts. Sometimes the part that regrew was not quite the same as normal. Yet something grew; this phenomenon Loeb called *heteromorphosis.* He set out to generate such results experimentally and found that the external conditions such as gravitational field generally did not affect the development. If he turned the crustacean *Antennularia antennia* upside down, for example, it grew upward. Thus, it seemed more likely to grow upward toward the light than to be directed by gravitational stimuli.

External chemical conditions held more interest and more promise for Loeb. Since botanist Julius Sachs had shown the importance of osmotic pressure for plant growth and differentiation, Loeb pursued the possibility that water might play a similar role for animals. He experimented with varying turgidity, for example, and with changing salt concentrations. Altered salt concentrations made a great deal of difference, he found, for several organisms. At Naples he had studied adults as well as developing lower organisms such as hydra. After his move to Bryn Mawr and his first summer at the MBL in 1892, he continued these experiments and extended them. Now he turned to sea urchin eggs as well, to look at the earliest stages of development. If the chemical conditions were truly crucial for directing development, changes should have some effect on eggs as well.

In fact, he discovered that soaking sea urchin eggs in more highly concentrated sea water did make a difference. While the eggs were in the salted solution, they appeared to rest. This occurred because they had less stimulus to change and thus experienced less "irritability," Loeb concluded, and he also held that the nuclei might meanwhile continue to divide and to prepare for cell division and differentiation. When returned to normal sea water, the eggs very rapidly divided into a large number of blastomeres, which demonstrated to Loeb that they must have been preparing. Despite the fact that he resisted the types of speculative theorizing which had characterized German *Naturphilosophie,* he nonetheless moved beyond his data at hand to conclude that "*the segmentation of the protoplasm is the effect of a stimulus which the nucleus applies to the protoplasm, and which makes the protoplasm close around the nucleus.*" Or that "segmentation of the protoplasm in the egg, and probably in every cell, is only the effect of a stimulus exercised as a rule by the nuclei."[60]

Yet Loeb did not like cytology. He did not prepare and fix and slice and stain his eggs. He put them in salt water and then back in normal water. He watched what happened and concluded that the chemical change must have had an effect. His physiological morphology brought together morphological problems and physiological approaches, but his theories remained more closely tied to the physiological tradition. He also strayed farther from the empirical evidence at hand than his more cautious American counterparts did. Undoubtedly, Morgan and others found Loeb tremendously stimulating. His work on "physiological morphology" and his related experimental work were, after all, excitingly suggestive. In addition, Loeb had a powerful personality. Whitman urged Loeb to join the MBL faculty as director of physiology and, in his capacity as first chairman of the new university, hired him at Chicago.[61] Thus, this creative "physiological morphologist" who found cytological work boring found a central place among a group of American biologists for whom cytology was central.

Certainly his new friends at the MBL and Bryn Mawr did not always agree with his conclusions. Nor did they always approve of his willingness to dispense with cytological studies and with more detailed and fuller examination of his organisms in their natural conditions.[62] European physiology as translated by Loeb and American embryology from Hopkins as influenced by Europe had not merged into one; they had instead become specialized parts of the same American biological enterprise. As Loeb moved to study development from the physiological tradition, his American counterparts moved to similar problems from the morphological tradition.

Morgan at Bryn Mawr

William Keith Brooks had said of Morgan when he was a graduate student that he tended to jump from one problem to another, that he was excited about and reacted to what went on around him, and that he responded to others and sometimes joined them in pursuing their problems and programs. This was certainly the case during the year that Loeb and Morgan shared at Bryn Mawr and the many summers thereafter at the MBL: the two worked together, and, even though they did not always agree about everything, Morgan's research program responded to Loeb's.

When Morgan first arrived at Bryn Mawr, he continued to pursue the kind of traditional morphological problems he had tackled as a graduate student. His first year he probably spent primarily in teaching and organizing the biology program. His first summer took him once again with the Chesapeake Zoological Laboratory group, this time to the Bahamas. Then he went on to the MBL, as he continued to do most summers thereafter. In both places, he pursued his study of the marine *Balanoglossus* and its larval form *Tor-*

Fig. 14. Morgan with students, Bryn Mawr College. From the Bryn Mawr College Archives.

naria, which naturalists had earlier regarded as a separate species. A short note and a longer paper a few years later recorded his observations of the differentiation process in this relatively simple form with its several life cycles.[63]

Morgan also looked at metamerism in the worm *Allolobophora foetida*.[64] The way in which worms form their regular, parallel segments had intrigued Wilson and had attracted more general attention at Hopkins and the MBL. Morgan took up the problem during his summer there, looking particularly at the unusual spiraling that occurs when a metamere splits at an early stage. Often such spiraling results as the developing worm regenerates a lost or injured metamere, for example.

This was a traditional problem of morphology that attracted Morgan, but also a problem of regeneration. Loeb had also pursued studies of regeneration and heteromorphosis during his summer stay at the MBL. Furthermore, instead of asking primarily, "What is the pattern of metamere formation?" as Hopkins students had before, Morgan had begun to look at abnormal situations and to ask what he could learn from those cases, as Loeb also tended to do. Comparing the normal and abnormal could yield new and valuable information about development, it seemed. Morgan was evidently inspired by Loeb in this direction.

By early 1893, Morgan had clearly found another line of research promising as well. Loeb had spent 1889–90 at Naples, overlapping with Boveri, and may have told Morgan about Boveri's work on chromosomes and heredity. We know that Wilson had also found Boveri's work attractive. Morgan therefore probably heard about Boveri and his cytological study of heredity from at least these two. He decided to translate Boveri's very suggestive paper of 1889, in which he had tried to hybridize cells by fertilizing with sperm from one species and eggs from another. The "larvae arising from the *enucleated* fragments of eggs have entirely the characteristics of the parent (*male*) species," Boveri had written. "Herewith is demonstrated the law that the nucleus alone is the bearer of hereditary qualities."[65]

Morgan decided to translate Boveri's paper, he said, because it was surely destined to become a classic. Also, he wished to "point out the new avenues of research that such work opens" since the results "touch the very heart of the question of Heredity." Further, "each advance in our knowledge gained by experimental work of this sort, carries forward rapidly our understanding of the most vital phenomena of life."[66]

By March 1893, when his translation appeared in print, and probably by 1892, Morgan had recognized the value of pursuing alternative approaches to questions of development and heredity. The use of unusual regenerative or heteromorphic phenomena in worms and the experimental manipulation of organisms to create new situations and to derive new data—Morgan felt that both had revealed their value for biological research. So had Boveri's cytological attack on problems of fertilization. The possibility that chromosomes might, in fact, act as *the* bearers of heredity also had therefore struck him as an idea worth pursuing. By this point, Morgan's work had begun to undergo change. Like Wilson, as he realized the value of exploring new approaches, using new organisms and perhaps new techniques, Morgan moved beyond the morphological tradition in which Brooks had trained him at Hopkins and into an expanded world. In this new world, changing convictions about how to do science prevailed; there the traditions had begun to change in fundamental ways. Once again, this was at heart a methodological and epistemological change rather than a shift in ontology or commitment to any new theory.

In the summer of 1893, Morgan worked at the MBL once again. This time he began his own experimental work, with studies of teleost and echinoderm eggs. With teleosts (or bony fish), he wrote in July, he had explored two problems. He had applied the experimental methods used by Pflüger, Roux, Driesch, and the Frenchman Laurent Chabry. Following their leads he had, for example, investigated the claim that the first cleavage plane corresponds to the lateral line of the embryonic and adult body. Using a mark on the egg to provide reference points, Morgan found that it does not, just as MBL researcher Cornelia Clapp had found that no such relation holds for the toadfish *Batrachus*.[67] The orientation of the embryo appears only after the head end is definitely established. With one species, *Ctenolabrus*, he shook apart blastomeres but could not get them to develop further. He did discover that putting the eggs in boiled sea water briefly seemed to affect cell cleavage, as Loeb had suggested, but he was not yet sure of the significance of such a result.[68]

He also worked with a second species, *Fundulus*, which Loeb also began to study at about the same time. Morgan succeeded in separating one cell of the two-blastomere stage with a needle. "This blastomere divides into two equal parts applied to one another, and is in all respects a miniature copy of the normal two-cell stage." Even the cleavage furrow appears in the same spot as it would have during normal cleavage. Removing the yolk of the otherwise normal egg allows abnormal cleavage but normal embryo development. "Does not this point to the conclusion," Morgan suggested, "that while the egg during development adapts itself to the necessities of the occasion by utilizing the mechanical means placed at its disposal, that it is a mistake to suppose the external conditions determine the series of phenomena."[69] The egg seemed already to have sufficient organization or direction to develop "properly" by itself, driven by internal causes.

Whitman had suggested that Morgan should also pursue a second set of questions, regarding concrescence, or the closing of the embryo along its lateral line. Morgan's studies showed that the body definitely begins at the head end and elongates posteriorly. He thereby disproved His's theory that the special series of cells making up the germ ring establishes the embryo, a theory endorsed at times by Roux, Oskar Hertwig, and others. Yet the proper interpretation remained open. Therefore, Morgan closed with words that came to characterize his own and also much of the American embryological work: "Perhaps I have stated my conclusion too positively. Any one working at such problems will realize and appreciate the difficulty of correct interpretation of such evasive and complicated phenomena. I should wish therefore to offer the explanation attempted above as an alternative view that may help as a working hypothesis and give the stimulus to further inquiry along the way."[70]

Perhaps Morgan had read a much discussed paper published in *Science* in 1890 or had heard discussion of it. There American geologist Thomas Chamberlin had published "The Method of Multiple Working Hypotheses," his presidential address before the Society of Western Naturalists.[71] Perhaps Morgan had heard of Weismann's call for bold theories, even if they proved inadequate in the face of future evidence. Scientists in America and abroad were self-consciously considering how best to proceed in doing science. In particular, how should one do good—by which they meant justified—science? Given that theorizing without empirical backing did not bring positive progress, according to the Americans, what would?[72]

Chamberlin called for multiple working hypotheses rather than the sort of single bold theory that Weismann or Roux favored. Having more is better than having only one working hypothesis, Chamberlin suggested, for putting forth only one makes its creator and followers too predisposed to accept it. Overall, what science, as well as other activities of society, needed was "greater care in ascertaining facts, and greater discrimination and caution in drawing conclusions."[73] Morgan, Wilson, and their Hopkins fellows agreed with that suggestion. Morgan continued his efforts to produce reliable facts and supportable working hypotheses.

Two months later, in September, he produced a second set of experimental studies. This time, he followed up on Boveri's and Driesch's work on sea urchins (*Arbacia punctatum*) and added a study of hybridized starfish and sea urchins. In the paper Morgan had recently translated, Boveri had held that fertilizing enucleated eggs of one species with spermatozoa from another results in embryos with the paternal type. Since they have only the paternal nucleus, Boveri had concluded that the nucleus must carry all hereditary information. Morgan called Boveri's conclusions into question because he found the experimental procedure inadequate. Boveri had claimed to have enucleated the eggs with the Hertwigs' method of shaking them apart, which produced many small pieces—some without nucleus, according to the Hertwigs and to Boveri. They determined which pieces had no maternal nucleus by size: some had only half-size nuclei and therefore presumably only the parental nucleus.

Morgan challenged this assumption by performing the same experiment with normal fertilization, that is, with both parents of the same species. Disagreeing with Boveri's claims, Morgan insisted, "Although I have studied large numbers of egg-fragments, I have never gotten any definite proof that the non-nucleated pieces segmented."[74] Spermatozoa enter the eggs, but they develop no further. The problem with the experiments is that one cannot really see with observations of living material either whether there is a nucleus at all or whether it is a whole or partial nucleus. Sometimes in sea urchins, the nuclear membrane breaks down and scatters the chromatin. Therefore, egg pieces may appear to have no nucleus when, in fact, all the nuclear material is really there. A smaller egg piece almost always has a smaller nucleus, but it may nonetheless have the same number of chromatin granules.

This meant that Boveri could not justify his conclusions. Proper comparisons with normal development undercut the significance of his experimental cases and called for the use of controls (though Morgan did not yet call them that). Instead of Boveri's conclusion that the nucleus bears all heredity, Morgan suspected that "a simple mechanical explanation is probably at the root of the matter, but I do not feel warranted in suggesting one."[75]

In other experiments, Morgan found that the sea urchin eggs are already cytoplasmically differentiated by the time of the two-cell stage and perhaps before. Therefore, the two-cell stage is *not* protoplasmically isotropic, as many of the German experimentalists had suggested. To show what he assumed was isotropy and the importance of external factors in influencing development, Driesch had subjected eggs to pressure by compressing them between glass plates and rotating them ninety degrees through their gravitational fields. He had then focused on the appearance of the micromeres, those smaller cells that occur in the upper hemisphere of those eggs that have a great deal of yolk (telolecithal eggs). These should appear at the ends of the eggs under pressure, according to Driesch's interpretation. Morgan repeated Driesch's experiments and extended them to gather further information. He found that the micromeres instead appear in the normal position, at the side. His results therefore brought into question Driesch's conclusions. Further experiments raised further questions and suggested that several alternative hypotheses could explain the experimental results. One seemed more probable at the moment, but he was not yet sure about its long-term success.[76]

A third set of experiments on echinoderms followed up Jacques

Loeb's use of sea water augmented with an additional 2 percent of sodium chloride. While in the salted sea water, the eggs remain unsegmented. When returned to normal sea water, they abruptly divide into many blastomeres. To Loeb, this suggested that eggs could develop normally with only one, maternal, set of materials plus external stimuli. This artificial production of cleavage became central to Loeb's research over a number of years, resulting in artificial parthenogenetic development all the way to the larval stage by the end of the decade. Morgan clearly found the work suggestive and fascinating. The technique was ingenious, but its significance less clear. "That this segmentation corresponds in any way to the normal stages," Morgan wrote, "I could not verify as the process seemed to me too irregular."[77] He was not at all convinced, without the sorts of cytological studies Loeb did not wish to pursue, that the nuclei were busily dividing and preparing for cell division. If not, what caused the cells to divide so quickly when returned to normal sea water? Only very few weak surviving larvae resulted from the procedure anyway. It remained to be seen how the experimental results should be interpreted.

A final set of experiments involved fertilizing starfish eggs (*Asterias Forbesii*) with sea urchin (*Arbacia*) sperm. Some embryos did form, different from normal *Asterias* embryos in both segmentation timing and size. Therefore hybridization must have occurred between "two forms belonging to entirely different 'Classes' of the animal kingdom."[78] This was an interesting surprise and one that might hold promise for later experimental research.

Morgan had run through the leading experimental work of the day, repeating and extending and questioning the experiments.[79] He reported results, taking care to record numbers of specimens used and failures as well as successes. His approach was very empirically based, always staying close to the procedures and results at hand. In offering interpretations he took great care not to step far beyond the data. He extended his occasional working hypothesis cautiously, always trying to give alternatives and to relate his results to other results and to other interpretations—as Wilson also had done.

During 1893 and 1894, Morgan turned to frog development, the study of which had begun the German move to experimental attacks on problems of development and heredity. There he once again considered the results offered by Roux and the Hertwigs and took exception to their interpretations. He once again rejected their bold hypotheses and speculative conclusions and offered instead careful working hypotheses tied closely to his own empirical results. Throughout his work to this point, he had asked morphological questions and used morphological approaches and techniques; like Wilson and Conklin and the other Hopkins graduates, he had also moved especially to embryology, which had become a new line of research lying outside traditional morphology but still within the modified morphological tradition, as informed by converts from the physiological tradition. In doing so, he joined Wilson and the Germans in helping to reorganize the boundaries between physiological and morphological research. He also began to embrace the promising techniques and approaches offered by some of the European experimental work.

Harrison's German Education

As the youngest of our four and the one with medical aspirations, Harrison was also the first to travel to Europe during his graduate training rather than after. Like his fellow Hopkins graduate students, he read the leading German morphological literature. This included some of the recent work on fertilization and cytology more generally as well as studies of later development. By Harrison's third year of graduate work, in 1891–92, Conklin and Wilson were already well under way with their cell lineage program. Boveri had performed his experiments that suggested that nuclei were the hereditary agents and that heredity and development might be treated as separate processes. Wilson was off to his year in Munich with Boveri and then to Naples. Morgan had gone to Bryn Mawr and begun to work with Loeb. Harrison had had a taste of marine work at the United States Fish Commission in Woods Hole in 1890 and the next summer in Jamaica. He had begun to look closely at teleost development (at the same time that Morgan and Loeb also did). Studying a higher organism than the marine invertebrates studied by Wilson and Conklin made sense for one with medical aspirations, though he had not yet firmly settled on an organism or a problem of research. Back at Hopkins he considered his options for his dissertation and decided on a trip to Europe first.

Off to Germany, Harrison planned to visit several cities, as Wilson had during his first trip abroad. Early in the schedule, he stopped at the University of Bonn and discussed with Moritz Nussbaum where he should settle for the year. Nussbaum argued that the young American should stay right there in Bonn and work with Nussbaum himself. He had received his M.D. at Bonn and had passed through the academic phases to become *ausserordentlicher* professor of physiology, so would be an appropriate advisor. Harrison did decide to stay and developed tremendous admiration for Nussbaum, describing him as "absolutely honest intellectually and in his dealings with men."[80]

Nussbaum, who had studied under Pflüger and Max Schultze in Bonn and had stayed to join the faculty, had written a series of articles concerning the growth and function of glands; he then had turned to problems of heredity, asking what is transmitted from one generation to the next. In 1893, while Harrison was in Bonn, Nussbaum published a paper dealing with the development of fish fins. During the time when he and Harrison were in most direct contact, Nussbaum produced a series of studies of muscle and nerve development. His papers suggest a primary concern with heredity and with delineating and describing histogenesis (the growth and differentiation of cells). Harrison's notebooks, with notes on Nussbaum's lectures about histogenesis of the nervous system in particular, demonstrate the deep influence that Nussbaum had on him.[81] Though it is difficult to assess how much Nussbaum caused and how much he reinforced Harrison's interests, it is clear that each profited through association with the other. One of Harrison's biographers pointed out that, though Harrison had originally remained undecided about where in Europe he wished to study, he never regretted having stayed in Bonn.[82]

After his year there, Harrison returned to Hopkins and received his degree in 1894. That year, after graduating, he applied for a Bruce Fellowship from Hopkins to allow him to continue his studies and research. As Brooks wrote in recommending him, Harrison was "a most promising student. He has marked ability for critical and literary work in zoology and for handling abstract problems. . . . He has fitness for research in the laboratory and technical skill as well as originality as an investigator."[83] Harrison received the fellowship. He then wrote to Daniel Coit Gilman and to Brooks that he wished to go to Kiel and to study with cytologist Walther Flemming. There he felt he could "best become acquainted with the latest methods in cytological research" and these he could "follow out to the greatest advantage."[84]

Brooks did not really favor the plan, since he would "prefer to see him use the methods at his command for the production of results, rather than to devote his time as Bruce Fellow to new methods." Further, Brooks worried that "his interest in Germany is not entirely educational, or scientific, but I hope he will put the three months to profitable use."[85] Brooks probably knew that Harrison had met a woman, Ida Lange, in Bonn during his first visit, whom he married in 1896. In fact, however, Harrison did not return to Bonn until later in 1895, and then again in 1898 and 1899. For the year 1894–95, he had received an invitation to join the Bryn Mawr faculty as lecturer in morphology while Morgan was off for a year in Naples.

In contrast with his fellow students, Harrison's first publication was not a short note in the *Johns Hopkins University Circulars.* Instead, he prepared a more lengthy study, in German, of the paired and unpaired fins in teleosts for a leading German journal. The typical short pieces followed the next year. Since he wrote his first major paper at Bonn, under the direction of Nussbaum, it was less in the Hopkins style than the others exhibited in their first papers. It employs typical techniques for fixing and preparing embryonic material, complete with staining cross sections with potassium gold chloride. Unlike the others, however, Harrison was looking at later development stages, at recognizable body parts that would grow into adult parts. He also was entering an ongoing German discussion, which influenced his style of presentation in some ways.

He found that to a notable extent, the skeletal parts developed independently, without intercellular connections.[86] Thereby he answered one of those questions that continued to attract the attention of the Hopkins group: is embryonic differentiation regulated by the whole interacting system of cells which make up the organism, or do the parts develop separately, either by some sort of cytoplasmic direction or (less likely, they felt) by nuclear direction? Separately, by cytoplasmic direction, he believed. The evidence also weighed on questions about how the teleosts are phylogenetically related to other organic forms. Thus, it fit the Hopkins program generally, though not the growing emphasis on earlier development or the style of presentation. In addition, since it appeared later, it reflected the considerable changes that had occurred in the short interim since the others had begun writing.

Harrison looked at later developmental stages, not at the very first cell divisions. Cell growth and cell divisions played an important part later, as well as for the earliest divisions, but it was masses of cells and different cell types that Harrison found interesting. He asked which cells became the muscles and which the nerves, for example, and where the cartilage that made up the skeleton and the fins came from. The problems and choice of organism diverged from that of his fellow students, but important overlaps remained. For in his early work, Harrison asked how metamerism occurs and what its significance is: how do the individual metameres arise and become differentiated as such different body parts as nerve and muscle? At root, then, Harrison's research focused on the problem of differentiation just as clearly as Wilson's, Conklin's, and Morgan's work had.

Wilson and Conklin had each asked how cells become differentiated out of the fertilized egg in a particular type of organism, gradually, step by step, while Morgan had begun by investigating in a variety of organisms how cells and body parts develop. Harrison looked at later stages in one set of organisms, focusing on the origins of parts out of the germ layers to a greater extent than the others did. He asked how the parts become differentiated. Not by external factors, he concluded, and evidently not by nuclear determinants or by cytoplasmic prelocalization either. It was not primarily the starting point in the fertilized egg or in some distant ancestral past but rather what happened thereafter in the interactive whole embryo that really mattered. For the cells' "position with regard to surrounding cells rather than their origin determines their ultimate fate."[87] Mechanical and structural rather than chemical conditions seemed primary to Harrison.

Harrison summarized and extended his work on teleost fin development in his dissertation, published in 1895.[88] Then, like his fellow Hopkins graduates, he turned to other research. Unlike the others, he moved to frogs rather than working with marine organisms, probably because he did not make the summer pilgrimage to the MBL each year. (This, according to his son, was at least partly because her violin was very important to Harrison's wife Ida Lange Harrison, and the humid Woods Hole summer kept it out of true.[89]) Like his fellows, however, he did adopt a new method in order to gain new information and reliable answers to age-old questions about embryonic development. Wilson and Conklin[90] had adopted cell lineage work and then experimental and cytological approaches to investigate the earliest stages of development. Morgan had explored experimental approaches, then frog development, then regeneration. Instead, Harrison embraced transplantation as his approach. The idea of experimentally transplanting tissue from one part of a body to another had occurred to Born and intrigued especially Harrison and Spemann. For each of these leading experimental embryologists, transplantation became the basis for his program through the late 1890s and into the twentieth century.

Traditions Transforming

During this period at the end of the nineteenth century, the morphological tradition experienced a shift, which brought relatively rapid change and ultimately transformed the tradition in fundamental ways. What we have seen is an initial outspoken enthusiasm among many morphologists for the sort of evolutionary program that Haeckel and others had outlined. Yet this enthusiasm came about against a strong background of interest in other aspects of morphological work as well, which continued to carry many researchers in other directions separate from Haeckel's or Gegenbaur's programs and toward other emphases.

Those historians who have emphasized the "revolt from morphology" have missed a central feature of late nineteenth-century biology and provide a misleading picture. First, the changes occurred too gradually and remained too nearly continuous with the past to constitute a revolutionary change; this was not really a *revolt.* Second, the changes were largely positive rather than negative. Thus, the changes were not so much *from* anything as they were toward something different, new, and exciting; a changing set of organisms, questions, and approaches enticed researchers in new directions. Third and most important, this was not a change away from *morphology* but a set of transformations within it. The Americans happily continued to call themselves morphologists and to see themselves as carrying on the morphological tradition. Rather than describing these events as a revolt from morphology, then, it is more accurate to characterize them as a transformation in the morphological tradition.

It is important to understand the nature of that transformation at this important time of change in biology. By 1900, some morphologists had expanded beyond the earlier set of assumptions of the tradition to look at new organisms, using new techniques and equipment, and adopting interventionist experimental approaches. Borrowing from the physiological tradition, morphology began to move into an area previously not occupied, an area focusing on questions about heredity and the earliest stages of development and their causes.

Above all, these morphologists embraced a new conception of what they wanted from science as well as how to get it. Specifically, they sought to move away from the sorts of grand explanatory but highly selective theories that had characterized some of the most highly publicized morphological work. Instead, they looked to working hypotheses to guide the production of knowledge and, in particular, the slow and careful production of definite results and therefore positive knowledge. These changes are not in the content of scientific belief; they are not changes in theory. Rather, they concern how to conduct science, what we should look for, and how to evaluate it thereafter. The changes are, in short, epistemic.

By 1900, the morphological tradition had expanded and was undergoing transformation, that is, though research remained historically continuous with, and strongly influenced by, the older tradition. Concern with structure, patterns of change, and careful descriptions of changes in the patterns—both within individuals and across species—prevailed throughout.

Yet the 1880s and 1890s also brought a stronger fragmentation along the various emerging divergent lines of research within the morphological tradition. The result was the beginning of establishment of concentrated specialty areas (in some cases disciplines, in others, fields or less well-defined, distinctive research clusters). This shift should not be surprising given that we should expect traditions to shift in response to progressive changes.

Until 1900, then, something that clearly remained a morphological tradition endured. After 1900, however, the tradition began to expand in such different directions that it began to diverge more significantly. The new research programs, fields, and even disciplines branched out so that soon researchers began to call themselves cytologists, embryologists, evolutionists, or geneticists, for example, instead of morphologists.

The problem for the rest of the book is to trace the diverging lines of research pursued by each of our four biological leaders. Part III therefore consists of four chapters that detail, in turn, the career and research progress of each of these men. The chapters discuss the problems they addressed, the work they did, and their changing views about how good scientific work ought to be done. They carry each career until roughly 1915, by which time each man had established the productive research program he pursued thereafter, had attracted students and gained institutional support, and had moved in his own direction. They also show shared values but diverging research interests. The question remains, in the face of this divergence, whether a continuous morphological tradition persisted or not. If so, what form did it take? If not, what replaced it, and more generally, what does this story tell us about the nature of scientific change? The concluding chapter of the book takes up that last set of central questions.

PART III

Diverging Research Programs

6

Edmund Beecher Wilson

MORPHOLOGY HAD BEEN led astray by speculative embryology, Edmund Beecher Wilson suggested in a theoretical lecture at the Marine Biological Laboratory (MBL) in 1894. The questions morphology addresses are "complex and difficult and must necessarily be attacked by means of inference and hypothesis." This was especially true for questions about the past such as phylogenetic questions. Yet precisely there, hypothesizing had gone too far, so that morphology had become "burdened with such a mass of phylogenetic speculations and hypotheses, many of them mutually exclusive, in the absence of any well-defined standard of value by which to estimate their relative probability. The truth is that the search after suggestive working hypotheses in embryological morphology has too often led to wild speculation unworthy of the name of science."[1] The modern student of morphology, as a result, must turn away from the "speculative pedantry" of phylogenizing and look instead to the solid empirically based facts and to the "brilliant discoveries of the cytologists and experimentalists." Therein seemed to lie the promise of progress that science requires.

In particular, Wilson felt that the "embryological method" must be abandoned as the ultimate source of morphological fact. That meant that in order to establish ancestral relationships or homologies of parts, morphologists could no longer automatically rely on embryological details as if those were the most primitive, or closest to the historical or "true" form of an organism. In fact, they could not appeal to any one developmental stage as more important than all others. In short, the morphologist must recognize that all embryological stages can undergo change and adaptation just as later adult forms change. It might even be the case that homologies can come into being in the course of development because of environ-

Fig. 15. Wilson with cello, 1889 or 1890. From the collection of Linda Timmons.

mental conditions. In order to assess relationships among organisms, to trace the ancestral form of a particular organism, or to identify homologies, therefore, the morphologist can assuredly look at embryological comparisons. But he must not take these as the ultimate data. Instead, "*we must primarily take anatomy as the key to embryology, and not the reverse.*"[2]

Wilson certainly never suggested giving up morphology. Nor embryology. Nor should morphologists turn away from interest in establishing homologies or evolutionary relationships. The changes Wilson urged in this lecture centered instead on how researchers should go about doing their science and what should count as reliable knowledge. Though not ostensibly the central focus of his lecture, this epistemological concern clearly mattered critically to Wilson.

Morphologists should move beyond the sorts of speculative work in which some had become bogged down. In particular, the exaggerated claims made for embryological comparisons had not yielded the kind of "positive results" that science needs but had instead produced scepticism and indifference. Morphologists must continue to set up hypotheses but should do so carefully and avoid speculative excesses. Start with the solid facts already assimilated. Then use reliable methods to achieve further facts and to move toward a "trustworthy basis of interpretation."[3] With such language, Wilson sought to purge contemporary morphology of its epistemological failings. In the same lecture, he also looked forward to currently emerging and future directions in morphological research.

Reliance on the embryological criterion had led researchers to see development of present-day organisms as directly caused by their historical past. Thus, they had come to regard "every operation of development as merely the result of 'inheritance.' " But that would not do. Yes, for normal development there exists something material called idioplasm. True also, this idioplasm is inherited from the past. Yet the past does not by itself directly *cause* present development. Nor can appeal to that inherited past in itself explain development. As a result, above all, "the normal operations of development are essentially physiological problems."[4] Morphologists should not despair of finding answers and of determining homologies, but Wilson felt that problems of development and inheritance remained inextricably interconnected. Reformed methods and further work were needed to carry their study further.

That further research is precisely what Wilson himself continued to pursue. Throughout the 1890s and until 1905, he directed all his work at the persistent problem of determining how development occurs. What are the precise morphological changes? Is development epigenetic or predeterministic in some way? These were problems shared widely by the morphological community and taught at Hopkins. Yet Wilson followed his earlier leads even farther into the cell and away from traditional morphologists. In order to learn about "embryonic physiology," he sought to understand increasingly detailed changes in embryonic structures. The structural changes in turn began to lead him to various tentative or working hypotheses about what physiological processes effected the changes. This chapter explores his various lines of research and hypotheses to about 1915: his close study of the cytoplasm, of the cell as a whole during all stages of inheritance and development, of embryological implications of the cytological study, of chromosome study, and of sex determination.

In two papers that appeared in 1895, for example, Wilson followed European suggestions by Hermann Fol and Theodor Boveri and looked in greater detail at changes during the fertilization process. What goes on in the nucleus and related parts? he asked. The first paper, reporting results of studies on echinoderm eggs by Wilson and his graduate student Albert Prescott Mathews, questioned Fol's assertion that fertilization brings a conjugation of two male and two female centrosomes, which Fol called a "quadrille of centers."[5] In particular, the two Columbia researchers focused on the sperm asters in sea urchin and starfish species, following them during each step from their initial formation to their role as guiding asters for the first cleavage of the fertilized egg. They found that the sperm aster has no centrosome, however, as Fol's interpretation required. Neither does the egg have asters except those that come from the sperm. This was a step in sorting out just what happens during fertilization. As such, it called into question some of the assorted and apparently conflicting information that researchers had begun to accumulate.

A few months later, further studies followed a paper of Boveri's and went farther. There Wilson attempted to make some sense out of the proliferating names for the various morphological parts of the cell during maturation and fertilization. Archoplasm, centrosome, amphiaster, aster, and spindle fibers, for example: all exist as identifiable parts of the cell with distinct roles in development. But where, when, and how do these many parts arise? Out of a preexisting reticulum of cellular stuff, Wilson concluded. The cytoplasm is not homogeneous stuff but is already a complex network of chemically differentiated material, which he called the cytoreticulum. The asters seem to form from this cytoreticular material, gathering around the spermatozoon's middle piece as it moves through the egg's cytoplasmic network. Similarly, the spindle fibers form from the nuclear reticulum, which may consist of chromatin.

Wilson was not confident on this last point, but he realized that it would provide an appealing connection between the otherwise separate chromatin and cytoplasmic developmental processes. He did not at this time believe that the nucleus and its chromosomes held any special uniqueness or significance. Like the cytoplasm, the nucleus must consist of a network of tiny fibers that have a chemical consistency and undergo mechanical changes. These together cause development. Functional cell organs form out of the reticular material so that "from the physiological point of view, therefore, a 'cell-organ' is a differentiated area of the cell-substance in which a specific form of chemical change occurs."[6] But how? Do these cell organs arise *de novo*, or do they always rise from a preexisting body that serves as a formative center? Are they therefore the result rather than the cause of the chemical differentiations that occur in the cytoreticulum?

Opinion had leaned heavily in favor of the continued existence of cell organs with modifications rather than new formation, Wilson acknowledged. He was not convinced, however. Even the chromosomes, which many researchers had begun to regard as the most clearly continuous individual structures, come and go so that "there is absolutely no proof that the chromosomes emerging from the network at the succeeding division are the same 'individuals' (i.e., the same group of individual molecules or other ultra-microscopical units) as before." Instead, "the constancy in their number may with equal reason be regarded as the outcome of formative process (i.e., at bottom chemical changes) affecting the chromatin-mass as a whole and causing it to crystallize, as it were, in a particular form at certain focal points."[7]

True, this view remained hypothetical, but so did the alternatives. And "even such a hypothetical suggestion may have some value, if only as a protest against the dogmatic assumption that each and every localized form of cell-activity must be referred to the agency of corresponding pre-formed material germs." The idea of cell organs as forming, by some process of chemical crystallization, out of the cellular reticulum had other advantages in its favor and seemed to Wilson "full of suggestions for further study."[8] What Wilson had begun to realize was the vital importance of formulating hypotheses carefully if he wanted to achieve reliable knowledge. He also saw the value of focusing on the specific chromosomal material to begin to make progress in solving problems about the nature of development and heredity.

The Cell

The further study that Wilson saw as needed was facilitated by a beautiful set of photographs of the fertilization process, the first such series ever produced successfully. It appeared in 1895 as *An Atlas of the Fertilization and Karyokinesis of the Ovum,* by Wilson and Columbia photography instructor Edward Leaming.[9] This volume and magnificently detailed drawings of the nuclear and other changes during fertilization and cell division have led historians and biologists to cite Wilson's work as fully "modern" and as "classic." It was, but perhaps not in the sense that that is often meant, for his interpretations of the phenomena did not all prevail in the long run. In both *An Atlas* and *The Cell in Development and Inheritance,*[10] Wilson continued to pursue the working hypothesis that the cell parts in fact crystallize during development out of the cytoreticulum. Some processes akin to Jacques Loeb's physiological changes in chemical composition or Wilhelm His's mechanical foldings seemed to cause the appearance of the cellular apparatus or organs. This interpretation soon gave way to revised alternatives, but the facts recorded in photographs and detailed descriptions did remain the standards and classics of quality.

In particular, Wilson moderated his preferred interpretation of cell development out of cytoplasm. Eschewing dogmatic proclamations in favor of either nuclear predetermination or extreme cytoplasmic isotropism, he instead sought a middle ground where nucleus and cytoplasm, heredity and development each retained a fundamental importance. He considered the collection of data that seemed to have been sufficiently well established as "facts" and looked for the best explanation that fit all those data. In *The Cell,* he worked through many purported facts and found some wanting. He examined theory after theory and found many of those inadequate also. Some failed to fit some of the accepted data; others went too far beyond the data.

The goal of science, for Wilson, was to achieve "positive knowledge." Observation and experiment yield careful descriptions of facts. These, in turn, add up to well-established empirical generalizations, which provide the ground for hypotheses. Some hypotheses follow as an "absolute logical necessity." Others are less certain and require inductive moves beyond the data. This was true, for example, for that central problem of embryology: explaining the innermost structure of the germ cell and its means of producing the appropriate fully differentiated adult. Wilson wrote in *The Cell* that:

> it should be clearly understood that when we attempt to approach these deeper problems we are compelled to advance beyond the solid ground of fact into a region of more or less doubtful and shifting hypothesis, where the point of view continually changes as we proceed. It would, however, be an error to conclude that modern hypotheses of inheritance and development are baseless speculations that attempt a merely formal solution of the problem, like those of the seventeenth and eighteenth centuries. They are a product of the inductive methods, a direct outcome of accurately determined fact, and they lend to the study of embryology a point and precision that it would largely lack if limited to a strictly objective description of phenomena.[11]

Thus, it is worth exploring those various alternative hypotheses that offer that "point and precision" to embryology.

Not all hypotheses would qualify as good scientific contributions, of course. In particular, Wilson felt that the Roux-Weismann theory failed in its assumption of initial cytoplasmic isotropy and subsequent qualitative nuclear division as directing cell differentiation. The theory suffered especially from its "*quasi*-metaphysical character." It had no facts actually in favor of it alone, and it had some against it. In contrast, good hypotheses must take into account the available data and must be able to explain the "well-known facts." They must also remain in accordance with all established facts. In addition, the good scientific hypothesis must be nonmetaphysical, meaning testable and capable of rejection in the face of new empirical evidence. Wilhelm Roux and August Weismann simply met objections by invoking auxiliary hypotheses, so that it was improbable for any appeal to fact to overturn the theory. That, for Wilson, made it an unacceptable theory, and not properly scientific.

Aside from its problematic nature, it seemed that the body of facts gathered by the 1890s had already ruled the Roux-Weismann theory out. Accumulated evidence showed that "the improbability of the hypothesis becomes so great that it loses all semblance of reality." Indeed, new work struck a "fatal blow to the entire Roux-Weismann theory" and "practically disproved" the idea of qualitative nuclear division central to it. Furthermore, even more "facts were determined that threw doubt on the hypothesis of cytoplasmic isotropy," which Roux also assumed.[12] As a result, it seemed definitely established that the initial cytoplasm of the egg already does have some differentiation, so that qualitative nuclear division is not the cause of, and does not explain, all of development.

Critics of the Roux-Weismann view offered an alternative explanation that the nucleus of each cell contains identical material because each contains the same idioplasm (that hypothetical material of inheritance tentatively identified with the chromosomes). Yet Wilson rejected that alternative as well since "there are, however, a multitude of well-known facts which cannot be explained, even approximately, under this assumption."[13]

Wilson instead offered his own alternative accounts of nuclear and cytoplasmic changes. He felt that these ideas fit the data. The nuclear chromatin did seem to specialize so that only the relevant chromatin goes to each cell, as Roux and Weismann had said. Yet, rather than casting out some of the material bodily, as Roux and Weismann insisted must occur, Wilson saw cell division as bringing the degeneration or dissolution of the part of the chromatin which was no longer needed. Or perhaps that unnecessary part undergoes transformation into nucleoli or the linin-substance that makes up the cellular reticulum. In addition, the chromatin might produce "nuclear ferments" that act as "formative substances" to direct development chemically. Though Wilson saw his ideas about how heredity of chromosomes gives way to development of body parts as "certainly full of suggestions," he also regarded them as put forward "only tentatively as a 'fiction' or working hypothesis."[14] He did "not desire to add more than is necessary to define some of the problems still to be solved; for I am mindful of Blumenbach's remark that while Drelincourt rejected two hundred and sixty-two 'groundless hypotheses' of development, 'nothing is more certain than that Drelincourt's own theory formed the two hundred and sixty-third.'"[15]

Both of his working hypotheses reflected what Wilson saw as the central need for a theory of development and heredity to explain how both the nucleus and the cytoplasm work together and how both internal and external factors shape the result. The "nucleus alone suffices for the *inheritance* of specific possibilities of development," but both nucleus and cytoplasm "are necessary to *development*." Any full "explanation of development is at present beyond our reach," but it is clearly a problem seeking a solution.[16] Tentative hypotheses can help guide the way if they are reasonable, that is, accessible to test against the facts. They must not be wild speculations or mere ad hoc creative exercises. Wilson concluded his first edition of *The Cell* with the view that:

> I can only express my conviction that the magnitude of the problem of development, whether ontogenetic or phylogenetic, has been underestimated; and that the progress of science is retarded rather than advanced by a premature attack upon its ultimate problems. Yet the splendid achievements of cell-research in the past twenty years stand as the promise of its possibilities for the future, and we need set no limit to its advance. . . . We cannot foretell its future triumphs, nor can we

> repress the hope that step by step the way may yet be opened to an understanding of inheritance and development.[17]

The Cell provided a brilliantly organized and reliable compendium of fact and theory of the day. It also presented the "state of the art" in such detail that it might very quickly have become out of date. Although new knowledge did require Wilson to begin almost immediately on a revised and expanded second edition, then to move after that to a third, the volume received rave reviews especially for its thoroughness and reliability, as well as for its readable narrative style. Yet Wilson's friend Edwin Grant Conklin objected to one aspect of the book.

In reviewing the volume, Conklin praised it for the abovementioned features. He outlined the main themes, then focused on the last chapter. Here he believed that Wilson had gone too far, and he noted that: "to me it seems that the feature which is most open to serious criticism is one which gives the work one of its particular charms, viz, its enthusiasm and, in some places, its controversial spirit. Professor Wilson frequently uses strong language, sometimes stronger than seems to be justified."[18] In particular, Conklin found Wilson's rejection of the Roux-Weismann theory unjust. That theory should not be so easily dismissed as a priori or as unscientific, Conklin felt. To him the primary difference between Roux and Wilson lay in the method by which each saw the nuclei as becoming qualitatively differentiated. The fundamental agreements went much farther than the disagreements, and Conklin felt that Wilson had stated the objections much more strongly than was really justified.

Today Wilson's presentation seems a masterpiece of diplomatic consideration of all sides. Yet surprisingly, Wilson himself agreed with Conklin's criticism. He wrote to his friend and acknowledged that he, too, had seen "the lack of sufficient restraint on some of the moot points that are not yet fully ripe for discussion" as the most serious defect of the book. "I know of nothing more difficult, than to be at once clear, brief, and critical; and you are entirely right in intimating that I have in some places allowed enthusiasm to run away with judgment."[19] As a result, in the second edition, Wilson moderated many of his statements and called again for further work to resolve the many as yet unanswerable questions about development and inheritance.

After *The Cell*

Having tied up the loose ends of fertilization and cell division as best he could for the moment, Wilson returned to more general questions about the significance of cell cleavage for development. He asked whether cleavage causes or results from differentiation of the egg and emerging embryo. By 1896, that had become even more difficult to answer because different organisms did different things in development. The experiments with isolated blastomeres that Roux and Hans Driesch had begun and others had since pursued had yielded apparently contradictory results for different organisms. Other researchers had put forth their theoretical explanations for their particular results, each assuming that his own results really represented all of development. A few, such as Conklin, had begun to suggest that several different patterns of development might actually occur, some involving determination from the beginning and others invoking interactive regulation until determination occurs at much later stages. Wilson disagreed with this suggestion. He believed that the various extremes, as well as positions in between, "must however be brought under a common point of view; for it is certain that development must be fundamentally of the same nature throughout the series, and the differences between the various forms must be of secondary moment."[20]

For Wilson, neither mechanical conditions nor inherited germs alone could explain the facts of development. Something like the type of localization of formative factors in the egg-cytoplasm (whatever those might be) that Thomas Hunt Morgan and Driesch had endorsed in a recent jointly authored paper from Naples held the most promise, he felt.[21] Yet he recognized that the causes might be complex and might vary in detail from case to case. The problem was becoming one of sorting out details of heredity and of development.

The remainder of Wilson's nineteenth-century research followed similar directions. In one line of work, he wanted to follow up that traditional morphological question he had addressed in his 1894 MBL lecture: how can one best assess phylogenetic relations? In that earlier paper he had concluded that embryological stages cannot serve as the ultimate reliable criterion of homologies and of ancestral relations. The adult form must remain the final source of data.

This might suggest that mechanical conditions acting on and within the individual direct development. Wilson rejected that conclusion. Current development is a double problem, he insisted, since each organism has a complicated mechanism with its own equilibrium but also has a past that influences that equilibrium.[22] Cell lineage provides clues. The precise way in which cell cleavages reflect the past was not clear, nor was the extent of ancestral reminiscence. Yet recent cell lineage studies, especially Conklin's work on *Crepidula,* had offered insights. Conklin responded to Wilson's "Cell-Lineage and Ancestral Reminiscence" with enthusiasm. He wrote to his friend that he had "no doubt that it will rank along with your Nereis paper as a classic on this subject, which seems to me the most important in modern Embryology. With every position which you have taken in your last paper my observations fully agree."[23]

The collected comparative evidence from a variety of organisms led Wilson to a "speculative working hypothesis." Perhaps an ancestral organism had acquired a particular "building pattern" in which certain cells give rise to certain body parts. The ultimate fates and details of development might then undergo changes while the pattern of cell division (or building pattern) remains constant. Such a speculative working hypothesis might eventually prove unsupportable, but for now it fit the facts and provided a "new point of attack upon some of the puzzling phylogenetic problems with which the study of cell-lineage has to grapple."[24] Any adequate theory had to account for both present mechanical conditions and ancestral reminiscences, and this one did.

In another line of research, he continued to develop details of the protoplasm, or cellular reticulum. By 1898, he had concluded that "a critical study of the living protoplasm shows that it is a liquid, or rather a mixture of liquids, in the form of a fine emulsion consisting of a continuous substance in which are suspended drops of two general orders of magnitude and of different chemical nature, as indicated by their staining reactions." The various parts, such as astral rays, grow out from a center along the chemical meshwork, Wilson believed in the summer of 1898. They seemed to grow by adding to the tip, taking up material from the surrounding reticulum rather than generating their own material at their center. Even that reticulum must have arisen secondarily, out of an initially apparently homogeneous matrix. Which left things just where they were as far as explaining development goes. For at the beginning of development lies "the invisible organization of a substance which seems to the eye homogeneous." Not a satisfying account, certainly. Yet, Wilson continued, there was "much in these conclusions to suggest, and nothing to contradict, the hypothesis that the 'homogeneous' or 'continuous' substance may be composed of ultramicroscopical bodies by the growth and differentiation of which the visible elements arise, and which differ among themselves chemically and otherwise."[25] These might be like Carl Nägeli's "micellae," which were molecular groups of protoplasm surrounded by layers of water. As such, they might retain an essentially chemical and physical nature but might combine in a way that could explain the apparent differences between living cellular and other, nonliving matter.

By putting forth such admittedly hypothetical units, Wilson could move toward explaining "some of the most puzzling questions of cytology," which included "the ultimate nature and origin of dividing cell-organs like the nucleus or the plastids, and especially such a contradiction as that presented by the centrosome which may apparently arise either *de novo* or by division of a preexisting body of the same kind." Yet he was not really happy with such a hypothesis, which strayed too far from established fact. Indeed, accepting it "leaves untouched the fundamental problem of division," and whether an answer would ever be forthcoming at all remained an open question.[26]

In a general paper considering recent research at the turn of the century, Wilson elaborated this sceptical point. He acknowledged that too many people had sought simplistic mechanical explanations of biological phenomena. Yet growth, reproduction, and cell division really raise very difficult problems for mechanical accounts, and biologists had at last begun to realize that fact. Unfortunately, the realization had led some to embrace a form of neovitalism which had its own problems. Driesch, in particular, had taken that step. Clearly, Wilson was not persuaded that the organic phenomena could not be explained in inorganic terms; he did not have any reason to embrace a revised ontological position himself. He felt that the current discussion actually helped biological research overall, because as yet biologists had not found any mechanical source and explanation of all "the adaptive and coordinating activities of the organism." Nor had they yet found any analogy between living processes and those of nonliving nature. In calling attention to these limitations of current biology, neovitalism was performing a great service and set "in a truer perspective the nature of the great problems still before us."[27]

Furthermore, biologists had finally begun to realize that they were a long way from answering all the fundamental questions about life and had accepted doubts and partial answers as a necessary part of scientific progress. Thus, "it is a happy augury for the future of biology that it is apparently entering upon a new constructive phase characterized by frank recognition of ignorance, by a decline of baseless theorizing, by steadily increasing thoroughness and range of observation, and above all by the extensive application in every direction of exact experimental methods to subjects which have, hitherto, hardly been approached along this path."[28]

So biology no longer pretended to have all the answers, and biologists must continue looking elsewhere with wider approaches. In keeping with his own suggestions, Wilson himself followed several lines of experimental work and then wrote at greater length about what he saw as the productive directions for biological research. He recognized that old problems and methods were giving way to a larger variety of divergent but related problems and approaches. Study of heredity diverged from embryology just as study of chromosomes pursued different methods than did embryology.

Experimental Embryology

With the new century, Wilson again embraced experimentation as an important additional approach to morphological problems. In his presidential address to the American Society of Naturalists, he took the opportunity to consider appropriate "aims and methods" for natural history. The traditional search for ancestral relations among organisms had produced vague conclusions, unsatisfactory solutions, "inflated speculation," and a general weariness with "wanderings through the scholastic maze." This had led researchers to new problems of cell theory and experimental and developmental physiology. With new problems came new methods, or actually a return to older experimental methods as supplements to observation and comparison.[29]

Some biologists had urged the "deliberately calculated and precise alteration in the conditions under which phenomena occur," or experimental manipulation, as the *only* valid biological approach. Nonexperimental methods, according to this view, are necessarily, by their very nature, not scientific. Wilson disagreed.[30]

In fact, all methods that advance and unify knowledge are properly scientific. Nature performs experiments, and man observes. The scientist also performs experiments. He then compares the results from the various observations. The accumulation of knowledge and the organization of the knowledge that results is science. A recognition of the value and validity of different approaches to nature advances science. So let us accept that the "bug-hunter," the "section-cutter," "worm-slicer," and "egg-shaker" are all scientists, each motivated by a love of nature and each seeking positive knowledge. Experimentation can bring exactness of method, so let scientists embrace but not demand experimentation.[31]

By that time, Wilson had identified the driving problems of his research. With this endorsement of experimentation as well as observation and comparison, he also established the full range of his methodology. By 1900, then, he had expanded his research program to address problems of what causes embryonic development through a variety of approaches.

By 1900, Wilson also had a research community in place at Columbia. As he had continued to revise *The Cell* in the period between its first presentation as lectures in 1892–93 and the publication of the second edition in 1900, so also had the Columbia program undergone revision. The Department of Biology had begun in 1891 with Henry Fairfield Osborn, Wilson, and paleontologist-zoologist Bashford Dean, an instructor in the old Columbia School of Mines (who also later served as founder and first director of the Cold Spring Harbor Biological Station). At first, these three set up shop in the School of Mines building on Forty-ninth Street. Joined by Oliver Strong, the first fellow in biology and a student of Osborn's, they taught only undergraduate courses at first. Actually, they did not teach anything until the second year since they were all off doing research and traveling. Then, as the number of assistants and junior staff members began to increase, the department added advanced work and a few graduate courses. This activity, in turn, attracted more graduate students, and very soon the department moved to larger quarters in the College of Physicians and Surgeons. There they remained until 1897, when they moved to the initially spacious-seeming Schermerhorn Hall on the new Columbia University site in Morningside Heights. They had also recently changed the department's name to reflect more realistically the actual interests and research of the group. They thus became the Department of Zoology in 1896, "by popular demand."[32]

With the increasing activity, Wilson put together a team of graduate students and assistants interested in participating in his research program. He had a few students at Bryn Mawr and had, of course, worked in close consultation with Conklin and others during summers at the MBL. Yet this year-round community of researchers, coming at a time when Wilson had settled into a research program of his own, brought new stimuli and new challenges. He had already published the paper with Mathews looking at the quadrille of centers and had found Mathews's interest in the physiological processes of development particularly promising. He worked with James H. McGregor on spermatogenesis of the amphibian *Amphiuma,* and he attracted McGregor to serve as instructor in zoology at the MBL from 1899 to 1906. McGregor also pursued his evolutionary field work, at Puget Sound (through an arrangement with Columbia), and turned increasingly to questions of vertebrate paleontology. Only after 1900 did a group of researchers begin to explore problems more directly related to Wilson's driving interests and to make suggestions that he decided to pursue. These associates soon led him more directly to chromosomes.

Also after 1900, Wilson himself turned to more explicitly experimental studies of development by applying Loeb's and Morgan's methods of artificial parthenogenesis to his own organisms.[33] In particular, the experimental production of artificially dividing eggs allowed him to begin determining what the chromosomes do in cell division. Artificial parthenogenesis allowed division with only half the normal number of chromosomes and also only two asters and centrosomes.[34] Along with Mathews, Wilson pursued the work further since these continued experiments produced information unattainable with normally developing organisms.

In particular, Wilson still wanted to determine how the centrosomes arise: whether *de novo* or with genetic continuity out of preexisting "promorphological" parts. Working with the sea urchin *Toxopneustes,* he concluded that it was "nearly certain" that the centrosomes, which direct cleavage, arise by division of the initial egg centrosome, which itself forms *de novo.* Comparing results from the artificially induced cleavages, from normal cases, and from egg fragments shaken apart provided bases for further comparison. It also yielded further information about normal changes in the centrosomes, asters, spindle fibers, and other pieces of mitotic apparatus.

Wilson concluded that the artificial exposure to magnesium chloride caused the eggs to begin dividing and to compensate for the lack of any spermatozoon by producing the complete mitotic division apparatus anyway. Yet at that point the egg can no longer compensate and "may manifest a multitude of aberrations which constitute a veritable carnival of development which one can hardly witness without a sense of amazement."[35] Comparing what the cells can do under these altered conditions with what they seem to do normally shows which parts of cell division require fertilization and which do not. He was left with only unsatisfying suggestive explanations and a call for further study, which led him onward.

A second series of experiments with eggs treated with ether instead of magnesium chloride produced similarly promising but incomplete results. These experiments showed that one could change the early conditions of cleavage and development without necessarily altering the end result; in Driesch's words, considerable regulation occurs. In Wilson's terms, fertilization and cell division exhibit considerable "plasticity." The action of centrosomes and other cell organs, and of nucleus and cytoplasm have an "essential relation" that directs cell cleavage and differentiation. His results showed Wilson that nuclear activity and division depend on cytoplasmic activity and suggested that the latter plays a fundamental role for the former.[36] Chromosome division seems to occur later, he felt, though the details were much less clear.

At this point, Wilson had certainly concluded that the cell division that is absolutely central to life processes is very complex. The chromosomes play a small part, but the intricate maneuvering of the other cell organs such as centrosomes, asters, and spindle fibers seemed key to him. These he regarded as cytoplasmic as well as nuclear phenomena; the nucleus remains closely tied to the cytoplasm, just as inheritance remains inextricably connected with development.[37]

Over the next few years, he applied experimental methods to a variety of organisms to extend the study of what morphological changes occur during development. Regeneration studies, undoubtedly stimulated in part by Morgan's and Loeb's extended discussions of the regenerative process, suggested that the developmental process does retain a great deal of plasticity.[38] The pincerlike claws or chelae of the snapping shrimp *Alpheus* can reverse their asymmetry, for example, so that if the larger claw is removed, the second grows and adapts to replace it while the new one replaces the functions of the old second one. The study certainly suggested that the initial differentiation process could have been switched and was not strictly and irrevocably determined. A good deal of cytoplasmic regulation even at relatively late developmental stages therefore seemed possible.[39] Apparently, an internal "mechanical regulation" made regeneration possible.[40]

The suggestion once again called for further studies to determine the extent of localization of later parts in early cleavage stages: to assess the extent to which differentiation was determined at the earliest cell divisions, or even because of the cell divisions. Returning to work with egg fragments which followed Boveri's and his own research from a decade before, Wilson now extended those studies to a variety of different organisms. Different organisms are differentially localized at early stages, he found. In nemertine eggs (*Cerebratulus lacteus*), for example, a fragment divides just as a whole, normal egg does, whereas an isolated blastomere after the first division acts as if it were still only a part of the whole. Such disparate results suggested to Wilson that the egg does not really exist as a prelocalized mosaic that is simply divided up throughout cell division. The particular morphological pattern of different egg materials is a primary factor in determining the character of cleavage, to be sure, but that morphological pattern is itself "a secondary result attained by a progressive (i.e., epigenetic) process."[41]

The question remained open whether obvious and demonstrable organization of the egg is caused by a priori organization of the egg or by a later regulatory, epigenetic process. It also left open the question of the role the nucleus plays in directing development. The best available conclusion still seemed to be that which Wilson had drawn in *The Cell*:

> Primarily the egg-cytoplasm is totipotent in the sense that its various regions stand in no fixed relation to thc parts to which they respectively give rise. Secondarily, however, development may assume more or less of a mosaic-like character through differentiations of the cytoplasmic substance involving local chemical and physical changes . . . The primary determining cause of development lies in the nucleus, which operates by setting up a continuous series of specific metabolic changes in the cytoplasm. . . . The cytoplasmic differentiations thus set up form as it were a framework within which the subsequent operations take place in a course which is more or less firmly fixed in different cases.[42]

The egg and subsequent cleavage stages do represent a mosaic, as Roux had insisted and as Wilson had been disinclined to accept earlier. The problem was to discover how and at what point the mosaic occurs. To what extent and at what point has heredity predetermined development? Once again, Wilson felt that comparative study of a variety of organisms could best answer that set of questions. He accordingly turned to the molluscs *Dentalium* and *Patella.*

From the earliest stages when eggs are discharged and hence become visible, they exhibit material differences that "foreshadow a corresponding distribution of these materials among the blastomeres during cleavage." The hypothesis that different formative stuffs exist in the visibly different regions seemed to fit the data best and to explain the phenomena at hand. It may well be, however, that the nucleus is directly concerned with every cytoplasmic differentiation as well. Some organisms have a greater degree of localized distribution of different materials at an early stage; others remain quite responsive to external conditions for a longer time. In some organisms, "the germ regions prelocalized in the unsegmented egg are, at least in the case of certain cells, accurately marked off by the subsequent lines of cleavage." In others they are not. Perhaps the nuclear organization translates to differentiated localized materials at different points. "But if the potentiality of the cytoplasmic system be primarily given in the nuclear organization, and if this be the primary determining source of the initial cytoplasmic localization in the unsegmented egg, this presents no insuperable difficulty. It is obvious, however, that this question is one not for speculation but for further experiment."[43] The same major questions had continued to drive Wilson's work up to this point, even while experimentation augmented the more traditional approaches. Those larger questions persisted, even as his research began to yield some answers about specific developmental phenomena.

The Move to Chromosomes

Throughout this period of grappling with the mosaic idea of development, Wilson was clearly also trying to assess what role the nucleus plays. He had largely treated the chromatin as just another cell organ, with chromosomes arising periodically but for uncertain durations. His emphasis on both cytoplasm and nucleus allowed him to set the one aside while exploring the other. Then his own students and associates provided research results that began to force the question: what, exactly, are the chromosomes doing? And to what effect?

The first indications of chromosomal importance had come by 1902: from Wilson's former student Clarence E. McClung, his graduate student Walter Sutton, and his MBL colleague Thomas H. Montgomery. In addition, people had rediscovered Mendel and had begun to ask whether hereditary units of some sort might exist, perhaps even along the chromosomes, as Weismann has suggested hypothetically. A short report of Wilson's from December 1902 reveals his budding interest in both subjects.[44]

Montgomery's work on insect chromosomes had strongly suggested that mitosis brings a synaptic stage during which the paternal and maternal chromosomes actually pair up and join together.[45] Reduction division thereafter separates the chromosomes from each parent into different germ cells. Montgomery's work of 1902 did not give definite confirmation for such an interpretation, in Wilson's view, but it extended the suggestions at which earlier results had only hinted.

In other work on chromosomes, Wilson felt that a student in his own laboratory, Walter Sutton, had obtained results that went much farther toward a "definite conclusion" about the role of chromosomes in development and even moved toward an explanation of "the Mendelian principle." Sutton's work on grasshoppers showed a set of eleven pairs of chromosomes of different sizes. The pairs separate and reappear in each synapsis, and Sutton concluded that each pair contains one chromosome from each parent. Furthermore, the work demonstrated that each chromosome does retain its individuality throughout the division process; it does not fade out of existence and reorganize at each division, as Wilson had been inclined to believe. Both Sutton's and Montgomery's studies showed that each germ cell contains a full, matched set of different types of chromosomes.[46]

Wilson's own work on parthenogenesis had shown that only one set of each type of chromosome is necessary for development; yet each cell in normal development had two full sets. All the evidence supported the conclusion that these sets are homologues deriving from the two parents. If so, Sutton explained that the subsequent separation of the members of the homologous sets during reduction division would, in fact, give a physical basis for Mendelian dominant-recessive results in each generation.[47]

Wilson was not yet fully convinced, but he was intrigued by the possibilities. He responded to a critic a year later that the accumulated results still showed "that these suggestions do not yet afford a full or positive explanation, but only, in my own former phrase, give a 'clue' which awaits further development and test. It is entirely possible that the clue may prove false, yet even so it may serve to illustrate that 'fertility of false theories' " which even his critic had recognized.[48] By two years later, in 1905, Wilson clearly regarded the suggestion as extremely fruitful and as much more than a simple "clue."

One might ask, as other historians certainly have, why it took Wilson so long to begin working actively on chromosomes. Why did he continue his experimental studies of cell cleavages and regeneration after his 1902 acceptance of chromosome individuality and their role in synapsis? Why was it not until 1905 that he adopted a fully developed research program to study chromosomes? The answer is that he really did just what one would have expected: he continued to follow what he saw as the most productive leads to answer developmental questions. There seemed to be no reason for him to jump at chromosomes any more than at the centrosomes or asters on which he concentrated, because with the latter apparently simpler cell organs he could carry out careful observations and more readily achieve what he regarded as definitive results. This success allowed him to address what he regarded as the key questions for biology.

As *The Cell* had so amply demonstrated, Wilson regarded actions of the whole cell as necessary for development. Chromosomes might have something to do with *heredity,* but he had spent over twenty years discovering the role of cytoplasm and mechanical actions and formative chemical stuffs for directing *development* and the way it responds to inheritance through the actions of the cell. As he had said before, the nucleus—perhaps through the action of the chromosomes—might serve as the "ultimate court of appeal" for inheritance. But the cytoplasm and the cell organs that arise within it also remain absolutely necessary for development.

In addition, Mendelism and chromosomes offered no major surprises. After all, Nägeli's idioplasm and Weismann's germ plasm both fit Mendelian ideas and emphasis on chromosomes perfectly well. There did not seem any call for radically new theories in the light of Sutton's and Montgomery's contributions. Nor did chromosomes alone explain anything much about development, which Wilson wanted his science to do.

Yet by 1905, Wilson had accepted that chromosomes do play a special role in heredity, which may follow Mendelian lines. For the first time, the accumulated evidence showed that the nuclear contribution is not just a chemical or molecular organization. Rather, it "represents beyond this some kind of definite material configuration of the nuclear substance."[49] Perhaps chromosomes and heredity deserved special attention for a while as a productive means to reliable knowledge.

How the unit characters are distributed to the chromosomes remained unclear, Wilson acknowledged, but he admitted that they somehow were. And "to just this extent have we admitted the principle of preformation as applied to the nuclear substance or idioplasm." It is not just substance but some kind of preformation that the chromosomes carry from the parent to the offspring. Therefore, the germ has two elements: one part undergoes epigenetic development while the other controls and determines the developmental process. That the individual morphological characters correspond in some way to chromosome characters now seemed inevitable. He and his associates had, unintentionally, helped to separate heredity from development. Wilson was, by December 1904, convinced of the chromosomal role in carrying hereditary units. "We can hardly imagine at present how this is possible; and it must be freely admitted that such a conclusion has an appearance of artificiality and crudeness that almost inevitably creates a certain feeling of scepticism. Nevertheless, to a conclusion similar in principle to this the facts seem to be pretty definitely pointing."[50]

The problem was to begin to pin down characteristics and to formulate a research program to tackle relations of chromosomal preformation and cytoplasmic response. This required no radical change of research direction. Nor did it bring a major change in theory. Because of the way Wilson thought science must work to achieve positive knowledge, he saw the discovery of individuality of chromosomes, their hereditary role, and their probable connection with morphological unit characters as a progressive step in experimental developmental morphology. Though a long way from Ernst Haeckel's morphological program, the new work on chromosomes was a logical next step in Wilson's own research. Because of the changes in epistemology, directing Wilson to look for different results in his science, his work lay historically in the same tradition as the older nineteenth-century morphology even though it had diverged significantly in problems attacked and in approaches used.

Sex Determination

A summer's research took Wilson to questions of sex determination in insects, particularly the Hemiptera. Montgomery had studied this group, and a number of other researchers had recently begun to tackle questions of sex determination as well, especially in insects.[51] Clarence E. McClung, a former student of Wilson's at Columbia who then taught at the University of Kansas, had pointed to the existence of an extra or "accessory" chromosome. McClung had, following H. Henking, postulated that this represented *the* determinant of sex: presence of the accessory made its possessor one sex, absence the other.[52] This certainly suggested a major role for chromosomes in directing development and indicated that the preformationist part of development might dominate, for sexual characteristics at least. Wilson acknowledged that his own studies on the Hemiptera revealed constant sets of chromosomes in each sex and that there could be "no doubt that a definite connection of some kind between the chromosomes and the determination of sex exists in these animals."[53]

The work of Bryn Mawr Ph.D. Nettie Stevens (who had been a student of Morgan's) on the mealworm *Tenebrio* and the aphid *Aphis Rosea* and *A. Oenotherae* supported that conclusion as well. She concluded that spermatozoa are "distinctly dispermic, forming two equal classes, one of which either contains one smaller chromosome or lacks one chromosome." Upon fertilization, one always produces the male, the other always the female. Yet Stevens, like Wilson and Morgan, remained cautious not to go too far beyond the data at hand, concluding, "Whether these heterochromosomes are to be regarded as sex chromosomes in the sense that they both represent sex characters and determine sex, one cannot determine without further evidence." For the sake of following the best available working hypothesis, she acknowledged that "we are not certain that we have a right to attribute the sex characters to these particular chromosomes or in fact to any chromosomes. It seems, however, a reasonable assumption in accordance with the observed conditions." Furthermore, the version "which brings the sex determination question under Mendel's Law in a modified form, seems most in accordance with the facts, and makes one hopeful that in the near future it may be possible to formulate a general theory of sex determination."[54]

Without question, some causal connection between chromosomes and sex determination must therefore exist. However, Wilson disagreed with McClung that the accessory *causes* sex determination. Nor was he persuaded by Stevens, though he also recognized the value of a promising working hypothesis. Instead, he concluded that the difference must be one of degree rather than of kind. It must at root be a metabolic or growth difference, directed by the different chromosomes that determine sex. For Wilson, the interaction of chemically differentiated cytoplasm and nuclear, chromosomal directors must cause development. Saying that a structural chromosome that resulted at fertilization could cause development and differentiation did not make sense. The regulatory environment must act alongside heredity, just as cytoplasm and nucleus must work together.

It does not make sense to ask of Wilson such questions as when he underwent his conversion to accept genetics or which piece of data proved to him the chromosomal and Mendelian theories of heredity. Wilson did not do science that way. He did not endorse a new theory because of one critical piece of evidence or after one "crucial" experiment. He looked instead for the weight of the evidence. He asked which of the alternative theories was better supported by the empirical data, which was most suggestive and productive for future research, and which was simplest. During 1905, he decided that the chromosome theory of heredity held the most promise, meaning that chromosomes held the material of heredity. He did not also accept that this meant that all subsequent development had been simply set at fertilization to be played out in a predetermined way. Such a view simply could not fit with the years of research he had carried out showing the importance of centrosomes and other cell organs and showing embryonic responses to experimentally induced changes. Development remained just as important as heredity, even if he was looking to chromosomes and heredity for positive knowledge for the present.

The chromosomes might well be "concerned with the transmission and development" of sexual and, by analogy, other characteristics.[55] Yet the precise nature of the "concern" remained unknown, and, at any rate, other factors clearly also played roles as well. "It is entirely possible," Wilson wrote in 1907, "that we are on a wrong track, that the so-called sex chromosomes are only associated in a definite way with the sexual characters, and have in themselves no causative influence on sex production. The whole chromosome theory of heredity, for that matter, stands unproved before the judgment seat." Nonetheless, he believed that "the chromosome theory as applied to the sex problem presents a sufficiently plausible force to be taken for a time as a guide to further examination of the facts. Perchance the true explanation may be found on the way, even should our working hypothesis prove a false leader."[56]

Wilson did not reject any older theories or evidence thereby. Instead, he embraced a new suggestion, an exciting hypothesis that could bring many phenomena under the same explanation. He also began to accept the separation of hereditary transmission through chromosomes, on the one hand, from the ongoing developmental process, on the other. This distinction, in turn, opened the way for a legitimate research program studying chromosomes, on the one hand, and exploring cytoplasmic developmental phenomena on the other.

For the years 1905 through 1912, Wilson carried out an extensive and lengthy set of "Studies on Chromosomes" in a variety of organisms. He also continued to look at sex chromosomes in various animals and to assess whether they really *determine* sex absolutely or just guide the development of sexual characteristics in some less rigid way. How, he continued to wonder, does hereditary transmission get translated into development? He worried about how sex determination can work in those species that have no different chromosomes rather than accessories. He recognized the problem of selective fertilization: namely, that a random pairing of chromosomes should result in a 1:2:1 Mendelian ratio, yielding a 3:1 with normal dominance. The ratio of males to females is, however, 1:1. Therefore, it seemed, some sort of selection must occur at fertilization to guarantee that equal distribution. Wilson did not like the idea, but he nonetheless considered various hypotheses about how it might happen.

Throughout the period to 1912, Wilson worked on chromosomes and on sex. By 1911 and 1912, his then colleague and friend Morgan had developed an explanation of how hereditary transmission might work. At that point, Wilson continued with his problems of how the cell works in inheritance and development and devoted himself to re-revising his opus for the last time. Since Morgan had moved to Columbia in 1904, the two and their students had maintained even closer contact than during the MBL summers. As Morgan's specialized program in genetics began to emerge after 1910, Wilson wandered in to talk and to help himself to the ubiquitous banana collection hanging in the lab to feed the *Drosophila*. Wilson himself continued to specialize in cytology, with his primary emphasis on the role of the cell and its parts in development. He continued to address questions that had concerned him since his introduction to morphology, using the approaches he had embraced during the 1880s and 1890s. He continued to call for a mechanistic or physicalistic approach to biology rather than a vitalistic alternative, since he felt mechanism offered the working hypothesis most likely to yield positive results.

Wilson's presidential lecture to the American Association for the Advancement of Science in 1914 expressed his views about how biological science had progressed in the previous twenty-five years. Among more empirical achievements, he said, science had learned the important lesson that it must keep moving. It seeks no final solution but follows "stepping stones to further progress" and moves on. Boundaries between fields become obliterated and re-form, for example. Throughout the changes, biology asks two fundamentally inseparable questions, Wilson explained: "What is the living organism," and "how came it to be?" By the 1890s, biologists had begun to tire of the latter, evolutionary, question as it was then approached and turned aside from the past to ask about present life processes: "They awoke to the insufficiency of their traditional methods of observation and comparison and they turned more and more to the methods by which all the great conquests of physico-chemical science had been achieved, that which undertakes the analysis of phenomena by deliberated control of the conditions under which they take place—*the method of experiment*. Its steadily increasing importance is the most salient feature of the new zoology." Yet experimentation was not new in itself, Wilson pointed out, and his reader should not make the mistake of thinking that he was rejecting older methods. Rather, the application of experimental methods to problems of embryology was new and brought progress because it reopened and reconceptualized old unsolved fundamental questions and, at the same time, provided a productive new means of attack. This experimental embryological work served as leaven for all of zoology. Rejoicing with a retrospective glow of success, Wilson saw that "it was a day of . . . revolt from speculative systems towards the concrete and empirical methods of the laboratory; of general and far-reaching extension of experimental methods in our science."[57] Again, the reader should not interpret this to mean that Wilson rejected the problems or the older methods of morphology; rather, he here recorded the sense of revolt from the unproductive speculation of an earlier approach. This is an epistemological claim. Indeed, his epistemological convictions, along with those of many of his contemporaries, had changed so that knowledge thereafter should be sought in the "concrete and empirical"—that is, in definitive, reliable, testable results and thus in positive knowledge, to use the terms that Wilson himself invoked over and over.

Great observational and experimental approaches and great fundamental problems of heredity and of development had brought great promise for biological success. Yet time had revealed that progress does not come instantly. Scientists must have patience and accept ignorance and error. Science must work more slowly and cumulatively than scientists might want. Little successes and accumulations of fact do not immediately or easily lead to solutions to the larger and tougher questions. History shows that "the remoter problems of science, like distant mountain-peaks, seem to recede before us even while our actual knowledge is rapidly advancing." In science, "we shall make lasting progress only by plodding along the old, hard beaten trail blazed by our scientific fathers—the way of observation, comparison, experiment, analysis, synthesis, prediction, verification. If it seems a prosaic program we may learn otherwise from great discoverers in every field of science who have demonstrated how free is the play that it gives to the constructive imagination and even to the faculty of artistic creation."[58] Science, like Wilson's own research career, must proceed sensibly and carefully, following suggestive working hypotheses and exploring promising experimental sources of new information but relying, ultimately, only on the positive knowledge of established and verified fact. Only then do we achieve the positive knowledge that Wilson believed made up science.

Wilson had grown beyond his Hopkins training in looking to different and more focused problems. He had turned to cytological study of heredity as well as to development to provide legitimate research problems. Above all, he had articulated his own research approach. Synthetic in incorporating what worked even from competing approaches, Wilson's research exhibited his carefully balanced judgment. Patience and plodding were necessary for good scientific work, Wilson insisted. He was, as Ross Granville

Harrison put it, "a classic" who "is more concerned with the perfection of his product, with setting his ideas in the proper relation to each other and to the main body of science. His impulse is to work over his subject so exhaustively and perfectly that no contemporary is able to improve upon it." The classic "resembles the legendary she-bear that licks her cub into shape with great patience and solicitude and does not let it go until she has done everything possible for it."[59]

7

Edwin Grant Conklin

FOLLOWING THREE short early papers and his graduation from Hopkins in 1891, Edwin Grant Conklin moved to Delaware, Ohio, and entered a five-year period of relatively little publication and professional activity. While he continued his research on cell lineage of the slipper snail *Crepidula* during summers at the MBL, serving as a course instructor there as well, he held a position as professor of biology at Ohio Wesleyan University devoted to teaching rather than to research. Though he sought to introduce a strong research component to the school and had even been hired with that intent, he evidently found it difficult to do much of his own work.

Conklin began his professional career with typical concerns. He wrote to the president of Ohio Wesleyan during negotiations that, frankly, he needed money to pay off his debts from graduate school. He also planned to spend each summer in original investigation at the seashore, which would create greater expenses than most faculty members incurred. Since he felt that such research would translate into better teaching, as well as bring honor to his university, it was reasonable that the university pay him for his summer research work in addition to the usual yearly salary. He had already received other teaching offers at higher pay, he told his old alma mater, but he was attracted by the possibilities for laboratory instruction. In the end, the university supported his request, and he went to Ohio Wesleyan.[1]

Conklin's position there was not atypical at the time in the United States, though he had more freedom to do summer research than most. Most colleges and universities in the 1890s expected very little original research of their faculty. One anonymous writer to *Science* had explained less than a decade earlier that American sci-

Fig. 16. Conklin, probably while at the University of Pennsylvania. From the Marine Biological Laboratory Archives.

ence lagged behind its European counterparts because its teachers were not also investigators. The small band of dedicated researchers coming from such new schools as the Johns Hopkins University had before them a difficult task, the writer explained. They want to carry out their own work, but "their first object is necessarily to render research more important in public estimation, and so to smooth the way for a corps of professional investigators."[2] Well into this century, leaders of biological societies continued to lament the poor regard for research found in many academic faculties. It seemed that a scientist was expected to teach what was known at the time and not to learn more.

Accordingly, Conklin spent much of his energy during his first five years of postgraduate employment in teaching and trying to persuade people of the value of research.[3] The effort paid off in quality, for many of his students commented on his excellent teaching. One biographer identified him as "a very great teacher. He was without question the best lecturer I heard as a student during eight years in two great universities. With a wealth of knowledge of science, literature, and philosophy, with a fine sense of humor and with a clarity that was almost unbelievable, he lectured in a manner that excited interest and admiration of generations" of students.[4] During those years of intensive teaching of a range of subjects, however, he published only one paper.

That one paper was actually the text of a lecture on fertilization presented at the Marine Biological Laboratory (MBL). As such, it represented a summary of the best evidence and current best available discussion of a problem of central interest to the MBL community and especially to the director, Charles Otis Whitman, who undoubtedly asked Conklin to prepare it and to couch it in the most accessible summary terms. Whitman's own work on fertilization demonstrates his continued interest in the subject, and he probably entered the discussion with enthusiasm.[5] Conklin's one publication during this time was therefore not a typical scientific article presenting research results but a public discussion. As a result, it may well have helped Conklin to crystallize his own convictions about such central embryological phenomena as fertilization.

Biologists including Oskar Hertwig and August Weismann, Conklin said in his lecture, had increasingly come to the view that fertilization involves the fusion of two nuclei, one coming from each parent. Indeed, nuclear materials were generally thought to be "the only essential substances upon whose union the act of fertilization depends." Conklin objected vigorously to this view. Far from being a purely nuclear phenomenon, fertilization must in fact be "a union of *all* the essential parts of the reproductive cells, cytoplasm as well as nuclei."[6] Here appeared in its first full form Conklin's commitment to remaining, as he often put it, a "friend of the cell—the whole cell." And of the whole egg.

It made no sense even on a priori grounds, Conklin insisted, to dismiss the nonnuclear cell parts as completely insignificant to the fertilization process. Yet a priori grounds remained all that the nucleophiles had, for they had not actually observed fertilization as a union of nuclei alone. They could not tell by simply looking that the rest played no role. All the evidence from experimental embryology did show that the cytoplasm cannot function without the nucleus for long, but neither can the nucleus operate without cytoplasm. "Both nucleus and cytoplasm are essential constituents of the cell, and one cannot be said to be more important than the other."[7] Nucleus, cytoplasm, and all the other cell parts act together.

"Of course," Conklin acknowledged, "it may be urged that there is some unknown and invisible influence emanating from the nucleus which controls all the processes of cell life. In the nature of the case such an assertion cannot be affirmed or denied on the ground of observation, and it seems to me sufficient to urge in reply that we should believe things are what they seem unless we are compelled to believe differently."[8] For the time being, then, the whole cell must remain the subject of inquiry in studying fertilization and all the other fundamental properties of organic structure and function.

This view clearly fit nicely with Edmund Beecher Wilson's ideas about the role of cellular form and activities at the same time. The survey of literature and the questions of interest which Conklin offered reflected the wider atmosphere in which he presented the work, since the MBL community as a whole had begun in the course of the 1890s to address a broad set of questions relating to all stages of early development.[9] It is interesting that Conklin decided to tackle this problem rather than to discuss some aspect of cell lineage, as so many others did, but he probably responded to Whitman's urging and perhaps to Wilson's encouragement to do so. Even with this lecture, he still had produced no real research papers aside from the very short notes from his Hopkins days.

The next year, in 1894, Conklin decided that it was "his duty" to leave Ohio Wesleyan. Deciding just where to move proved more difficult, however, for Stanford and Northwestern Universities each tried to lure him. The choice was not at all obvious to him. He liked Stanford's president, biologist David Starr Jordan, and was attracted by the climate and the prospects for specializing at Stanford. Yet the Chicago area had already become established as an intellectual center, was closer to the east-coast concentration of biologists, and offered a much better established and larger university in the form of Northwestern. Finally, after several months of negotiations, he resolved to move to Northwestern University, where he began a two year stay as its first professor of zoology and director of a new zoological laboratory. There, as before, he continued his cell lineage work during the summers at the MBL but evidently still found little time for further research and publication. Several years later, in 1897, he acknowledged to Jordan that he did "not underestimate what I have missed in not casting in my lot with you. I heartily congratulate you upon the magnificent prospects which are now opening before your university."[10] He also found the Northwestern community less than perfectly supportive. When a clerical group attacked his teaching of evolution at the end of his two years, he resolved to leave.[11]

Only after Conklin accepted a position as professor of comparative anatomy (changed later to professor of zoology) at the University of Pennsylvania in 1896 did his research career really begin. Almost surely, his new location in the center of east-coast activity and in the intellectually stimulating Philadelphia environment made possible Conklin's blossoming as a professional biologist and his rise to international distinction. That year the MBL appointed him as trustee for the first time, for example. His earliest papers at Pennsylvania also came during his first year and were based on presentations at the annual American Morphological Society meeting and at a special invited conference at the American Philosophical Society (APS) in Philadelphia. Being in Philadelphia and taking the opportunities offered him made all the difference to Conklin's career and to his scientific work.

Pennsylvania

The short report of his American Morphological Society paper shows Conklin following up suggestions only hinted at in his earlier lecture at the MBL. It is not cell size that determines an individual's body size, he said; rather, the number of cells is what matters. This conclusion had clear bearing on the half-embryo experiments and their interpretation, where researchers sometimes obtained smaller and sometimes full-sized but half-formed embryos. Some embryos recovered normal size but not shape, or shape but not size. Part of the explanation of the differences might lie with differences in numbers of cells in each case. Yet Conklin simply reported his observations and did not draw that conclusion in print at this point, though he did so later. He probably did discuss the idea with Wilson, who was concurrently exploring similar questions and similar results.

In his other major paper of 1896, Conklin explored another theme central to Wilson's work. In a conference organized by Edward Drinker Cope and Liberty Hyde Bailey, participants addressed various factors of organic evolution for the APS audience. Conklin focused on embryology and evolution. In agreement with Wilson, he regretted that embryology had so long been asked to perform the impossible service of determining phylogenies. With Wilson, he agreed that only comparative anatomy of all stages could serve as a reasonably safe and reliable guide to past and present relationships. Yet embryology retains a central role for studying evolution, for "more than any other discipline, embryology holds the keys to the *method of evolution.*" That is, "the study of the causes of development will go far to determine the factors of phylogeny." This is true because in order to understand evolution, it is necessary to know the causes of all basic organic phenomena, including growth, differentiation, reproduction, and variation, which make up the evolutionary process.[12]

To make his case, Conklin offered six propositions and discussed each at some length. Examining evolution generally and exploring differences between Lamarckian and Darwinian evolution more specifically, Conklin sought to show that study of embryology holds a fundamental key to understanding evolution; that an individual's embryonic development results from a combination of intrinsic and extrinsic factors that provide the material for both heredity and the capability for adaptive change; and that experimental embryology produces the strongest and best evidence to demonstrate just how those factors operate on particular organisms. On the last point, he insisted that without experimentation we end up with the alternative theories of Lamarck, Darwin, and Weismann and a basic "deadlock of opinion, each challenging the other to produce indubitable proof. This can never be furnished by observation alone. Possibly even experimentation may fail in it, but at least it is the only hope."[13] Even without full experimentation to "prove" it, Conklin nonetheless believed that Darwinian theory would prove best able to explain the causes of evolution.

The tone of this first major paper is somewhat different from his tone later. Conklin had long been particularly interested in evolution and how it works, but his real concern was to show that evolution made sense, that Darwinian evolution offered the best account of evolutionary causes, and that evolution remained perfectly compatible with religion. Nowhere else did he put so much emphasis on the relative differences among the alternative versions of evolutionary theory. Perhaps this was because this APS conference had been co-organized by Cope, a committed neo-Lamarckian. Conklin may well have felt it incumbent upon him to address questions that would not otherwise have captured his primary attention. The way that *Time* reported the event years later, after an interview with Conklin, suggests that the conference had made a great impact on him. The magazine article depicted the rather bashful young (at age thirty-three) biologist as appearing "before his elders." "He was pitted in debate against a booming bigwig, Professor Edward Drinker Cope of University of Pennsylvania, who advanced the Lamarckian view that acquired characteristics (e.g., muscular development or manual skill) can be inherited. Conklin defended the opposite view, boldly stated that inherited characteristics are determined solely by the germ plasm." Time eventually awarded the win to Conklin.[14]

The paper and the entry into the professional biological world which it brought show that Conklin had begun to join the biological elite as he did not do and probably could not have done at Ohio Wesleyan or Northwestern. A flurry of activity, with publications, invited lectures, active membership in the APS, and scores of other professional and social undertakings characterized the rest of Conklin's life.

With his arrival at Pennsylvania, Conklin also began to publish reviews of leading writings of the day, as he continued to do throughout his career. His first such review considered a recent paper of Weismann's, in which the German biologist had forsaken his commitment to a strict mosaic sort of development and advocated an extension of selection theory into the embryo. Recently translated and republished in English, Weismann's paper, "On Germinal Selection," acknowledged that his earlier work had failed to convince everyone and might even have employed an inadequate methodology in appealing to inductive logic. We cannot explain all organic phenomena through combinations and recombinations of biophores, he admitted. Now he appealed also to the principle of selection, which operates at all levels and all stages of life. Thus, a struggle for existence among those theoretical determinants and biophores (which were the units of heredity and development for Weismann) joined the Darwinian struggle among individual organisms to effect evolutionary change. This shift in view amounted to nothing less than a "revolution of opinion," Conklin felt, and it was "scarcely less sudden and wonderful than that manifested in a certain historical conversion on the way to Damascus." The new Weismannian theory was intriguing and important, Conklin conceded, but as yet it remained unconvincing. "Evidence should be the crucial test for this and any theory," and on precisely these grounds, Weismann's failed. As yet "not a particle of evidence is adduced in proof of a single proposition named."[15] Further research was needed, and Conklin took up the challenge.

Once he was fully established at Pennsylvania, Conklin settled into serious and energetic pursuit of three major sets of problems. The first was strictly biological, in which he continued to explore the patterns and processes of fertilization and cell division. The second was a concern with the relations of science and his other great commitment, religion: what are the relative domains and the respective limitations of each? He also explored the role of education in science, particularly in biological and evolutionary science.

The first concern assumed primary importance, and 1897 brought Conklin's first major opus on the subject, as well as several smaller papers. A second MBL lecture assured his listeners that accounting for the development of an individual organism was the most important problem of biology. Further, he agreed with MBL colleagues, including Wilson, that this study ought to be undertaken using both experimental and observational methods, for "there is no such sharp distinction between observation and experiment in biology as is sometimes assumed."[16] This epistemological point was important to all four of our Hopkins graduates. While some biologists, in conflict with rejections of any experimentation by other researchers, insisted that experimentation was *the only* way to do science, these four each called for a coordinated embracing of a range of methods and approaches. This eclecticism fit their Hopkins training. They also endorsed Whitman's desire for a community adoption of a range of problems and methods, even while individual researchers or researches adopted one or another specialized choice.

Conklin clearly thought about questions of how best to do science, just as Wilson did. Yet Conklin initially paid less explicit attention than Wilson to epistemological questions. Like Wilson, he assumed that biologists should adopt a careful, empirically based epistemology that would take them toward well-founded factual statements and better explanations of the causes of natural phenomena. Unlike Wilson, he spent more time worrying about which interpretation of developmental phenomena seemed to fit the facts best. Currently, Conklin explained, Wilhelm His's theory of organ-forming germ regions had gone out of style, but empirical evidence demanded that it be taken seriously. Alternative accounts that insisted on strict epigenetic development of initially homogeneous stuff clearly had proven inadequate. What seemed to occur, in fact, was some degree of initial organization followed by more adaptive development. The earliest cell divisions are most important, Conklin concluded, because they are the most constant and the most predetermined by heredity; they also most often serve to separate different later body parts. Thus, the earliest stages are morphologically the most important, he insisted, thereby disagreeing with Wilson's emphasis on the later physiological and mechanical processes of developmental change.[17]

Conklin's emerging view that early development is more mosaiclike and only later gives way to significant adaptations led him to comment on Wilson's lack of substantial justification for rejecting the Roux-Weismann mosaic interpretation of development. In his review of Wilson's *Cell*, Conklin noted that Wilson at times fell into inconsistencies. In some places, he asserted that Wilhelm Roux and Weismann were wrong and that the very type of theory they offered was necessarily problematic. In other places, Wilson agreed with the basic assertions of the very theory he had earlier rejected. Unfair, Conklin insisted. While he agreed very substantially with Wilson's conclusions and respected the quality of his friend's work, Conklin wanted a more serious look at various versions of the mosaic theory and at His's ideas. The relative roles of heredity and development remained very unclear, he felt. Accordingly, he called for further careful study of the earliest stages of development in a wide range of different organisms.[18]

At about the same time, his own lengthy paper on the development of *Crepidula* development finally appeared in print. This represented his dissertation research, augmented by all those summers spent at the MBL expanding the project. Whitman had agreed back in 1892 to publish the work in his journal, despite the ominous length of the paper, the number of expensive quality color plates, and the consequent high cost.[19] He kept his word, even though it took far longer than Conklin had hoped to complete the project. In part, the delay reflects Conklin's lifelong tendency to keep working at a project and perfecting it until the editor insisted on having it in hand.[20] He had, however, finished the work by 1893. Yet the publisher held up the volume, saying that it would lose too much money (two thousand dollars) on printing it. Conklin said that he tried to increase subscriptions and to work with the company, but Whitman took little interest in such practical matters.

Even though this one large paper was held up in press for so long, Conklin continued to pursue the same research problems with the same approaches. His move to the research environment in Pennsylvania clearly stimulated his activity. Also, as he continued his study and began to compare details in different species of the same type of organism, he found differences. These raised questions of interpretation, amplified further by the conflicting results arising from the various other researches on cell lineage carried out simultaneously by a dozen or so people at the MBL. At long last, Conklin had determined to draw his evidence together anyway and to present the research results as best he could.

In what had become the traditional manner for modern work

in morphology, Conklin began with an outline of the project and its methods, then progressed to introducing the species he had studied. Thus, he considered the natural history and breeding habits of *Crepidula* in a way that reflected his morphological upbringing at Hopkins. He then moved on to a detailed examination of the history of cleavage, meaning the exact patterns and significances of division at each stage of cell division. This paralleled Wilson's work on *Nereis* and Whitman's on *Clepsine.* A major question for all three was, What is the significance of cleavage?

> Is it an orderly sifting of materials, a "mosaic work," or, as Driesch ('93) has maintained in the case of the echinids, a mere quantitative division of homogeneous material? Can the cells of cleaving eggs be compared with each other as the organs of adult animals can? Can one properly speak of the homology of blastomeres? Are the chief axes and regions of the egg or embryo homologous in different animals? And finally, are the causes of the various forms of cleavage to be found primarily in the constitution of the egg itself, in other words, in the internal conditions, or rather in the external conditions, such as pressure, surface tension, gravity, etc?[21]

Even in these days when "'all the world shakes eggs,'" he said, researchers must depend on careful observations of normal conditions, as well as on experimentally derived data. It is more important to understand how nature works with normal eggs, he insisted, than to know how abnormal changes are handled. Interventionist experimentation was not necessary for the purpose at hand, namely, providing further information, and it would distract from the effort to understand normal development.

To that end, Conklin produced a chart and a system of nomenclature to indicate precisely which cells give rise to which others in the course of division. Unfortunately, as he acknowledged, he had trouble obtaining perfect results. When he removed the cluster of eggs from the mantle cavity of the mother, they soon stopped developing. As a result, he could not follow any one cluster through more than two or three days of development and consequently had to rely on numerous different clusters of eggs from different mothers and exposed to differing conditions. He therefore had no perfect uniform control throughout the entire process. Fortunately, enough females became fertile at about the same time to provide sufficient material. With many egg clusters he could therefore put together the lineage with considerable confidence in his results.

Up to the fifty-two-cell stage, Conklin discovered, the cleavages are virtually the same in the different species he studied. After that they diverge increasingly, probably for a variety of reasons, each stage influencing the following divisions as well. A slight rotation in a spiral direction in one stage translates into a significant spiraling in later stages, for example.

Spiral cleavages appeared in the cytoplasmic division more than in the nuclear division, which suggested that the division was mechanically caused. Yet it was not external mechanical conditions that caused the cleavage but the complex internal mechanical conditions, "which in our ignorance we call the coördinating force, or hereditary tendency." Indeed, "how anyone can follow the history of the blastomeres of an ovum like that of *Crepidula,* and still maintain that the peculiarities of each cell are due entirely to external conditions or to intercellular relations, is more than I can understand. To me it seems absolutely necessary to believe that *between cells with such different histories there must be some internal or constitutional difference.*" The internal direction even appeared sufficiently strong and coordinated to be purposeful, so that "the end seems to be in view from the beginning, and the building materials are sorted and arranged with reference to this end result."[22] Yet it is only that complex internal mechanical condition, inherited from the past and therefore already highly adapted, which provides the apparent purpose. Similarly, the later organization and orientation along the central body axis arise because of mechanical necessity, given the initial conditions (or "the structure of the germinal protoplasm") and subsequent epigenetic responses. The body axis therefore arises because of internal conditions and not as a product of the first cleavage, as many researchers had suggested.

Such a mechanical explanation of development and other organic phenomena is not something to be feared, Conklin insisted. Instead, it is the proper goal of biology, which is, after all, a causal science. Cell lineage study helps to provide a causal account of development, at least. But a full "mechanical explanation of vital phenomena is a great task, and one not to be accomplished in a year or a century." Thus, we ought not to pretend that we have even gotten very far as yet. We should instead keep working toward such a goal, however distant.[23]

By this point, then, Conklin had laid out his basic program of study. He sought to understand the mechanical proximate causes of development beginning with the maturation of the inherited egg and sperm cells. This would necessarily involve study of the whole egg and the interactions of the parts that make up the whole. The study must be materialistic, since that is all that exists in the world, and it would focus on parts of the organism but would not be reductionistic in any epistemological sense. Observation of normal development should provide the starting point, but experimental manipulations would provide additional information, sometimes even yielding crucial information making it possible to determine which of several theories best fit the whole range of available facts. Finally, generation of hypotheses should proceed only very carefully, with the primary emphasis on understanding facts about normal development.

Public Science

Wilson had remained committed to his own research and to building professional biology through the development of journals, the MBL, the program at Columbia, and his own laboratory. Conklin held similar professional commitments, but he also felt a call to deal with broader questions and to address a larger and different audience. Because of his religious commitments, which his fellow Hopkins students did not share, he felt a need to help interpret how biological science relates to religious questions, especially in the light of public concern about the implications of evolution. In addition, his experiences in a one-room schoolhouse, then as a missionary teacher before graduate school, and his work in teaching- rather than research-oriented universities before Pennsylvania made him particularly interested in matters of science education. He devoted considerable energy and time to both interests.

The year after his major research summary and manifesto for mechanical explanation appeared, for example, Conklin addressed the Methodist Episcopal Church Congress in Pittsburgh. He argued there that we must learn about science from nature directly and not from the Bible, or, as Galileo had put it, "the intention of the Holy Ghost is to teach us how one goes to Heaven, not how heaven goes."[24] People think that science, and especially evolutionary science, necessarily opposes religion because they think that Darwin denies creation and thereby implies that there is no God. Not true, Conklin pointed out to his fellow Methodists. Instead, science deals with secondary causes as we see them manifested in nature. In principle, it cannot deal with the invisible and inaccessible first cause. Science "traces effects to causes and these to pre-existing causes, and so on until the process must stop, hanging in mid-air as it were, without finding the first cause. . . . Where science ends faith begins and, like the child or the savage, the profoundest philosopher or scientist must say: 'In the beginning—God.'" Though evolution is a theory, science accepts it with the "utmost probability," and it has achieved "almost universal acceptance." Therefore, not only are science and religion compatible, but indeed "it is the duty of a progressive theology to relate this new knowledge to the old faith."[25] Conklin continued to take that as a goal for his own work throughout his life.

He often had to deal with people like his undergraduate professor of mental and moral philosophy at Ohio Wesleyan. The professor often railed against Darwin in class, explaining the horrors to be found there. Conklin had decided to look at Darwin himself, so he had obtained *The Origin of Species* from the library and read it. As he recorded later, he could not understand very much of it, but he could tell that it offered a great deal of evidence in favor of the idea of organic evolution and that it ended on a "quite idealistic note." So when the professor next indulged in a tirade against Darwin, Conklin asked whether he had read any of Darwin's books. The professor said "No, I wouldn't touch them with a ten foot pole!" Conklin then read the last sentence of *The Origin* to the class: "There is grandeur in this view of life, with its several powers, having been originally breathed [by the Creator] into a few forms or into one; and that, whilst this planet has gone cycling on according to the fixed law of gravity, from so simple a beginning endless forms most beautiful and most wonderful have been, and are being, evolved."[26] "When the professor heard this, he said in amazement, 'Did Darwin write that?' "[27]

The third area of Conklin's professional concern, science and education, emerged in 1898 when he presented a syllabus for six lectures on evolution theory. The series comprised a set presented for the price of ten cents through the American Association for the Extension of University Teaching in Philadelphia, and the published syllabus made the course more generally accessible.[28]

The next year, Conklin considered recent advances in teaching zoology. Such advances must necessarily be based on advances in research, he maintained, and "the interest and value of teaching is directly proportional to the teacher's acquaintance with original sources of knowledge." Though certainly not the only scientist to put forth such an opinion, Conklin became one of the leading biologists advocating that the zoology teacher must quit relying on textbooks as the sole source of information and on antique bottles of preserved specimens that "he wearily exhibits . . . before his suffering class."[29] Instead, Conklin lobbied, let us recognize that zoology consists of much more than classification or morphology, of tired and dead old forms. Zoology teachers must embrace field work and laboratory experiences as vital parts of the education process. The cooperative research carried out each summer at the

MBL was the ideal that Conklin offered to other educators.

The following year, he followed up on his praise of the MBL with an article assessing the institution as the "center of biological instruction and investigation in this country." He felt that the work done there represented "substantial contributions toward the solution of some of the most fundamental problems of biology."[30] For Conklin, the MBL was a very special place, where he could combine his research interests and his commitment to and love for teaching. While his fellow Hopkins graduates also attended summer sessions there and also served actively on the board of trustees throughout their lives, they did not continue to serve as instructors in the courses, as Conklin did.

Studies of Development

At the MBL and at Pennsylvania, Conklin continued to pursue his studies of fertilization and early development. Like Wilson, he was concerned to establish precisely what happens beginning at the time that the two germ cells meet. What do they bring with them in the way of inheritance? What processes of change do they undergo because of the fertilization process itself? And what other factors influence the course of differentiation? Like Wilson also, Conklin turned to the most advanced available cytological methods to study cell changes. And like Wilson, he recognized the value of examining as many individuals as possible, comparing results among different species, and introducing experimental manipulations to produce additional information where observation of normal cases could not suffice. For the early years of his career, then, Conklin followed a research path quite similar to Wilson's and very like what their Hopkins training would have suggested. During the later 1890s and increasingly thereafter, however, the two began to diverge.

While Wilson eventually moved farther and farther into exploration of the nucleus and its role in heredity and development, Conklin, in a lecture to the MBL in 1898, revealed his concern with the whole cell and with the role that the cell's protoplasm plays. He also made it clear that he saw heredity and development as fundamentally interconnected and not as separable as Wilson had begun to suggest. As he said:

> The fundamental problems of development and inheritance are in the last analysis questions of differentiation. Development is progressive differentiation coördinated as to time and place; hereditary likeness consists in the repetition by the offspring, at certain stages of its life cycle, of definite differentiations of the parent; and hereditary unlike-

> ness, or variation, is a modification of these differentiations either as to their character or as to the time of their appearance. The phenomena of differentiation are therefore of the greatest interest, and their causes one of the most important problems of biology.[31]

At bottom, those phenomena of differentiation depend on movements of the protoplasm, since "in this, as in other phenomena, the cell acts as a whole, and in the interaction of its various parts are to be found the causes of all vital phenomena."[32] In this emphasis on the cell, the whole cell, Conklin carried a traditional morphological concern to one reasonable conclusion, while Wilson moved in another direction, also reasonable. Neither was preferable to the other in any tightly logical way. Rather, the choice of research direction followed underlying concerns with different problems.

Conklin also continued to emphasize the importance of the earliest developmental stages. There the product of inheritance acts most strongly to set the mechanical conditions, which, in turn, dictate what the later stages may do.[33] Inheritance therefore exerts a very powerful force, Conklin suggested, a force that environmental factors have little success in changing. In contrast, Wilson had emphasized that the mechanical physiology of later development does respond significantly to environmental changes as well as to heredity. The differences really were more a matter of degree than of kind, however, with fundamental agreement even while the two were specializing in their emphases. Conklin did agree with his MBL and former Hopkins colleagues in unequivocally rejecting the popular idea of inheritance of acquired characteristics, for example. Such an idea simply makes no sense, he often insisted, saying that "wooden legs do not run in families, although wooden heads do."[34] Even as they diverged in emphasis, therefore, Conklin and Wilson remained tied to their shared roots in the Hopkins training.

A Friend of the Cytoplasm—and the Cell

By 1902, Conklin had embarked on what amounted to a campaign to demonstrate the importance of the cytoplasm. Theodor Boveri, Roux, Weismann, Oskar and Richard Hertwig, and others were delving deeper and deeper into the nucleus. The MBL lectures, discussions, and research projects increasingly reflected the view that the nuclei play the most important roles in development. Though not yet completely convinced either that the nucleus is crucial or that the chromosomes retain their autonomy and therefore their opportunity to serve as the fundamental units of heredity, Wilson had become increasingly inclined in that direction by 1902. Conklin looked at the role of nuclei also, but then even more intensely at the protoplasm and at the whole cell.

In a second massive study of *Crepidula*, published in 1902, Conklin tackled the phenomena of both karyo- (or nuclear) and cytokinesis. Though cell divisions at later stages often simply serve to produce new material in a rhythmic and nondifferential way, in the earliest stages they can carry great significance for later differentiation. For these divisions, "the minutest details of unequal, bilateral or qualitatively dissimilar division of cells may be of great importance. The forms and peculiarities of such cleavage are inherited quite as certainly as are any adult features, and when the problem of inheritance may be reduced to a certain peculiarity of a certain cell division it is evident that we have this problem reduced to relatively simple terms."[35]

So Conklin set out to perform such a reduction. He examined in detail every aspect of cell division that he could imagine. In assessing the nature and significance of chromosomes in cell division, he studied their size, shape, and changes. So much remained unobservable, however, that, with even the most careful attempts and most advanced techniques, many questions remained. He could not, for example, determine whether chromosomal division during maturation of the germ cells occurs transversely across the narrow center of the chromosome, or longitudinally down the entire length. Centrosomes, polar rays, spindle fibers, and spheres also undergo complex changes, all details of which could not be followed perfectly. Yet all seemed regular and therefore probably of significance for later development. Conklin was certain enough of this to refute Boveri's contention, for example, that the centrosomes come from the spermatozoon alone and that they bring about the division of the egg. Neither is true in all cases, Conklin's empirical observations showed.[36]

After describing in detail what happens during maturation of the germ cells and fertilization, Conklin moved on to cleavage. Here the chromosomes undergo division and aggregate to form new nuclei, he found. Each nucleus is initially the same size, but they grow in accordance with the growth of the cytoplasm of the cell in which they are found. In some way the nuclear material extends outward to reach the centrosomes and to combine with cytoplasm to form the spindle fibers and other cell parts characteristic of cell division.

Turning then to movements of the cell body, or cytoplasm, during division (or cytokinesis), Conklin sought to show the way in which they were the "immediate cause of many important differentiations." Cytoplasm is a thick fluid through which movement of cell parts can at times take place, he felt, even though it also exhibits a system of stable fibers and subsystems that do not move. The way in which the cytoplasm moves during cell division therefore sheds much light on the causes of cellular differentiation. Movements of fluid produce various unlike substances that become aggregated to form differentiated cell parts at a very early stage. As was apparent with careful observation, these become localized in different regions of the egg or later cells and, in turn, determine the nature and significance of later divisions.[37] The structure of the initial egg cytoplasm therefore determines how cells, and their nuclei, divide rather than the nuclei's determining how cells divide. Cytoplasmic development guides and directs development, but we do not as yet know how.

This conclusion certainly does not mean that the nucleus is not important. Indeed, it is. The nucleus may well remain the bearer of heredity and, as such, the prime mover for all processes of differentiation. The results of Conklin's study suggested that there was "good reason to believe that the structure of the [cytoplasm] is influenced by the nucleus through the large amount of nuclear material which escapes into the cytoplasm at every mitosis. Certainly many features of later development are derived from the father and the conclusions as to the part which the nucleus has in hereditary transmission, founded as they are upon the remarkable apparatus for such transmission afforded by the nuclei, cannot be lightly cast aside.[38] Nuclear inheritance was one part of the process of differentiation of the egg, but only one. Cytoplasm and the interaction of nucleus and cytoplasm also remained key.

In 1905, as Wilson moved even farther toward embracing nuclei and chromosomes as basic to heredity and probably also development, Conklin produced a major study of ascidian eggs to support his emphasis on the whole egg. He saw organization of the egg, in the form of differentiated regions already marked out in the maturing germ cell, as the starting point for development. The current controversy, he realized, focused on the "nature and contents of the germ cells." Did they contain a collection of some sort of inherited determinants that correspond to the morphological characters of the organism, as Weismann maintained? This view represented the modern version of preformationism, or evolution as it was often called since it involved an unfolding of predetermined structure. Alternatively, do the germ cells consist instead of a complex of chemical substances that become transformed in the course of development into organismal structures, in a manner parallel to crystal formation? This view represented epigenesis. Conklin followed his fellow MBL researchers in believing that the truth must lie somewhere between the extreme positions.[39] At precisely what point remained unclear.

Wilson advocated a more epigenetic view, in which the ongoing physiological processes of development played a fundamental role in causing differentiation. Thomas Hunt Morgan agreed and gave factors external to the egg even greater influence. Conklin, on the other hand, followed Whitman more closely in expressing his strong attraction to His's idea of very early organ forming germ regions. Why would it not make sense that much, at least, of the cause of later form could lie in the differentiated regions of the maturing germ cell and in the mechanical processes of cell division which followed as a result of the particular organization?

Ascidian Studies

Ascidians were particularly valuable organisms to study, Conklin pointed out, because their cell cleavages are so wonderfully regular and easy to observe. In addition, they have a relatively small number of cells during gastrulation and organ formation, and it is reasonably easy to determine the egg and embryonic axis. These advantages, he insisted, make the ascidian egg "the most favorable in the whole phylum of the chordata for an exact study of the early development."[40] Their advantageous nature also lent some authority to his conclusions. When others disagreed on the basis of their studies of less perfect species, Conklin could reply that his results were more informative.

Conklin used *Cynthia* and *Ciona* primarily, thanks to a suggestion from Morgan one summer at Woods Hole. Morgan reminded Conklin, upon receiving a copy of the latter's lengthy ascidian paper, "Little did I dream that day when I first showed you the eggs with the red cheeks that you would make so much out of them."[41] His ascidians showed Conklin that, once the yolk forms in the oocyte, the chief axis of the egg and later the gastrula are fixed for all later cleavages; the axis corresponds to that passing through the centrosome and nucleus. At least on this point, predetermination of the later organism therefore occurs. It was "interesting to observe how recent studies of development have led to the recognition of morphogenetic differentiations at earlier and earlier stages in the ontogeny; a dozen years ago the germ layers were the earliest differentiations of this sort which were generally recognized." Now differentiations could be traced back even to the unsegmented egg, thus leading to the conclusion that "the cleavage cells and even the unsegmented egg must be organized with reference to the parts and axes of the future animal."[42]

Roux had believed that the unfertilized egg remained essentially undifferentiated. He concluded that the action of sperm penetration then caused the median plane of the embryo to form along the penetration path, since the two paths corresponded in the frog species he studied. Conklin found the same phenomenon but insisted instead that the structure of the egg caused the sperm to penetrate and the egg's cleavage plane to develop where they did. So even here, in Conklin's interpretation, some preorganization of the egg directed the course of later development.

The question still remained open as to the extent to which the nucleus influences this organization and the extent to which it is purely cytoplasmic. Evidence had accumulated to show that the nucleus plays a central role, Conklin recognized, but he did not believe that the nucleus or the chromosomes could thereby be said to control development. Instead, as he had suggested earlier, the best evidence showed a considerable amount of nuclear material "escaping" into the cytoplasm, where it directed the formation of numerous specialized cell body parts. The nuclear substance in effect used protoplasmic material to produce all the apparatus necessary for cell division to occur. In this way, the nucleus exerted control over development, as the nuclear advocates demanded, but the cytoplasm retained an essential role as well. With the escape of nuclear substance we have a possible mechanism for nuclear control of the cytoplasm, Conklin said, and "when, as in the case of the ascidians and fresh water gasteropods, these substances are definitely localized in the egg, and can be traced throughout the development until they enter into the formation of particular portions of the embryo, a specific mechanism for the nuclear control of development is at hand, and the manner of harmonizing the facts of cytoplasmic organization with the nuclear inheritance theory is clearly indicated."[43] The escaping of substance and localization of parts in the cytoplasm takes place gradually, he continued, so that the developmental process is partially epigenetic even though it begins with preorganization. Throughout the process, cytoplasmic movements continue to provide the immediate cause of differentiation.

In another paper of 1905, this time directly addressing the mosaic theory of development, Conklin applied what he had learned from his cell lineage studies to Roux's theories. The ascidian is a mosaic work, he concluded, "because individual blastomeres are composed of different kinds of oöplasmic material."[44] Thus, the mosaic is of germinal substances and not just a cleavage mosaic that arises as development goes along, as the Weismann-Roux theory insisted. In effect, Conklin took Roux's term and redefined it more in line with His's idea of organ-forming germ regions, so that he was not really endorsing Roux's mosaic theories at all. In agreement with Roux, however, he rejected Hans Driesch's interpretation of development as a process of regulation beginning with an essentially totipotent mass of undifferentiated material. His own study of cell lineage and experiments on fragments of ascidian eggs showed no regulation and considerable preorganization.

Morgan, who like Wilson was more inclined toward Driesch's sort of epigenesis than to Roux's sort of preformationism, disagreed with Conklin on this interpretation. "You have smitten Driesch on the hip," Morgan acknowledged, "*yet I am not sure* but that even what you describe for the half (lateral) larvae does not involve *something* of a regulation towards a whole."[45] Furthermore, Morgan objected to what he took to be Conklin's claim that those who did not also know normal development through cell lineage studies should not be allowed to do any experimental work either. More on this later, he promised. Conklin responded with enjoyment: "Oh! what a beautiful rise I got out of you. Morgan you are impaled on a bare hook. I never said that experimental work should be in the hands of cell-lineage cranks." Nor did he deny the importance of work by experimentalists such as Roux and Driesch, but he did nonetheless believe that Driesch was in error with respect to the regulative ability of ascidians.[46]

Additional study of ascidians extended his conclusions and reinforced the view that some sort of preorganization exists to guide later development and that regulation is limited. While Wilson continued to oppose Roux's mosaic view of development, and Morgan continued to favor Driesch's emphasis on external conditions, Conklin became increasingly attracted to his version of preorganization. His look at a range of ascidians complemented other studies to demonstrate that "*all the principal organs of the larva in their definite positions and proportions are here marked out in the 2-cell stage by distinct kinds of protoplasm.*"[47] Cell divisions cut across these differentiated materials in different and irregular ways, so that a cleavage will not neatly separate the yellow crescent from surrounding material, for example. Yet the organization is there. Slight modifications in early differentiation or cleavage chop up the material differently and can cause all the subsequent divisions to follow a different pattern as well. Ascidians proved extremely strongly fixed in their development from an early stage, with each cell and part unable to adapt to changed conditions.[48] Thus, very simple early changes or mutations could effect great differences later, as Hugo de Vries had suggested with his mutation theory. This suggested to Conklin that evolutionary changes in phyla might occur not by transmutation of the adult forms, as was so generally assumed, but perhaps "by relatively simple alterations of the type of germinal organization."[49]

All these ideas he continued to pursue in further research with additional organisms and using experimental as well as descriptive cell lineage studies of normal development. Sometimes experimentation did not work well for a particular organism or a certain question; in these cases, the researcher should continue to use cell lineage or other available descriptive approaches to gain information. Thus, for example, the study of the gastropod *Fulgur* (later reidentified as *Busycon*) required detailed descriptive work. Indeed, Conklin "regarded the case of *Fulgur* as a triumph for the method and doctrine of cell-lineage." Critics had misunderstood the value of careful cell lineage study of normal development, he charged. "Those who see in this method only 'the counting of cells,' 'mitotic bookkeeping,' 'the drudgery of dull minds,' have missed the whole point and significance of this method, which is not to name every cleavage cell, but to determine in what areas of the egg certain morphogenetic processes are located." Study of various organisms then reveals which patterns are variable and which hold throughout a group of organisms. This knowledge, he insisted, served to illuminate important questions of the day.[50] Though Conklin insisted that science need not be experimental, experimentation did have its place as well.

Experimental Study of Development

While Conklin had long advocated the use of experimentation as an appropriate method for zoology, and indeed as occasionally the only way to adjudicate among conflicting theoretical interpretations, he did not at first actually perform experiments himself. His methods through the 1890s remained largely observational and descriptive. After 1902, he began to add experimental studies to that more traditional descriptive work.

In 1902, he reviewed Morgan's volume on *Regeneration* and took the opportunity to reflect on the place of the book and its approach in recent history of biology. The past decade or so had brought a period of revolution in biology, he suggested. Researchers had reacted strenuously against the building of speculative theories about evolution, against the construction of speculative phylogenies, and against the idea of inheritance of acquired characteristics. Conklin found these reactions entirely wholesome and felt that experimental morphology had played a central part; "in fact, it was the attempt to make biology an experimental science which first aroused interest in this subject, and while at times some of these experimental morphologists have illustrated the uncritical methods which they have denounced, while their conclusions have often been open to the criticism of having been hasty and ephemeral, no one can deny the fact that their work has introduced a new spirit into the study of zoology."[51] Morgan's work typified the best of such experimental study, Conklin felt, and his look at regeneration as a process parallel to normal development had shed much light on the nature of the latter.

Morgan had earlier carried out experiments parallel to Roux's, Driesch's, and Wilson's on egg fragments. He had initially been inclined to agree with Driesch that eggs must from a very early stage be essentially homogeneous matter with relatively little differentiation. Yet his study of regeneration had led him to revise that view. Already at the earliest stages, the protoplasm seems to exhibit some differentiation and is hence capable of regenerating the correct parts to replace the missing ones. This differentiation does not arise out of invisible biophores or other units of heredity, however. Instead, there is some sort of cell differentiation in the form of "tensions" that guide development and regeneration. Conklin found particularly intriguing the "pregnant suggestion" that a system of tensions exists in the living protoplasm and serves as the cause of differentiation in both normal and regenerative cases.

In a set of what are essentially nature's experiments, Conklin looked at the occasional occurrences of inverse symmetry. Sometimes, he noted, the plane of symmetry is inverted, even in man. The cause seems to lie in the reversal of polarity in the egg. This occurs occasionally for different reasons in different cases, but the reversal always occurs during maturation of the egg cell. The appearance of this abnormal circumstance provides new information, just as that acquired through manipulative experimentation. In this case, Conklin concluded that since reversing polarity at maturation can effect a total reversal of all subsequent development across the axis of symmetry, then "there must be a definite localization of germinal primordia or anlagen in the egg before maturation, e.g., the substance out of which the kidney of the snail will ultimately form must be definitely localized on one side of the chief axis, and so for every other part."[52] The egg, as Morgan had also come to conclude from his regeneration work, cannot be homogeneous at the beginning but must experience considerable organization resulting from the maturation process.

Further experiments followed. For example, he used the hypertonic sea water that Jacques Loeb, Morgan, and others had shown effective in producing parthenogenetic development, in order to stimulate formation of centrosomes and spindles.[53] Conklin used recently fertilized eggs instead of enucleated eggs but sought to determine what effect the altered conditions would have. He found that instead of one unchanging pattern by which cleavage centrosomes arise, he could produce several. This suggested that the centrosome might very well arise differently in different animals or even in the same animal under different conditions. Obviously, such a suggestion raised many possibilities for further study, which Conklin pursued in part with the help of research and collecting trips to the Marine Laboratory of the Dry Tortugas (Florida), run by the Carnegie Institution, and to Nassau or Bermuda, as well as the MBL.

Experimental examination of half-embryos extended Conklin's cell lineage study of ascidians. In 1906, he reviewed the literature on the subject, then pursued further the suggestion that Morgan had made the year before. Morgan had questioned Conklin's conclusion that, since only partial embryos result from single blastulas, ascidian blastomeres do not undergo regulation and instead exhibit a strict preorganization. Driesch had also attacked Conklin's conclusions and his results. Conklin returned to examine the question further and concluded that, indeed, some regulation does occur, but never so far as to produce a complete larval form from isolated blastulas. Instead, differences in types of cells and in egg substances are sufficiently "great that they can give rise to no other types of structures than those which they form under normal conditions. . . . These visibly different ooplasmic substances are therefore 'organ-forming substances' and the areas in which they are located are 'organ-forming germ regions.' "[54] Experimental as well as descriptive studies revealed the same thing.

Experimental study had become quite exciting and a major innovation in biology, Conklin pointed out later in reviewing Morgan's *Experimental Zoology*. He felt that Morgan had set forth the various alternative hypotheses and then considered them fairly and very clearly. While Conklin did not always find Morgan's interpretations convincing, he was nonetheless certain that the discussion would stimulate research. "And after all this last is perhaps the greatest service which any book can render."[55] Experimental biology had brought experimental manipulations and consideration of various theories to the attention and discussion of biologists in a way that helped to advance the science. Experimentation builds upon the descriptive work that precedes it, as Morgan's did, thereby also advancing the goal of embryology: "to trace to their origins the principal differentiations of organisms." Experimentation also provides control and hence can yield valuable information.[56]

The Research Program

As Conklin became more and more firmly established within the professional biological world, with his appointment at Pennsylvania and memberships and even presidencies or chairmanships of numerous scientific organizations, his research program also became firmly established and successful. Clearly, cell lineage and similar descriptive study of early stages of development remained the central approach to acquiring new information, though experimental manipulations provided legitimate supplementary information. Studying a variety of marine invertebrates also yielded additional information since comparisons allowed inferences about the ancestral conditions of development and therefore about how divergences from the primitive form, especially in the earliest stages, effected changes in differentiation. His work, with its continuity of questions and approaches, provided a solid research program for Conklin to follow. In addition, he built a small research community of students and colleagues following the program. At least eleven doctoral students adopted his research program at Pennsylvania.[57]

Even as he established himself and his program at Pennsylvania, however, Conklin began to consider moving elsewhere. He had become effectively the head of the zoology group within the biology program and had been instrumental in building a vivarium in 1900 to provide important facilities for living organisms for research. Yet the separation of biology in 1896 into zoology and botany, along with other specialties, left the organization of biology a bit unclear. In addition, the biologists were in serious need of a new building. When the university acted too slowly in providing one, Conklin looked elsewhere for a position.[58]

As early as 1897, the year after he had arrived at Pennsylvania, Jordan had again inquired whether Conklin might be willing to move to Stanford. He had replied in the negative and had instead recommended a former student for a position. Other possible positions appeared along the way. Ross Granville Harrison worked hard to get Conklin to Yale, for example, including an offer shortly after his own arrival in 1907. Then late in 1907, Princeton offered Conklin a professorship and the chairmanship of the biology program. He might well have preferred to have had Pennsylvania offer better resources and incentive for him to stay in Philadelphia, which he had come to love. He had become sufficiently discouraged about the prospects, however, that he determined to leave. Somewhat reluctantly, he accepted Princeton's offer. In letters to his best friend, Harrison, he expressed his nervousness and concerns about the move. "Cheer up!" Harrison urged him, "Even if Princeton can't play football, it may become a good place to study biology." Harrison had himself found it difficult to move from Hopkins to Yale the year before and had felt homesick at times, but he had gotten over it. He did sympathize with Conklin's having to endure so long a process of preparing to move, though, since Conklin had received and accepted his offer in the fall for the next year. Nonetheless, Harrison assured his friend, "when you are snugly settled in your new berth you will be better satisfied."[59]

The same year that Conklin moved to Princeton to head his own department, where he remained for the rest of his life, he also became a member of the National Academy of Sciences. This therefore began the final phase of his career. At Princeton, as in his earlier positions, Conklin sought to develop a community of biologists. He maintained a personal relationship with his students and staff and created something of a family feeling among the group. One former student recalled that students at Princeton felt comfortable visiting professors' houses, perhaps for Sunday tea or for rarer formal occasions. These were pleasant and important times for all. In addition, the Conklins invited any graduate students for Thanksgiving dinner and a visit with the family. "We came," the student reported, "to regard both Professor and Mrs. Conklin as genuine friends and as personalities who were in a real sense a part of our lives."[60]

At Princeton also, Conklin continued his work in embryology but also began to address problems of heredity more directly. He saw that the early organization of the egg must have some cause, presumably in heredity. Therefore he took the subject as the focus for his lecture to the American Association for the Advancement of Science in 1907–8. The mechanism of heredity necessarily begins with the fact that the two germ cells are not equal, "*all the early development, including the polarity, symmetry, type of cleavage, and the relative positions and proportions of future organs being predetermined in the cytoplasm of the egg cell, while only the differentiations of later development are influenced by the sperm.*" The nucleus also played an important role in heredity, but not because there is any special, unique hereditary substance there. In fact, "the evidence in favor of an inheritance material, which is distinct from the general protoplasm of the germ and whose function is the reproduction of hereditary characters, is not convincing."[61] Heredity is, Conklin maintained, just another manifestation of the basic process of development.

Only after 1910 did he begin to consider that heredity might, in fact, be separable from development, that heredity might be controllable through an enlightened program of eugenics, and that evolution might even be directed by controlling heredity. He also began to endorse a Mendelian view of hereditary units, influenced in part by Wilson's and Morgan's move to do so and the accumulated evidence in favor of it. This work, which led to his *Heredity and Environment in the Development of Man* of 1915, represented a natural extension as well as a modification of his views of development. The bulk of his research program continued along the same lines as before, with his doctoral students at Princeton following lines similar to those at Pennsylvania.[62]

After his move to Princeton, Conklin also began to travel more. He finally went to the Naples Zoological Station for the first time in 1910, to work on development of ascidians and other invertebrates. He worked at the Columbia University table there, thanks to Wilson, since Princeton had not subscribed. Fortunately, the Conklins escaped from Naples shortly before a serious cholera outbreak. They went on to visit other parts of Europe and then on to Australia and New Zealand. Conklin had visited much of the Caribbean in connection with his marine research, but he began to travel more widely. Obviously, he took great pleasure in these travels, collecting brochures, maps, and a host of other souvenirs to remind him and the family of all details of the adventures.[63] With such opportunities, with a new science laboratory building, and with other encouragements of various sorts, Conklin stayed at Princeton despite continued efforts by Wisconsin, Yale, and other places to lure him away.

Though progress in science, in the form of new data and better theories to explain them, carried Conklin somewhat beyond his early work, his research program remained dedicated to solving the same set of embryological and evolutionary problems with similar sets of approaches. His underlying convictions about the way science ought to work remained the same. His research also remained relatively close to its roots and initial approaches. While Wilson and the others diverged increasingly to address more specialized sets of problems, Conklin stayed closest to an updated version of Hopkins morphology.

What distinguished Conklin's career and insured him a place in this group of four outstanding biologists was the combination of his solid, productive research results and his public science. Influential through the various societies to which he belonged, through his role on funding panels, through his high status at Princeton and the MBL, Conklin held considerable political influence. As a powerful and persuasive orator and teacher, he commanded public attention. His continued emphasis on development and the cytoplasm guaranteed continued support for that work throughout his long and active career, which really continued after his retirement in 1933.

8

Thomas Hunt Morgan

AS COAUTHOR of the standard American textbook on general biology, as Charles Otis Whitman's primary confidant at the Marine Biological Laboratory (MBL), and as head of a major new research program in biology at Columbia University, Edmund Beecher Wilson had ample opportunity and stimulus to reflect on the nature of biology. What counted as biological, and more generally as scientific, and how to do good scientific work in biology were questions he addressed in various ways in his writings, lectures, and more general theoretical addresses. Edwin Grant Conklin also considered such questions and offered answers in his public capacities. The American Philosophical Society and his missionary work encouraged him to develop general views about science rather than just carrying out his own specialized research. As a result, Conklin and Wilson had articulated many epistemological, metaphysical, and other convictions early in their careers.

Thomas Hunt Morgan, in contrast, developed his views more gradually. When he stepped into his first professional job at Bryn Mawr, the biology program and course outlines were already in place, established largely by Wilson. Morgan joined the MBL slightly after Wilson and Conklin, when its mission had already largely been set. Since he was not a churchgoer or involved in other public activities, he had no reason to espouse general views about the nature of science. So he did not, except insofar as he developed such views in the course of doing his own scientific research. In that capacity, he wondered about how to deal with alternative interpretations of phenomena, for example, and about how to judge whether an experiment reveals anything about normal conditions. He also considered the proper role of hypotheses in science. His views gradually began to emerge explicitly, following largely

Fig. 17. Morgan in the Biological Laboratory at Bryn Mawr, probably about 1900. From the Bryn Mawr College Archives.

pragmatic lines similar in most fundamental respects to Wilson's and Conklin's.

Morgan's work fell into several different research programs, which on the surface may seem quite distinct. Study of frog development, experimental zoology, regeneration, and sex determination provided the main focuses. What remains central to all these studies is a set of core questions and approaches. Thus, in reality, the several sets of research fit together in ways that reveal Morgan's major concerns.

Frog Studies

Morgan had become interested in frogs during graduate school when he wandered around the hills of Maryland in search of prize specimens. Presumably that occupation was stimulated by the enthusiasm for frog embryology in Germany during the 1880s. By the mid 1890s, he had begun a systematic survey of all the various frog studies to demonstrate what the collection as a whole revealed about the nature of development. Unfortunately, he had trouble keeping up with the increasing volume of these studies. As a result, the book he planned suffered delays. Only in 1893, while spending part of the summer working at the University of Berlin, was he able to concentrate sufficiently on the work to make real progress.[1]

The summer of 1893 ended Morgan's second year at Bryn Mawr and the first without Jacques Loeb, though Joseph Warren (a physiologist from Harvard who had arrived as a lecturer when Loeb did) remained. With a stable job and permission to take a year's leave of absence, Morgan sought his turn at experiencing the European biological world. He also looked forward to contact with the leading research centers, since, at Bryn Mawr, he felt somewhat peripheral to research activity.[2]

Morgan had presumably heard from Loeb about the excellent medical and biological traditions at Berlin, for Loeb had attended his *Gymnasium* and had begun his university study of medicine in that city. In addition, Loeb must have told him about the advanced work in physiology and physicochemistry emerging there. Loeb may also have had connections to help introduce Morgan around since he had relatives on the faculty. Morgan made a reasonable choice, then, when he began his European adventure with a visit to the Biologische Anstalt at Helgoland and then a stay at the Zoologisches Institut at the University of Berlin.

At Helgoland, a North Sea island off the German coast first identified in midcentury by Johannes Müller as a choice location for marine research, Morgan explored the process of fertilization in sea urchin fragments which Theodor Boveri had recently made popular. In Berlin, he worked on his frog book into the fall of 1894. Then in late October he made his way to Naples. From Wilson and other American friends, Morgan had heard of the great advantages of Naples for experimental marine research. Unlike the MBL, which was only open in the summer and only had a short breeding period for many marine animals in early summer, Naples remained open all year round. Most Americans chose to stay away during the hot and cholera-prone late summer months, but they could work throughout the academic year. Morgan spent 1894–95 at the Stazione Zoologica in Naples. There he concentrated on experimental embryological studies, but he also continued with the survey of studies of frog development. After his return the next year to Bryn Mawr, he spent the summer of 1896 in Zurich and finished *The Development of the Frog's Egg.*[3]

Historians and biographers who have looked closely at Morgan's work in genetics and have ignored the developmental studies have missed a vital part of what he held as important. *The Frog's Egg* provides a fine example. This first book of Morgan's was, he explained, intended as a survey of the "general problems of development," in particular the early developmental stages. The frog provides a particularly interesting case "owing to the ease with which the frog's egg can be obtained, and its tenacity of life in a confined space, as well as its suitability for experimental work."[4]

As a result, many researchers had examined the frog's egg through a variety of methods, variously poking, heating, cooling, compressing, centrifuging, and otherwise manipulating its conditions. They had generated alternative theoretical explanations of the assorted results. Morgan's presentation of the situation reveals him as widely read, as open-minded, and as having understood with considerable insight the significance of the work done to date. He clearly assumed that good scientific discussion must weigh the evidence carefully and move to an interpretation only very cautiously, and only tentatively. In addition, this book demonstrates that Morgan had not been seduced by the call to experimentation. He ignored Roux's bold claims that biological work *must* be experimental in order to make any real scientific progress. Instead, Morgan joined his fellow Americans and Johns Hopkins University graduates Wilson and Conklin in proclaiming the value of combining traditional descriptive studies based on careful observations of normal development with experimental investigations.

Experimental work, Morgan felt, had proven itself "most instructive for an interpretation of the development." But the juxtaposition of the experimental results with normal studies made possible the most "definite conclusions" of all. In any case, he believed that it was important to stick as closely as possible to the actual evidence at hand, avoiding theoretical discussions whenever he could.[5] When he did consider theories, he tried to include all the alternatives, all sides of the issue, and discussion of evidence opposed as well as that in favor of each particular theory. The result was an excellent textbook summary of the best research to date, though, as Morgan later admitted of texts in general, the good ones are inevitably out of date just about as soon as they appear in print.[6]

Morgan's book began in the way any classical embryological text would begin: with discussion of the production of sex, or germ, cells and what happens during fertilization. Then he proceeded to questions of hybridization, or cross-species fertilization. Such hybridization most often occurred during the height of the breeding season, according to Eduard Pflüger and Gustav Born. Yet Oskar Hertwig had disagreed, maintaining that Pflüger's eggs had been kept under artificial conditions that had caused deterioration of both eggs and spermatozoa. They had lost their normal "irritability," he concluded, and therefore only appeared to hybridize best at that stage. Normally the eggs resist penetration by foreign sperm, but when they begin to break down or to "lose their irritability," they lose the power to resist. Hertwig's objection showed that when subjecting eggs to experimental conditions, the researcher must be extremely careful to avoid producing artificial results that really reveal nothing about normal processes. Morgan reported the disagreement and the evidence for each point of view, then concluded that many complicated factors enter into the fertilization process and suggested that experimenters must be careful in their studies.

Cleavage follows fertilization—in nature and the *Frog's Egg*. The many German and American studies of cleavage had raised questions. Two held particular interest for Morgan: What relation does the first cleavage furrow hold to the later body axis (and similarly, what relation does nuclear division hold to cytoplasmic)? How does the embryo form within the egg: by concrescence or by cell migration?

The first cleavage plane and the later embryonic axis do coincide, Morgan acknowledged, but not because the cleavage actually causes the axis to form in that location. In an earlier paper, he had noted the identity of the two furrows and had held that "it has been almost conclusively proved by previous experiments and observation that the plane of the first furrow in the case of the frog divides the egg into halves corresponding to the right and left sides of the embryo."[7] At the same time, the dorsal lip of the blastopore always corresponds to the same point with respect to the primary body axis; furthermore, what Morgan saw as the overgrowth of the blastopore lip seemed to have a great deal to do with later organization of the embryo. Therefore placement of the emerging blastopore lip might somehow cause the placement of both cleavage furrow and body axis. It was not at all evident what caused what, though the relation of furrow to axis was clear.

In his earlier experimental attempts to clarify how cytoplasmic organization acts, Morgan had given up and resorted to observation of normal processes alone. Experimentation had proven "impossible," he said, and the results "valueless." He had failed to watch the developing eggs attentively during every single moment after fertilization. This is absolutely necessary, he realized too late, because the eggs could have undergone changes that the unsuspecting experimenter might not realize had occurred. The egg could heal itself after an injury, for example, and therefore appear to have developed normally from a single blastomere (at the two cell stages) when it had really regenerated the supposedly missing blastomere. This, Morgan surmised, had happened in some of Roux's cases of supposed "postgeneration." As a result, one blastomere had not really developed a whole organism; two blastomeres had.[8]

Experimental results must remain completely useless, Morgan insisted, if the researcher fails to keep track of each cell division. It is also necessary to identify which cells are involved. The observer must record which cell is injured, and what its normal relation to the later embryo would be, in order to make sense of the experimental results.[9] Though he did not argue for cell lineage research as such, his emphasis on the particular cells and their positions in the egg clearly reflects his conviction that cell lineages and other details of cytoplasmic organization matter.

In addition, Morgan believed that the orientation of the egg was vitally important since the egg has such differences in its white and black regions. If an egg was rotated so that the white side was up, it could generate a whole but small embryo even after the other blastomere had been pricked with a hot needle. Those with the black side up could not. Morgan concluded that the internal organization and localization of cytoplasmic materials both direct embryonic development and differentiation. As he said,

> The results show, I think, that the phenomena of half or whole development of an embryo from one of the first two blastomeres is entirely

> a protoplasmic phenomenon. The results have nothing whatsoever to do with a qualitative division of the egg at the first cleavage or with a later postgeneration. Whether we get a half- or a whole-embryo will depend upon the subsequent arrangement of the protoplasm in the uninjured blastomere and upon the relation of the protoplasm of the uninjured and injured halves. If the egg is turned after one blastomere has been injured, so that a rotation of the contents of the uninjured blastomere takes place then a whole embryo tends to develop. The completeness of the development will depend upon the extent of the rotation.[10]

Since the internal conditions of the egg seemed central to determining the later developmental stages, Wilhelm Roux had explored what would happen under different sets of mechanical conditions. Most significantly, he had looked at a set of oil drops under various conditions of surface tension. They behave very similarly to the normal cells just after the early cleavages. This showed that internal conditions such as surface tension may play a vital role in determining the shape of the blastomeres. The individual blastomeres themselves may not be completely self-differentiating, as Roux had suggested. Instead, the regulatory action of the whole organism seemed important. Yet the answer to exactly how that action acts was not clear. Alternative theories still existed, with different evidence in favor of each. With respect to the questions about what causes the cleavage plane and the body axis as well as nuclear division, "the plain answer is, we do not know."[11]

The egg is not isotropic, Morgan was certain, but it experiences considerable organization from the beginning. In addition, the cytoplasm, not the germ nuclei, has the greatest organization. Some researchers had suggested that the nuclei are critical since the male contributes only the nucleus and plays a significant hereditary role. Morgan once again found alternative evidence, however. One line of work "seems to throw some light" on the question, "although the interpretation is extremely difficult and hazardous."[12] When some of the cytoplasm is cut out of a ctenophore egg, the resulting part gives rise to an imperfect embryo—even though the nucleus remains intact. Thus, the nucleus seems unable to compensate, and the cytoplasm seems to control normal development. Furthermore, by the time of the *Frog's Egg,* Morgan had concluded that the nuclear spindles form in response to internal conditions within the egg. The spindles then determine which way the egg will divide in each division. This division in turn correlates with the orientation of the blastopore lip. At the same time, the body axis is set as well, also in response to internal conditions of the egg. At the root of the various developmental steps, then, is the initial cytoplasmic organization. Yet, as with the question of what directs the cleavage patterns, the answer to questions about the relative role of nucleus and cytoplasm must remain, "We do not know."

In addressing the other major question, about how the embryo forms in the egg, Morgan also relied on a combination of observations of normal development and experiments. After following in detail the actions of the blastopore, he concluded that, as its dorsal lip closes, the material behind it also closes in along the median line. Thus, embryo formation occurs by concrescence. It does not occur by growth outward from the blastopore or by slipping of material backward along the body. Instead, the process brings together two sides of the germ ring along the median line. Experimental work included analysis of the case of spina bifida, which the large yolky mass remains in the center of the two sides, which fail to close properly along the median. This condition occurs in nature, in frogs and other vertebrates. Researchers could also make it happen in the laboratory by affecting the actions of the dorsal lip of the blastopore, as Morgan himself had shown in an earlier series of experiments. Those studies showed that frog embryos normally form by concrescence. Surprisingly, however, similar studies showed that teleost fish embryos do not.[13] Once again, Morgan responded to the plethora of complex data by emphasizing the data rather than any particular interpretation of it.

After his summary presentation on the *Frog's Egg*, Morgan moved on to other related problems. He returned to frogs and to the various problems of vertebrate development a number of times over the next decades. But his research carried him to different organisms and to different questions and different ways of framing the same questions as well.

On to Naples

Late October 1894 to July 1895 Morgan spent at the Naples Zoological Station. In addition to working on the frog book, he carried out numerous experimental embryological studies and became a friend of the Naples station's director, Anton Dohrn. The "chief aim and work of the station is original investigation," Morgan wrote, and its primary virtues included "absolute freedom to work on any subject desired, a plentiful and never-failing supply of fresh material and a well-filled library always at hand." In addition, the laboratory provided a center for biological activity, especially precious to a young American, for, "isolated, as we are in America, from much of the newer, current feeling, we are able at Naples, as in no other laboratory in the world, to get in touch with the best modern work."[14]

When Alexander Agassiz, who had supported a table at the Naples Zoological Station, decided to terminate that support and began publicly criticizing the place, Morgan and other Americans came to the defense. Articles in such widely read publications as *Science*, lobbying at national meetings, and other efforts resulted in increased support for the *stazione*. Three Americans tables, including a special women's table and the ongoing Smithsonian Institution table, guaranteed a warm welcome for any qualified American researcher who wished to spend some time in the stimulating international cultural and intellectual environment at Naples.[15]

Morgan enjoyed his European visit and extended it as long as possible into the summer. With Hans Driesch and Curt Herbst, he continued on to Zurich in August, then he went to Paris, London, and back to work at Bryn Mawr. He returned for summer research in at least 1896, 1898, 1900, and 1902 and had nostalgic visits with Driesch and Herbst at Vincenzo's, the favorite local tavern.[16]

Amidst the great wealth of materials, equipment, knowledgeable researchers, library, and other perfections of the *stazione*, Morgan pursued the sort of work that had gained so much attention in recent years. The Hertwigs had shown the particular value of sea urchins for embryological study. Boveri had demonstrated the possibilities of shaking sea urchin eggs to produce fragments, which then go on to experience some further development. Driesch, Roux, and others had raised questions about the fates of isolated blastomeres. And Loeb had suggested that altering external conditions such as the salt content of water can influence the course of development. All of these studies had stimulated considerable discussion. All had also raised new questions. During his first visit to Naples, Morgan began to explore some of those questions with the same local organisms that others had used.[17]

With sea urchins, especially *Sphaerechinus*, he shook eggs apart in test tubes to produce fragments, then subjected them to a variety of procedures. In one set of studies, he dropped the fragments about a foot into a dish of water below. This served to separate the bits most effectively. Other studies explored the respective effects of hard and gentle shaking. Still others chopped the eggs into bits with scissors. A further set of explorations looked at the effects of shaking the embryos at the blastula or gastrula stages instead of prior to fertilization. All the studies combined to demonstrate, in agreement with Wilson, that the isolated blastomere or egg fragment that divides and becomes a smaller normal embryo is not just the same as the normal case. It has fewer cells than normal, and it does not regulate development in the same way. Furthermore, a certain minimum size is necessary for sea urchin cells, with different sizes at different developmental stages; a minimum number of cells is necessary for each stage to progress, but the full normal complement is not required; and a fixed number of cells invaginates during gastrulation, even when there is a smaller total number to start with.

The upshot was that the Roux-Weismann theory of qualitative division and mosaic development of independent cellular units had been refuted. So had the Driesch-Hertwig view that the whole organism undergoes regulation in a way specific to that type of organism. Since the same number of cells (about fifty) invaginates in the smaller and in the normal embryo, the same kind of mechanical regulation could not explain the facts of development. It seemed to Morgan in the midst of his studies that Whitman had been right after all when he had said that "the formation of the embryo is not controlled by the form of the cleavage."[18] Instead, something about the internal structure and constitution of the cytoplasm seemed crucial.

A series of two investigations carried out jointly with Driesch on the ctenophores (or comb jellies) led to the same conclusions.[19] So did another study, of *Amphioxus.* Inspired by Wilson's work on *Amphioxus* and following similar experimental manipulations, Morgan asked how many cells those larvae that are formed from isolated blastomeres have. Fewer than usual, he found once again, concluding that "despite the physical constraints that have been brought to bear on the developing egg, there remains always a tendency for the egg or part of the egg to reach its prescribed goal despite, even inspite of the modifications impressed on the egg from outside." This seemed to Morgan one of "the most important results of the experimental work of the last few years."[20]

The results showed that a piece of an egg, which under normal conditions would never have done so, could become a normal embryo. Unfortunately, "we can find no chemical or physical explanation for any of these phenomena." The processes might well be normal physicochemical processes, but that was simply not known yet. The evidence remained inconclusive. Thus, "so far as we can see at present the vital factors that control the development do make use of many known chemical and physical properties of matter, but it seems to me that it is very rash at present to conclude therefore that the vital processes of living things are necessarily only the complex of known physical and chemical processes."[21] Perhaps the piece acts like an adult that has been injured and somehow regenerates the missing parts to make itself whole. It would therefore make sense to study the vital processes involved in regeneration.

After a set of experiments in the winter of 1894–95 following up on the effects of Loeb's altered salt conditions on the production of artificial centrosomes and nuclear parts, Morgan turned to a full-scale and sustained study of regeneration. He looked at a wide range of organisms under a variety of experimental conditions to try to get at the heart of the regeneration process.

Regeneration

Regeneration, Morgan believed, is essentially similar to normal development. It also parallels regulatory production of a whole larva from isolated embryonic blastomeres. As a result, a study of regeneration could help to explain those other phenomena. The large number of normally occurring cases of regeneration made it a promising line of study. In addition, Morgan's friend Loeb and other MBL researchers had begun to study the physiological factors involved in regeneration. When August Weismann had devoted a section of his *Germ Plasm* to problems of regeneration and helped to stimulate renewed interest in the subject, it was almost overdetermined that Morgan would also turn to this set of phenomena.[22]

His first major paper on the subject appeared in 1897 and reported the results of studies on oligochaete worms (terrestrial hermaphroditic earthworms). In particular, Morgan wanted to determine how small a piece may be and still regenerate, just as he had asked how small a sea urchin fragment may be and still grow into a whole sea urchin larva. He recognized the need for working with large samples since the worms exhibited such wide individual variation; this in turn led him to report some of his results in the form of tables and charts, which he had not typically used for his embryological studies.

The major thrust of that paper and of several others, however, was its attack on Weismann's theoretical interpretation of regeneration. Weismann's appeal to hypothetical hereditary units failed to persuade Morgan, as did his discussion of the efficacy of natural selection. Morgan saw facts that he felt did not conform to such a theory. For example, after a worm is sliced through its midsection, it begins to regenerate in one direction almost immediately but takes some time to regenerate in the other direction (the exact details depending on the exact place and size of the cut). To

Morgan this showed "that we are dealing here with something that is connected with the organization of the worm itself." Perhaps the organism possesses some head "stuff" that diminishes as one moves posteriorly, some tail "stuff" that diminishes in the other direction. In the central areas where there is less, it might take longer to produce enough more to initiate regeneration, which, at any rate, generally produces just the parts lost and in just the proper places. "I do not pretend that this explains anything at all," Morgan readily acknowledged, "but the statement covers the results as they stand."[23] In contrast, the theory of reserve cells or reserve idioplasm favored by Weismann and Roux could not explain the precision of the process. As Morgan put it later, we should stick with the production of facts, so that

> the work is not likely to come to a standstill, but when we leave the analytical method and attempt to construct injudicious theories that make the pretense of explaining a complicated process without attempting to resolve the process itself into factors, then progress stops. Such, I believe, to be the case in the attempt to explain the process of regeneration by a theory of preformed imaginary germs. Any theory of this kind is only a pretense; imagination takes the place of verifiable hypothesis, and the process that we set out to study is explained by saying that there are "germs" present that have been set aside to bring about the result![24]

Weismann's appeal to natural selection had similar problems and also represented poor science. He had said that the power of regeneration must depend on the ancestral conditions of the species in question and could not arise in this generation. In the past, according to his view, individual organisms that had the power to regenerate particular body parts that were most liable to injury had a selective advantage. Those individuals developed their power by the presence of an auxiliary set of germs held in reserve for such injury conditions. Thus, the species acquired the set of germs and the ability to regenerate those parts most liable to injury. This was Weismann's theory, at any rate.

Morgan ridiculed it. How could even an infinite number of injuries to a part create in it the power to regenerate, he objected. Would not injured animals necessarily be at a competitive disadvantage, so that those with the power to regenerate would never really have a chance to use it, he asked.

> In my experiments, for instance, I find that only rarely do posterior ends of worms cut in the middle regenerate anteriorly, and even in those cases where this happens, the regeneration is almost always im-

> perfect. Does this mean that as yet an insufficient number of these worms have been injured in this way? In the course of time if more worms have been injured in accidents, will Allolobophora foetida acquire a capacity for regenerating the anterior end? Are we to consider seriously this interpretation of the selection theory?[25]

Besides showing Morgan's disinclination to accept natural selection as the mechanism of evolutionary change, the paper reveals his recognition that further work on regeneration would be required before any major progress could be made in understanding it. As a result, from 1897 through 1904, he wrote over thirty papers and a book on regeneration in a variety of animals, explored with a variety of experimental and descriptive approaches. By the end of that time, he had developed his own working hypothesis, decided that it was no longer productive to pursue this line of research, and had moved on to other, related questions about development and inheritance.

Throughout the studies, Morgan continued to expand on his idea of organization and to explore theories based on the existence of formative stuffs in the body. In planarians, he found that almost every part of the animal can form a new animal as long as it is sufficiently large. Central pieces do that by adding new tissue at the ends to generate a head and tail, then the whole undergoes a transformation to act as a new whole organism. Thus, the planarians exhibit not only the power to regenerate missing parts but also the ability of self-regulation, to bring about normal relations of the various parts as they develop rather than simply adding new material.[26]

This theme of organization and self-regulation received greater emphasis as Morgan became more deeply involved with the research. When the departure of his boat from England back to the United States in September 1898 was delayed, he spent his extra time in the British Museum, reading earlier work on regeneration. He found Trembley and Bonnet especially interesting, indeed more than anything his contemporaries offered.[27] As a result of that visit, he probably revised the lecture he had presented at the MBL that summer, for it begins with an introduction to the questions Trembley and Bonnet had raised and goes on to consider their theories in some detail.

That lecture and another the next summer outline Morgan's thinking on the problems of regeneration to that point. The first surveys the phenomena of regeneration, demonstrates that questions abound, and asks, "Is there not here a problem that we can hope to solve by means of further experiment?"[28] Indeed there were several: is there a correlation between liability of a part to injury and its regenerative ability?—a question he had raised in an earlier paper in direct response to the assertions by Weismann and others that this was the case. Also, is a certain minimal size required before an organism can regenerate, as is true for egg fragments? Most important of all, what are the relative internal and external conditions in regeneration? Finally, do old cells form new tissue directly or are there "reserve" cells that come into action when needed? Morgan had a response to each question but also recognized that the whole picture remained far from clear.

At this stage, he supported the view that regeneration was an "intelligent process" because

> something more is included in these phenomena, I think, than can be explained by simple physical interaction or by chemical influences. The process that takes place suggests that something like an intelligent process must be at work—I mean that what we call correlation of the parts seems here to belong rather to the category of phenomena that we call intelligent, than to physical and chemical processes as known in the physical sciences. The action seems, however, to be intelligent only so far as concerns the internal relations of the part, i.e., it acts rather as a "perfecting principle" than as a process of adaptation to external needs.[29]

Morgan was not a vitalist, but he felt that developmental processes exhibited something more than just physics and chemistry. That something he referred to over and over as *organization*. Something about the organism made it more than the sum of its parts, he felt sure. This theme he spelled out in 1898, as well as the questions to be addressed in future research.

Once again, after outlining the different theories that had been proposed and assessing each, Morgan attacked Weismann. Once again he stressed that Weismann's approach was unscientific. His appeal to preformed reserve germs invokes an unnecessary complicating hypothesis. Even more important, he charged that Weismann simply did not understand how to do good science. While reasonably restrained in his criticisms of other authors, Morgan said of Weismann:

> I find in this whole argument only an attempt to shift the difficulties of the problem back to the unknown ancestors of present forms, just as the difficulties of other parts of the problem are also shifted back upon the unknown germs that exist preformed in the egg or in the parts that can regenerate. Weismann does not, perhaps, realize the difference between himself and those whom he somewhat scornfully calls "the younger investigators." The problems that they are trying to solve are those that Weismann also tries to answer, but "the younger investi-

gators" base their interpretations on the assumption that when a change takes place a sufficient cause for the change is to be sought in the organ itself and in the external conditions surrounding that organ. They are not content to rest their "explanations" on "the phyletic origins" of the changes. It is not necessary to deny the theory of descent, but it is unsafe and in many cases unscientific to base "causal explanations" on an imaginary line of ancestors.[30]

Instead of any of the available theories, Morgan believed that the best hypothesis was that the whole protoplasm undergoes a change, probably a molecular one. The subsequent processes, he had decided by 1899, are probably "only the expression of the physical, molecular structure that has been assumed by the piece." He recognized that the hypothesis remained provisional, as a "working hypothesis" and a "possible point of view," but he felt that it alone "can bring under one point of view many isolated observations."[31] A major problem, obviously, was to clarify how the physical, molecular structure acts to produce what seemed like action of a whole organism. This quest occupied the rest of Morgan's work on regeneration.

From 1900 to 1904, when he reported his last major regeneration results, Morgan studied such assorted species as the earthworm, planarian, teleost fish, sea urchin, hydromedusa, tubularian, antennularian, stentor, hermit crab and crayfish, and the terrestrial triclad bipalium. He asked well-defined questions about each animal and used each to address different aspects of regeneration. For example, with some species such as frogs, he followed Ross Granville Harrison's lead in transplanting part of one embryo onto a host embryo of a different species. He then cut off part of the tail and observed which cells went to make up the regenerated material at the end: how did the parts from two animals join to act as a whole? This experiment he found difficult because he could not observe directly what was going on inside the developing tissue. He had to rely on serial sections of different animals and different stages of development instead of on living specimens, as he had been able to do for most of the other studies.

With hydromedusae, he showed that the regeneration of parts could not be the result of mechanical action alone but required some active reshaping of the living tissues as well, a process he termed morphallaxis. Hermit crabs and crayfish showed that those parts that most often are injured are not, in fact, the most likely to be able to regenerate. Terrestrial triclad worms (which occur commonly in greenhouses and which Conklin sent him) gave further evidence about how the piece of the worm transforms itself into a new, whole, smaller worm. Planarians also demonstrated their morphallactic abilities. Teleost fish exhibited the ability to undergo unsymmetrical development, so that the new part actually becomes unsymmetrical "but completes an unsymmetrical old part in such a way, that in the end the old and the new supplement each other to form a symmetrical whole."[32] He did not know how that transformation takes place, Morgan reminded his readers, but he felt it better to wait for "the discovery of new facts" before putting forth any further hypothesis.

In addition to looking at regeneration in adults or in embryos, Morgan also looked at the process in egg fragments, making clear that he held the processes as fundamentally similar. He also examined the role of different germ layers in effecting regeneration and found that sometimes the regenerated part arises from a different germ layer than the normal part does. This certainly undercut the germ layer doctrine, which held a strong determinism of different parts from different layers.

While stressing over and over that the internal organization in some way directs regeneration, that simple mechanical action is not sufficient, and that the processes must involve some kind of formative stuff, Morgan also looked at the effects of external conditions. Loeb had suggested that geotropism, or the differential effects of gravity on different parts of the organism, explained regeneration as well as much of normal development. Morgan tested the hypothesis in antennularia and did not rule it out, but he felt that the evidence failed to support the efficacy of external conditions. Other studies followed.

By January 1900, Morgan had carried out most of his preliminary studies; later research followed up the same basic questions and tested the same set of theories, refining the working hypothesis in each case. In that month, Morgan presented his work in a series of five lectures at Columbia University that were extended and revised to become *Regeneration.* Like *The Frog's Egg,* this book provided a summary of the questions, theories, observations, and experiments carried out to that time. In *Regeneration* he also developed his best version of his working hypothesis, on which he had been working for some time and which he had revised in various ways in light of new evidence. He also articulated in some detail what he felt the proper scientific method should be, again a theme that he had pursued earlier but clarified here.

His own theory of how regeneration, like normal development, occurs was based on the conviction that the organic system is a coordinated system of integrated parts, and not a set of separate units. Even cells are not separate, isolated parts, for they are all connected and intercommunicate through their protoplasm. Regeneration and development occur because of the organization of the integrated parts that make up the whole organism and the system of tensions which exists between parts to effect the whole. Regeneration can best be understood in terms of internal factors, namely "some sort of tension." Or, as he put it later, we can refer the polarity "to a gradation in the pressure relations, since these are the dynamical expression of the gradation of the materials, as shown in their differentiation. These differences can be traced to the egg where the differences in the pressure relations of the cells give rise to the later differentiation."[33]

This remained rather vague, Morgan realized, but he felt that it fit with all the assorted data from different organisms and different types of experiments. It avoided hypothesizing into existence any invisible units. It was legitimately scientific in that it was subject to proof or disproof, he concluded, and it pointed the way to further study. As such it was a good working hypothesis, the best available after he had ruled out all the existing alternatives.

For Morgan was very concerned to do what he considered good science. He had become alert to the kind of work he was doing and self-conscious about following a legitimate scientific approach that did not lapse into imagination or speculative philosophizing, as those of too many others, including his friend Driesch, often did. Morgan agreed with Conklin and Wilson that metaphysics in particular has no place in scientific work.

This did not mean, however, that science had no place for proper hypotheses:

> The preceding hypotheses that have been advanced to account for the phenomena of regeneration, draw attention to some of the most fundamental problems of regeneration and, even in those cases in which the hypotheses have not given a satisfactory solution of the problems, some of them have served the good purpose, both of directing attention to important questions and of leading biologists to make experiments to test the new points of view. We should not underrate their value, even if they have sometimes failed to give a solution of problems, for they have been useful if only in eliminating certain possibilities, and this simplifies all future work. So long as an hypothesis is of a sort that is within the range of observational and experimental test, it may be of service, even if it prove erroneous: for our advance through the tangled thread of phenomena is not only assisted by advances in the right direction, but all possibilities must be tested before we can be certain that we have discovered the whole truth. The value of a scientific hypothesis depends, it seems to me, first, on the possi-

> bility of testing it by direct observation, or by experiment; second, on whether it leads to advance; and, lastly, on its elimination of certain possibilities.[34]

Though not ordinarily inclined to redundancy within any given paper, Morgan emphasized and reemphasized what he saw as the value of hypotheses throughout his work of this period, as well as the value of both experimental and observational study. He was clearly articulating for himself just what counted as legitimate scientific work, particularly biological work.

After completing his regeneration volume and tidying up the remaining studies of regeneration in various organisms by early 1904, Morgan made several major changes in his life. Probably most importantly, he married Lilian Vaughan Sampson in June 1904. The daughter of an upper-middle-class family in Philadelphia, Sampson had entered graduate school at Bryn Mawr in 1891 with a specialty in embryology and morphology. Morgan wrote to Driesch that: "I am engaged to Miss Lilian Sampson of whom you may have heard since she studied one year with Lang in Zurich. I need not tell you that I am the luckiest fellow in Christendom."[35] Lilian Morgan tried to keep her own scientific career active and contributed several important papers.[36]

In the same year, Morgan accepted Wilson's offer to move to Columbia University as professor of experimental zoology. His students at Bryn Mawr were indignant at his leaving and felt "almost as though Dr. Morgan belonged to us, and that when he departed there would be no department left."[37] Yet going to Columbia made sense. There he would have the advantages of being in a major department and a large city, with significantly greater resources than Bryn Mawr could offer. He would also have his old friend Wilson as a close colleague. Before taking up that post, Morgan and his new bride traveled to California for sightseeing and collecting, with visits to the marine laboratories of Stanford and Berkeley. Several final regeneration studies followed including a look at polarity in tubularia. This work resulted in the provisional hypothesis that hydranth-forming material decreases from the apical to the basal end of the animal, but Morgan evidently did not see how to pursue that hypothesis further with experimental tests. After a few more attempts, he gave it up in 1909.[38]

After that summer, Morgan and his wife returned eastward with Loeb and the Dutch mutationist Hugo de Vries, both of whom were also going to meetings in Saint Louis. The Morgans then made their way back to New York, to take up the new position at Columbia. Wilson later said of this move that it was his great claim to fame in genetics. In his welcoming address to the Genetics and Eugenics Congresses in New York in 1932, he said, "I am not and never was a geneticist." Yet in 1904, "I was able to effect a cross between T. H. Morgan and the Department of Zoology at Columbia, of which I happened to be at that time the Executive officer. The experiment succeeded beyond my wildest expectations."[39]

Experimental Embryology

After pursuing regeneration through a variety of animals and a wide range of different kinds of studies, Morgan had reached a provisional working hypothesis but no clear productive direction in which to go with it. It was no longer evident how to test the hypothesis, useful as it had been, and so he moved on to other problems. Since he had explored regeneration in order to illuminate the working of normal development, he returned to that set of problems. Actually he had never left those questions but had just focused more sharply on regeneration. The period after completion of *The Frog's Egg*, from 1898 to 1910, therefore included numerous studies in experimental embryology as well as new explorations of heredity and evolution.

Morgan continued to ask what he could learn from experimental work in particular. The traditional questions of embryology remained: is development essentially self-directed or externally directed, and how? What are the relative roles of nucleus and cytoplasm? To what extent does the embryo lie preformed (or at least prelocalized) in the egg? To these he added: to what extent do the results of experimental, or abnormal, manipulations reflect the normal process of development?

In principle, Morgan reiterated on various occasions, experimentation makes biology scientific. This was not to say that doing experiments was new to biology, as some historians have suggested. Rather, biologists were newly recognizing that they needed to add experimentation to traditional descriptive and observational methods to help achieve scientific progress and, in fact, to place biology on a "more scientific basis."[40]

The essence of the experimental method, Morgan clarified in 1907, requires "that every suggestion (or hypothesis) be put to the test of experiment before it is admitted to a scientific status." Each hypothesis must do more than just bring together under one account a number of phenomena, though that function may have heuristic value. A good hypothesis must do more, however, and must be useful in leading to new facts in the form of clear answers. It must not be simply a nice or elegant fiction but must have "practical bearing" as well. Such a proper hypothesis serves as a working guide to further study. Yet it carries with it dangers as well since it may blind us to reality. "Therefore the investigator must not only be an inventor of working hypotheses, but cultivate also a skeptical state of mind toward all hypotheses—especially his own—and be ready to abandon them the moment the evidence points the other way."[41]

Once one has the hypothesis in place to guide research and to point the direction in which to begin testing, then it is necessary to construct a proper experiment. What, Morgan asked next, is an experiment? Is it an experiment if someone chops an earthworm into two to see what will happen? Well, that depends. It depends on the purpose of the chopping.

> If the worm is cut in two in order to study the physiological behavior of the two ends, as had been done in fact, with some interesting results, or in order to see what regenerates at the two cut ends, we have a distinct purpose in view, although no formulated problem. If we proceed farther and remove a definite number of segments in order to see how many come back, and then try to determine what conditions are involved in the results, we are clearly carrying out an experiment with a more definite aim. This illustration will serve to show that the most essential feature of an experiment is the anticipation of the results of a test.[42]

Such experiments are what Morgan had attempted to carry out in embryology, seeking always to address the traditional questions about what directs development and how. Beginning with the study of isolated blastomeres of frogs and sea urchins, as so many others were doing, Morgan moved on to other questions and other animals. He did not pursue the cytological details as far as Wilson or Conklin had, but he learned from their work and incorporated cytological with other techniques and questions.

Several leading techniques allowed control of at least some factors of development and therefore facilitated testing of some hypotheses. These included the temporary exposure of eggs, especially sea urchin eggs, to altered conditions such as salt concentrations or strychnine in order to assess the effects of these external factors. Morgan concluded that the division that occurs thereafter is not normal and that the additional sodium chloride somehow acts as a stimulus on the nucleus in a nonnormal way.[43] He observed the details of changes in the nuclear and cytoplasmic parts but regretted that the study of those internal parts required the killing, fixing, and preserving of specimens rather than allowing the researcher to watch their living actions. Fortunately, Wilson made some helpful suggestions as to cytological techniques.

In earlier work, Morgan and others, including Wilson and Conklin, had wondered whether the size of isolated blastomeres or egg fragments was crucial in determining whether they could develop normally; a minimal size, which varied in different species, did seem to be required. Now Morgan wondered whether the embryo had to have a certain number of cells before it could develop normally, which would explain why whole embryos did not result from one blastomere out of thirty-two, for example. Driesch had asserted that all developing embryos have the same number of cells, at least prior to gastrulation. Morgan found that not to be the case and raised further questions about how the smaller embryo with fewer cells accommodates.

Other experiments considered the effect of gravity on development. Pflüger had argued that gravity was crucial for determining the embryo's orientation. Roux had carried out similar experiments, also with frogs, and had devised a wheel for spinning the embryo within a gravitational field to determine whether a constant force is necessary and what effects changes would bring. Morgan extended that study to toads. He agreed with Roux and against Pflüger that a constant gravitational effect is not necessary. He also agreed with Roux in opposing Pflüger's idea of the virtual homogeneity of the dividing blastomeres and arguing for an internal direction of development. It is the internal organization of the egg, or "heterotropism," and its accompanying near totipotency rather than any sort of qualitative division which makes up the critical internal factor.[44]

Another related technique gaining currency was centrifuging. Morgan carried out a number of studies using relatively stronger or weaker centrifugal force, culminating in a long paper in 1910 which surveyed much of his work on various species.[45] Centrifuging did have an effect in moving the materials of the egg around. For species such as sea urchins which seemed truly totipotent and relatively homogeneous, centrifuging made relatively little difference to the course of development. To such relatively organized and differentiated eggs as frogs' eggs, in contrast, the experimental centrifuging did make a difference. This effect Morgan explored in an impressive series of ten papers published in *Roux's Archiv* from 1902 through 1905. There he examined that basic question of experimental zoology: what is the relation of experimentally produced abnormal cases to normal cases of development?

In these studies, Morgan proceeded systematically to attack questions and techniques that had gained attention in recent years. First, with injury to the frog's egg yolk: what happens and what does this tell us about normal development? On the latter point, Morgan recognized that two opposing schools of thought coexisted: one held "that the evidence from abnormal development can be applied directly towards the elucidation of the normal development. The other school maintains that just so far as the embryo is abnormal it departs from the normal method of development, and the evidence from the one source has no application to the other."[46] In fact, the proper use of experimental cases lies somewhere between. They can produce valuable information if carefully used, but they can also produce cases sufficiently abnormal as to obscure the normal processes. It is the experimental researcher's problem to determine exactly how far information derived from abnormal cases bears on normal development.

The other studies in the series included a look at the effects of lithium chloride, abnormal forms of development, injuring blastomeres as Roux had done (but he had not killed them, Morgan determined), removal of the upper cells from the black hemisphere (since that might be where the embryo appears), incompletely injuring blastomeres (as Roux had apparently done), injuring the top of the egg, low temperatures, insufficient aeration, and movement of cells in the early stages of development (where the cells remained able to move extensively because of what Morgan saw as their contractile powers, like the tension he had hypothesized as allowing regeneration). The whole set of studies supported his conclusion that the frog has organ-forming materials from the beginning. Not organ-forming germ regions, as Wilhelm His had maintained, but differentiated materials. These could be moved around experimentally and their changed effects observed.

As a whole, then, the frog studies and the numerous other, related experimental embryological studies of the time brought progress in the form of useful working hypotheses and the generation of new facts about development. The results from all the lines of study to date pointed back to the internal organization of the egg as central. Somehow the vital action of the whole differentiated heterotropic but totipotent material effected development—under a wide variety of external normal and artificial conditions. The next obvious set of questions related to the nature of the organization and to just how it exerts its effect. We have seen that Morgan was not convinced by the natural selection hypothesis, especially not Weismann's version of it. However, his embryological studies did not yet tell him where else to turn for further productive exploration of the questions at hand. So, by 1903, he had also begun to look at problems of heredity and evolution, alongside his questions of embryonic development.

Evolution and Heredity

During his regeneration work, Morgan had found it difficult to explain the origin of the ability to regenerate. He regarded Weismann's natural selection interpretation as annoying, unscientific, and unsupportable. It seemed clear to Morgan that no liability to injury—that is, no need—could possibly cause the regenerative ability to arise. As he wrote in the introduction to his next book, in 1903, the problem of how to explain the origin of regeneration stimulated his wider look at the nature of evolution and adaptation. For "the conclusion that I have reached is that the theory is entirely inadequate to account for the *origin* of the power to regenerate; and it seemed to me, therefore, desirable to reëxamine the whole question of adaptation, for might it not prove true here, also, that the theory of natural selection was inapplicable? This was my starting-point."[47]

The regeneration study also provided a starting point for Morgan's concern with problems of heredity. For simple inheritance could not explain regeneration either. Simply asserting that the organism had inherited some "reserve idioplasm" or other unobserved something that initiated regeneration did not appeal to Morgan. Theoretical units were not subject to scientific observation, and theories involving them therefore could not be properly tested by experiments, as any legitimate scientific theory must be for Morgan. The regenerative ability must result from the internal constitution and structure of the organism in some way, but exactly how that organization arises remained a question far in the background during Morgan's years of intensive exploration of regeneration and experimental embryology.

By 1903, however, he had declared interests in new areas, interests that grew along with the availability of new evidence over the next decade. That year brought publication both of Morgan's third book, *Evolution and Adaptation*, and of his survey of studies of sex determination. As with so many of Morgan's writings, these two provided excellent surveys of both historical and contemporary thinking on the problems in question. They both outlined the available alternative theories and discussed in detail the evidence offered to support each. In fact, another paper of 1903 in the popular magazine *Harper's* also discussed Morgan's views on Darwinism and natural selection in a very accessible way. This paper most nearly parallels his popular survey of sex determination in *Popular Science Monthly* and warrants attention as an outline of Morgan's key concerns.

First of all, Morgan pointed out, we must carefully separate the two views: evolution and natural selection. One does not absolutely require the other. The former idea, that species have come into being through a process of evolution, Morgan regarded as the "most fruitful of all modern philosophical conceptions."[48] This theory had brought many facts into the same explanatory framework and pointed the way to future work in a provocative way. Further, it suffered no real problems. As such, it was an excellent scientific hypothesis.

Natural selection, according to which natural forces select among individuals with different variations to change the population as a whole, did not qualify, however. On the face of it, the idea might seem convincing, but serious objections remained. In particular, Morgan worried about swamping: when a variation arises in one individual, that individual would seem to have to mate with another with the same variation in order to make a difference and be perpetuated within the species as a whole. In addition, experimental embryology provided cases that he felt could not be explained on a hypothesis of natural selection. Isolated blastomeres of embryos can regenerate the whole organism even though they could not possibly have ever experienced such conditions themselves in the past. Thus, they could not have adapted to the conditions, and the variation allowing regeneration could not explain the ability to regenerate—as Weismann insisted it did. To Morgan such an objection showed that natural selection could not, in fact, explain how evolution occurs in all cases, though it might possibly act in some cases. One had to look elsewhere instead. Biology needed another, properly testable scientific theory to explain evolution.

De Vries had provided such an alternative with his mutation theory. Morgan explained that he had seen de Vries's experimental garden and had been impressed. Here was evolution at work, in the form of rapid mutations that produced new species in a single generation. Thus, "the important fact, from the point of view of the theory of evolution, is that the new species have sprung full-armed from the old one like Minerva from the head of Jove."[49] Evolution is not gradual but sudden, Morgan concluded. The popular objection to evolution that we cannot see it in action is met by the fact that we *can* do so in this case. Morgan was not convinced that de Vriesian mutation provided the conclusive answer to how evolution always works, but he was certain that it served as a better working hypothesis than natural selection. The case, like so much of biology and of science generally, called for continued investigation in the form of testing the hypothesis.

It is important to recognize that Morgan was at no time a confirmed de Vriesian. He primarily found the speed and visibility of evolution which de Vries supplied attractive, even while he questioned other aspects of the view. After his opportunity to travel with and talk with de Vries at greater length in the summer of 1904, as the Morgans, Loeb, and de Vries made their way from California to Saint Louis, Morgan continued to explore the possibilities of this alternative hypothesis throughout the first decade of the twentieth century.

Sex determination provided another puzzle for Morgan. Again, the coexistence of different working hypotheses served to stimulate work and helped to make the discussion productive. By 1903, he had ruled out the importance of external factors in causing an individual to become determined as one sex or the other and had concluded that, as with most processes of differentiation, sex determination was a product of internal factors instead.

This raised the question of whether an individual's sex is inherited and hence predetermined, or whether it becomes determined gradually, epigenetically, in the course of the individual's development. William Castle offered an account whereby the inheritance of one or another Mendelian factor would cause the production of one sex or the other. Morgan found Castle's suggestions fascinating but not convincing. At the root of his objections was that he found epigenetic explanations more promising than preformationist alternatives. He also saw Mendelism and heredity in terms of predeterminism.

Wilson and Morgan agreed (at least until 1905) that the egg exists in a balanced condition, with the ability to become either male or female. It is the conditions to which the egg is exposed that determine which sex the individual will actually become. It might very well be the case that different factors have the deciding influence in different cases, for "here, as elsewhere in organic nature, different stimuli may determine in different species which of the possibilities that exist shall become realized."[50] Yet precisely how those conditions effect their result remained a question for this confirmed epigenesist.

Morgan resolved to explore this question in the usual scientific way: by taking the existing alternative hypotheses and holding them up to the experimental evidence to test them. He began with a look at the ascidian *Ciona intestinalis*, a hermaphroditic species that

Castle had declared incapable of self-fertilization. Why should this selective infertility occur? Morgan asked.

He found no evidence for Castle's suggestion that the phenomenon is caused by the existence of different male and female types of eggs and sperm. He therefore explored the idea that the selective self-sterility is related in some way to the selective fertilization that Castle had said is necessary for sex determination. Instead, Morgan found that self-fertilization is actually possible in some experimental cases. He concluded that the egg has some substance that brings the spermatozoa to rest and makes fertilization possible. This internal substance was therefore key to understanding the process, but he concluded that it had nothing to do with sex determination and that Castle's linking of the two phenomena was therefore unjustified.[51]

In another set of studies, Morgan looked at the gynandromorphous condition in insects, defined as possession of some male and some female characteristics by the same individual. Boveri had put forth the hypothesis that the condition resulted when the spermatozoon failed to unite properly with the female nucleus at fertilization and then united with one of the egg nucleus products at a later stage. This would produce some nuclei with only the female nucleus and some with both. He generally assumed that male nuclei would result in male characteristics, while normal fertilized nuclei would produce normal embryos. Morgan saw no evidence for this delayed fertilization, since he had never seen such a union occurring at later divisions. He proposed an alternative explanation: that two spermatozoa enter the egg simultaneously; one unites with the egg nucleus and the other acts on its own. Thus, there is one normal and one male nucleus in each of the cells but no difference among the cells themselves.[52] This theory allowed for the individual epigenetic development of the cells in response to the whole.

It was not to advocate his own hypothesis instead of Boveri's that he put forth his suggestion, Morgan wrote. Nor did he have any particular interest in weighing "their relative merits on the grounds of probability. I have raised the question not to invite discussion, but to appeal to those who may have an opportunity to examine gynandromorphs from mixed hives."[53] Once again, he called for empirical fact and experimental test as the ultimate scientific court of appeal.

Once again also, Morgan lobbied for an epigenetic rather than a preformationist interpretation. "Which of these general points of view, preformation or epigenesis, we may think more profitable as a working hypothesis is, I believe, the question of the hour. My own preference—or prejudice, perhaps—is for the epigenetic interpretation, but the whole truth may lie somewhere between these two forms of thought that are the Scylla and Charybdis of biological speculation."[54]

The possibility of a preformationist account raised questions for Morgan largely because preformation at that time generally implied the existence of preexistent—that is, inherited—particles. Yet he saw no evidence for such particulate theories of inheritance. He saw no particles. Similarly Mendelian factors, he pointed out to the American Breeders Association, are only hypothetical units. The assumption that such factors exist and that they segregate in the way that Mendelism suggested remained a hypothesis, and a preformationist hypothesis at that.

Such a hypothesis could be productive for scientific inquiry. Indeed, Mendelism offers a ready explanation of heredity: if the characters already exist preformed (in the indirect sense of predetermined) in the inherited factors, then they have only to exert their present influence in order to effect proper development. According to this interpretation, the factors remain segregated on the respective male and female nuclear contributions, and they remain independent in that they sort separately; they "may be shuffled like cards in a pack but cannot become mixed." This was a traditional preformationist view, Morgan saw, with its attendant advantages and disadvantages. It gave apparent success in explaining heredity, but that "does not, in my opinion, justify the procedure; for the preformation idea has always led to immediate, if temporary successes; while the epigenetic conception, although laborious, and uncertain, has, I believe, one great advantage, it keeps open the door for further examination and re-examination. Scientific advance has most often taken place in this way."[55] Epigenesis was just better science.

This did not, however, mean that Morgan rejected Mendelism as a whole. Indeed, he accepted "Mendel's Law," by which he meant the empirical claim that, within a large population of individual pea plants, certain ratios obtain and that these result because of the persistence of both parental contributions in hybrids. He even considered it plausible that the germ cells contain factors that behave in the way that Mendelism suggested. Mendel had offered a valuable generalization, which put zoology on a firm basis and which Morgan was surprised had not been observed before.[56]

Mendelism does not, however, adequately explain development. The Mendelian factors are not the same as the characters of the adult. The idea that characters are transmitted from one generation to the next is metaphorical rather than explanatory, Morgan felt.[57]

Instead, the egg and the sperm each contain "a particular material which *in the course of the development produces* in some unknown way the character of the adult." The fertilized egg is a balanced hybrid of the substance that can go to produce either of two characters, but it is the local conditions internal and external to the egg that actually determine which the adult will become.[58] Experimental embryology certainly supported such a view. Though Morgan recognized that the issue remained far from resolved, he favored an epigenetic view, in terms of local conditions acting on the inherited material, as much the preferred sort of scientific theory.

Furthermore, Morgan questioned the attempt to establish a Mendelian-chromosome hypothesis. He recognized that the chromosomes offered a very seductive opportunity to explain the apparent segregation of characters from the male and female parents. He certainly knew of the work by his colleague Wilson's student Walter Sutton along those lines. Such an association of chromosomes and Mendelian heredity had begun with Boveri, he pointed out, when Boveri argued for the persistent individuality of the chromosomes. This made them plausible candidates for the position of carrier of hereditary particles and therefore placed them as the "chief bone of contention" in explaining heredity and development. Sutton had gone on to develop a Mendelian-chromosomal theory of heredity, whereby the particulate determinants for characters reside on those nuclear units and the presence or absence of a particular determinant would decide whether the individual had a particular character.[59]

Morgan disagreed. Indeed, the theory suffered what he saw as a fatal objection. Namely, if the chromosomes carry all hereditary factors and factors determine characters, then a large number of characters must be inherited together on the same chromosome and should therefore be shared by many individuals. This does not happen. Therefore the hypothesis fails as presented and demands significant revision, at the least. Instead, Morgan favored the view that chromosomes may well be the bearers of heredity, or the vehicles of transmission: however, "we must be prepared to admit that the evidence is entirely in favor of the view that the differentiation of the body is due to other factors that modify the cells in one way or in another." We have two factors determining characters: heredity and the modification during development.[60]

As a result, Morgan inclined to the same sort of epigenetic view he had consistently preferred, according to which internal factors work together to bring development of the whole organism. Such an interpretation made sense to an embryologist so familiar with the complicated working of experimental embryology, regeneration, and such. Yet, true to form, Morgan left the door open for alternative theories. For "science advances by carefully weighing all of the evidence at her command. When a decision is not warranted by the facts, experience teaches that it is wise to suspend judgment, until the evidence can be put to further test. This is the position we are in to-day concerning the interpretation of the mechanism that we have found by means of which sex is determined."[61]

The next year, of course, he observed the white-eyed male *Drosophila,* which suggested with compelling simplicity and consistency that sex and eye color are always related. This offered just the sort of support for the Mendelian-chromosome hypothesis that Morgan had required in order to overcome the otherwise "fatal objection." The story of his move to work on fruit fly genetics is well known and need not be reiterated here.

The important thing to remember about Morgan's move to genetics is that it did not involve a rejection of embryology. Nor did it represent a radical conversion to something new and completely different. Instead, Morgan was practicing what he had always argued for: he was following the most promising and provocative hypothesis, testing it, and exploring its implications. He did not believe before or after beginning his work on fruit flies that heredity determines all of development and thus provides a sufficient explanation of that complex phenomenon. The data from white-eyed males and other studies convinced him that it was worth testing the hypothesis that chromosomes are the bearers of heredity, that they can explain Mendelian inheritance, and that they may be more direct determinants of differentiation than he had previously thought. As he cautioned his readers in 1913 in *Heredity in Sex,* "I beg to remind the reader and possible critic that the writer holds all conclusions in science relative, and subject to change, for change in science does not mean so much that what has gone before was wrong as the discovery of a better strategic position than the one last held."[62]

During the period between Morgan's graduation and 1915, when he was firmly established as an independent researcher, he published nearly eighty articles, six books, and a host of reviews and abstracts. These pursued several different lines of exploration, following different sets of working hypotheses with different techniques and different sets of organisms. On the surface, they may appear to be very different research programs indeed. They do represent a divergence of research projects. Yet a careful look shows the interrelatedness of the lines of research within the larger tradition, as well as the consistency of Morgan's epistemological convictions.

Throughout all these lines of research, Morgan also held to the common set of epistemologic concerns which he developed early in his career: seek positive facts by following productive working hypotheses. He seemed to jump from one problem to another because he followed what seemed to be the most promising line of research given the methods, equipment, and knowledge at hand. This led Harrison to consider him the "romantic" counterpart to the classic Wilson. For Harrison, the romantic's "ideas come thick and fast; they must find quick expression. His first care is to get a problem off his hands to make room for the next."[63] Perhaps it might be more appropriate to see Morgan as an American entrepreneur, following opportunity when the resources are abundant and productive and abandoning projects with diminishing returns.

9

Ross Granville Harrison

AFTER HIS FIRST visit to Bonn in 1893 and his final year in the Johns Hopkins University graduate program completing his dissertation research on fin development in teleosts, Ross Granville Harrison spent a year teaching at Bryn Mawr. There he replaced Thomas Hunt Morgan, who was visiting in Europe, and taught courses in general biology, general zoology, and theoretical biology.[1] President Martha Carey Thomas of Bryn Mawr said when she lost Edmund Beecher Wilson to Columbia, "never mind, I can always find a better one." She got Morgan, who teased Wilson about Thomas's claim. Harrison could have done the same when Morgan left for Columbia and it was Harrison's turn to be the "better one."[2] The following year, 1895–96, he returned to Europe himself to continue his work in Bonn with Moritz Nussbaum. Harrison also got married in Germany in January 1896, after which he traveled to Naples for a honeymoon visit to the aquaria and a few months of research at the Zoological Station.[3] Morgan had visited Naples the previous year and had undoubtedly talked enthusiastically about the opportunities there, so Harrison joined the pilgrimage even though he was not primarily interested in the local marine invertebrates, as Wilson and Morgan had been.

After returning from Europe, Harrison stopped at the Marine Biological Laboratory (MBL) for the summer and then began teaching at the Johns Hopkins University Medical School in the fall. While Wilson, Edwin Grant Conklin, and Morgan had each gone on to positions in biological programs, Harrison found his place first in the Anatomy Department at Hopkins. This was admittedly a unique anatomy department, headed by someone who had neither studied human anatomy nor done the traditional dissections as his major field of research. In fact, Franklin Paine Mall recalled that,

Fig. 18. Harrison at work at Yale. From the Marine Biological Laboratory Archives.

when he agreed to direct the department when the medical school opened in 1893, he had not done human dissections for a decade.[4]

Mall was nonetheless an ideal head for this anatomy program, which aspired to take its place alongside leading European programs. While virtually all American medical schools offered primarily (or exclusively) a traditional course in gross human anatomy, Hopkins sought to include such newer fields as histology and more general anatomy as well. As director of the program, Mall addressed his first tasks, namely teaching gross anatomy and designing a new building for the department, in unorthodox ways.

He designed the building so that the dissecting and laboratory spaces were accessible to students and faculty alike. And he invested considerable effort in guaranteeing that the building would be modern, spacious, well planned, and clean—all quite unusual for anatomy, which normally involved a good deal of mess and crowdedness. With an expensive refrigeration system, Mall could even hold the requisite human cadavers during the hot Baltimore summer months and still maintain sanitary conditions. Further, the facilities allowed the sort of teaching that he had experienced in his European studies, in which the students carry out their own research under the professor's direction.

Instead of holding formal lectures, with anatomical dissections carried out for the group by assistants, Mall taught through what he called the inductive method. He led the students into the dissecting room instead of the lecture hall, pointed them in the direction of their cadavers, and put them to work—to learn by doing. He walked through the laboratory and looked over students' shoulders, passing out rare praise and frequent caustic criticisms in a way that did not always succeed. The result was that "many of the students likewise were not sympathetic with Mall's method of teaching and did not enjoy his visits to their dissecting table, perhaps because of his cryptic questions and answers. As a result he soon became a legend in the School and generated not a little fear on the part of the students, which perhaps was not altogether a good thing, either for the students or for Mall himself."[5] Given that Mall reportedly "had no ability as a lecturer," it was nonetheless probably just as well that he stayed with his particular version of "elbow-teaching."

Mall's great strength lay in his ability to inspire those advanced students, younger colleagues, and members of his staff like Harrison who understood and accepted his sometimes abrasive and rigidly perfectionist demands. He worked hard for these people. "For them apparently nothing was too much trouble, and in them he had his greatest reward."[6] Harrison undoubtedly profited from such attention, and he evidently liked the environment, for he always held Mall in the highest regard and generally patterned his own teaching approach after Mall's.[7]

Rejecting the traditional view that anatomy consists primarily of human gross anatomy, Mall insisted that study of basic biological subjects, including cytology, embryology, and neurology, for example, should also provide a central part of any anatomical program. Therefore, he hired people, including Harrison, from biological rather than medical programs to build up his group. Harrison began as an instructor in the department in 1896, after his return from Europe, with the understanding that he would teach embryology and histology and that he would be given time to return to Germany to complete his M.D. degree at Bonn.

Transplantation and Regeneration

The first year in the Hopkins medical school, Harrison carried out his teaching duties and in the spring began a new series of experiments. For this work he turned to a different set of problems, using a new technique, and with a different emphasis from that of his earlier studies. Inspired by the rich possibilities for Gustav Born's technique of transplanting pieces of tissue from one animal to another, Harrison applied that grafting method to address his own questions.[8] This involved cutting a piece off of a host frog embryo and replacing that piece of tissue with material from another organism of the same general type. The resulting hybrid would survive and continue to grow and develop.

Specifically, Harrison used heteroplastic grafting, in which the host and the donor material looked sufficiently different in coloration for the researcher to tell easily which parts of the developing hybrid organism came from which original. More particularly, Harrison studied embryonic frogs, in which he sought to determine what would happen to the tail after experimental grafting. By taking a piece from a donor and placing it on a different place on the host, and by observing how the transplant and host cells and tissues changed with respect to the original graft plane, he hoped to illuminate how cells normally move during development. This technique would also allow study of different groups of cells in later stages of development, the point at which nearly all studies stopped for lack of evidence.

In one set of experiments, Harrison observed the frog embryo's tail and reported a surprising shifting of the two types of tissue layers. He particularly followed what happened to the distribution of nerves. In a second set of experiments reported in the same paper, Harrison studied regeneration in the embryonic frog's tail.[9] In both cases, he relied heavily on daily observations of the living specimens, though he also made some traditional histological preparations as well.

Harrison's friend Morgan had turned to frogs in his work of 1893 and after and had used frog embryology in his classes and laboratories at Bryn Mawr. Frogs had also figured so centrally in the work of Born and the other German experimentalists that it made sense for Harrison to use the same experimental material.

It is also clear that Harrison felt the attraction of the sort of experimental work that Born did, with its promise of definite results. He referred to the success in obtaining "decisive" conclusions through experimentation. He also saw great value in experimentally testing one theory or another, writing on numerous occasions of the "crucial" test that a particular experiment could offer for deciding between two alternative theories. Thus, Harrison was much more bullish on the efficacy of an experimental approach to biology than Wilson, Conklin, or Morgan, each of whom embraced its value but did not award it a superior status. Harrison seems to have been influenced by his German study in this view, especially by the ideas of Wilhelm Roux and Born.

Born, in turn, was pleased with Harrison's application of his technique to useful purpose. He wrote to Harrison at the end of 1898 and thanked him for the friendly references to his own work. Further, he expressed his delight that Harrison had not gone wild in using transplantation, simply sticking together any random uninteresting bits of organisms for the sake of adventure to see what happens. Rather, Harrison had used the technique to bring "sharply focused formulation of questions." In addition, Harrison had taken what was really only an exploratory method and had made it much more precise and useful. Born was clearly delighted with the results of Harrison's work and with his general experimental approach. He also sent his personal greetings to Mall and expressed the hope that he would see Harrison in Naples in the near future.[10] With this experimental study, then, Harrison solidified his position as a respected researcher within the German experimental embryological community.

Yet Harrison was not one to wave the flag flamboyantly for any one method or approach. Nor did he discuss his philosophy of science openly or easily. He just did science and wrote relatively little about *how* to do it. He made assumptions about how to proceed in science without apparent self-conscious reflection about whether some other approach might prove more successful. Perhaps because he never really liked to write anyway, he generally stuck to presentation and restrained discussion of facts.[11] Most of his papers began as talks delivered to national meetings or as invited lectures, which he then published. For whatever reasons, he tended to present the research as reasonably straightforward. It was only later in his career that he began to consider questions about the validity of results or approaches more explicitly.

In his first set of experiments with transplantation, also reported in his first major paper from Hopkins medical school, Harrison sought to test the received view that embryonic frog tails grow by adding new material at the end of the tail. By obliquely cutting off a chunk at the end of the tail of a *Rana virescens* embryo and replacing it with the tip from a *Rana palustris*, Harrison made a surprising discovery. The epidermal tissue of the host moved outward to cover the donor tissue, which produced a very uneven edge between the two. Lest any sceptics question whether it might be the pigment rather than the cells themselves that migrated, Harrison squelched their doubts. With a strong light, he could see each cell and could tell that each remained either highly pigmented (as with *palustris*) or not (as with *virescens*). There was no mixture or changing pigmentation within any one cell.[12]

While the skin tissue wanders backward, so do the attached muscles. But the associated nerves do not. On the contrary, the major nerve cells seem to have become established in their connections quite early in the developmental process, before the embryonic stage used for Harrison's experiments. Thereafter, growth draws the nerve fibers out farther but generally does not alter the basic pathway. Harrison did not pursue this immediately but instead turned to other questions. Yet this brief consideration of nerve connections became a subject of central interest in Harrison's research by 1904. The work also led him to his best-known scientific contribution, concerning the tissue culturing of nerve fibers. The interest in neural connections and growth also reflects an interest shared by his associates Nussbaum and Mall, as well as others.[13]

With his second set of experiments reported in the same long paper of 1898, Harrison took up regeneration, the subject that had come to occupy Morgan's central interest by 1897. Morgan had begun by studying the way isolated half-embryos regenerate their missing half. Then on to planarians, earthworms, crabs, and other familiar regenerating organisms. Morgan's primary motivation was to use experimental regeneration to discover the normal processes of tissue differentiation and development. Because of the phenomenon he called morphallaxis (in which existing tissue undergoes transformation into a new type of tissue as needed by a whole organism in response to its injury), Morgan saw regeneration as lending critical support to his preferred epigenetic interpretation of development.[14]

In contrast, Harrison was interested in the very specific question of how the peripheral nervous system regenerates in the tail of the embryonic frog. When the tail (with its neural connections already established) is cut, what determines how the nerves grow out? And how do they reestablish functional pathways—given that they apparently do so, since he could directly observe the nerve fibers growing out of the severed ends of the tail. These fibers appeared to grow out of ganglion cells that had not made up the original neural cord, so that the organism proved itself capable of accommodating to the artificial, injured conditions. Yet the manner of development remained very close to normal, Harrison concluded.[15]

Another question that interested many embryologists, including Morgan, concerned the "polarity," or asymmetry, which seemed to exist in injured organisms. The head position can produce a new tail, but isolated tails do not generate new heads. Why did such an asymmetry exist in the system? Experiments showed that, in fact, the existing portion takes over and reshapes the tissue to regenerate the missing parts. The tail has difficulty in effecting a sufficient reorganization, probably because of the lack of available nutrition to support the necessary functional activity. The question really was "whether a tail will be produced under the influence of the position of the regenerating center with regard to the whole organism, or whether the elements in the transplanted stump retain their original orientation and strive to reproduce the lost body." What was needed was to test the role of functional activity and the impact of nutritional availability. In fact, Harrison concluded that "these experiments establish beyond a doubt the fact that the regenerative power of the tissues of the tail is very considerable in both directions."[16]

Yet the result was not the sort of heteromorphosis that Jacques Loeb saw as typical, in which a part takes on the form and functional role of another part.[17] Rather, what Harrison decided had occurred was incomplete regeneration. Tissues seemed to him to exhibit a remarkable degree of self-differentiation. Functional activity alone cannot reshape or reorient any body parts; it only

> brings about a higher degree of efficiency in a structure already capable of considerable independent development in a given direction, regard-

> less of surrounding conditions. The present experiments do not justify us, however, in going further, in the conclusion that unusual relations imposed upon a regenerating part can call forth out of material which would normally be used otherwise, an entirely new heteromorphic structure, as a functional adaptation to new surroundings, or as the result of a striving to complete the mutilated organism.[18]

Generally, Harrison saw each tissue as preserving its characteristics, which had been established at some very early stage of development. Yet rather than pursuing that differentiation back to earlier and earlier stages of cell division, as Wilson and Conklin continued to do, Harrison sought to understand the degree and nature of differentiation at later stages in normal organisms and the way in which experimentally injured individuals compensate.

After his 1898 report, Harrison continued to work on frog embryos and to ask similar questions. He sought to refine the questions and to apply additional techniques in order to obtain more and more definite results and to test the available theories. In 1897, he had advanced to the position of associate in anatomy. Then, after some time in Bonn in 1898 and 1899, he finally received his M.D. degree and became associate professor in the Johns Hopkins University Anatomy Department.

Also during this time, Harrison began in earnest his work on the development of the nervous system. This research occupied his attention for the next decade. His son later explained that this period was a very busy time in his father's life since he spent long hours at his work: "My Father left home early during the Baltimore days and came home late in the evening, he had little time for family life."[19]

Nerve Development

The next set of papers, from 1901 through 1906, should be considered as a group since with them Harrison offered a sustained attack on basic questions about nerve development. The way the work evolved tells us a great deal about Harrison's own approach to biology and the way in which his medical training and interests pulled him in directions away from the Hopkins experience.

His first publication since 1898 indicated that it came from both the anatomical institute in Bonn and the anatomical laboratory at Hopkins, whereas the next mentioned only Hopkins. Both followed up on earlier work, the first pursuing some of the same problems with additional techniques and the second applying Born's experimental techniques to new problems.

The first paper, published in 1901, looked directly at the development of the nervous system. In an effort to place his work within the broader context of ongoing discussion, Harrison offered an extensive bibliography. As he pointed out, much work had been done on nerve development. Important results by several leading researchers had offered morphological "proof" (*Nachweis*) that individual nerve fibers develop by an outgrowth from a single cell. At least they do in higher organisms, and the majority of researchers believed that the same was true of lower vertebrates as well. Yet most of the previous work remained morphological, presumably looking at the general structural changes and patterns of growth, rather than histogenetic. Harrison promised the latter sort of study, which would describe in detail, "step by step," the developments of individual cells and their relations to developing nerves and connections between central and peripheral nerves.[20]

Harrison did not begin without a guiding theory to interpret his results; he adopted one of the existing alternatives, which he felt was far better supported than the others. At least with respect to his particular work on salmon embryos, he felt that the outgrowth model of nerve development held true. According to this interpretation, each individual nerve fiber develops by growing out from a single cell. He found absolutely no evidence to support the alternative suggestion that nerve fibers emerge out of cell chains. These two interpretations had gained considerable attention by 1901, as each had attracted its own leading researchers with their different methods and different organisms. By taking up the subject of nerve histogenesis and by endorsing the outgrowth view, Harrison had entered a lively field of discussion, one in which he remained until he generated what he regarded as a compelling argument in support of outgrowth by 1910.

The debates about the development of nerve fibers were actually embedded in a larger set of debates about the nature of the nervous system more generally. Here also two major theories existed by 1900: the neuron and the reticular theories. The neuron theory held that each neuron is an individual cell, which serves as the basic unit out of which the system is made up. Wilhelm His had provided the major arguments in favor of such an analytic and reductionist view, with his carefully prepared histological observations. He felt that these showed that each cellular unit remains structurally and functionally independent and communicates with other cells only by indirect contact across a synapse. He saw no material interconnections between the autonomous cells.[21] Others did.

Those others argued for a reticular interpretation, according to which the nervous system is a system first and foremost. The interconnected functioning network remains the relevant unit for this group. They insisted on the existence of material connections between any two parts of the system. Though begun by anatomist Joseph von Gerlach and supported strongly by anatomist Oskar Schultze (son of Max Schultze, who had taught Nussbaum), this view soon boasted Camillo Golgi as the leading protagonist. As professor of histology in Pavia, this Italian researcher produced outstanding technical innovations such as silver staining to make the neural components more visible.[22] With the help of arguments in favor of the organism-as-a-whole and with the commonsense idea that such a complex working system as the nervous system must be more than a collection of independent nerve cells, Golgi gained considerable favor. As a result, the reticularist view appears to have become the majority position by the early part of this century.

Yet the neuronists also found a powerful advocate in Spanish anatomist Santiago Ramon y Cajal.[23] Ramon y Cajal's own studies and his interpretations of Golgi's illustrations told him that nerve fibers actually grow out from the neuroblast cells that reside at the end of an area called the growth cone. By 1900, the debate about the structure of the nervous system had accordingly begun to focus less on static structure and more on nerve development. If only researchers could determine precisely how the elements of the system arise and the nature and origin of their connections, they could then answer the structural questions. So alternative theories of nerve development competed during the first decade of the twentieth century. Unfortunately, all researchers relied on study of prepared material for their evidence. They fixed, stained, sliced, and microscopically observed their specimens. As a result, what they saw was far removed from the living, growing nerves themselves and required a great deal of interpretation. Not surprisingly, different people sustained their different interpretations even in the face of conflicting evidence.

The three major theories of nerve development included the cell chain, the protoplasmic bridge, and the outgrowth theories. The first, elaborated especially by physiologist Victor Hensen, held that there exists from the earliest developmental stages a series of cells making up a path from central ganglion cells to the organism's periphery. These cells begin to coalesce, according to this view, and a nerve axon forms along the path, elongating until it reaches the ends.[24] This means that the nerve fiber never arises by itself but always appears as a secondary product of noncellular crystallization of material from the Schwann cells, which then serve as a sheath around the emerging axon. In addition, the axon only arises because of the functioning of the Schwann cells, so this was essentially a physiological theory where function determined the form.

The second, protoplasmic bridge, theory also offered a preformationist explanation of how properly functioning neural connections become established. Bridge advocates held that from the very earliest developmental stages, the egg must have tiny bridges set up so that the fibers can find their way. When the Hungarian Stephan Apáthy argued that he had seen neurofibrils materially passing through ganglion cells during development, he thereby suggested that those ganglia cannot be the source of the neurons, as the neuronists argued. Instead, the ganglia might contribute a neural "force" to the developing nerve fiber, which also suggested a physiological primacy in which the form arises after and because of the functioning. This work also supported the reticularist view of the structure of the nervous system.

The climate by 1903 was heated. As Ramon y Cajal put it in his rather colorful and reconstructionist autobiography, "the contagion of reticularism" and the bridge theory had infected researchers by 1903, so that even leading neuronists had been shaken in their faith.[25] Indeed, the reticularists could offer an apparently quite satisfying explanation of the complexity of nervous action, and the bridge theorists could explain the complexity of neural connections. It was much more difficult to maintain an epigenetic view. Yet the neuronists' conviction that each cell remains autonomous and grows out independently by outgrowth of the protoplasm from its ganglion cell origin did remain essentially epigenetic.

Harrison clearly inclined toward this interpretation and sought to make the case in its favor. With his earlier work of 1898, he had offered hints about his observations of outgrowing fibers from the transplanted materials in frog embryos. With his paper of 1901, he extended the discussion to other observations of living as well as prepared material. His focus on individual cells and their patterns of growth and their role in establishing neural connections reveals his attempt to use embryological methods to contribute to the discussion. Rather than emphasizing static morphology, as he felt most of the other disputants had done, he relied on histogenesis for new information, a focus he retained in later works.

Harrison's next paper, though it appeared in print only in 1903, actually reported on research begun by 1901. Indeed, Harrison had presented a preliminary discussion of the experiments on the development of the lateral line sense organs of amphibians at the American Morphological Society's annual meeting in December

1901. The lateral line sense system exists in aquatic vertebrate fishes and amphibian larvae and consists of a series of organs lying along the head and sides. Clusters of sensory cells develop at these points and are stimulated by currents in the water; thus the lateral line organs give the organism a highly sensitive system for detecting changes in the aqueous environment. Looking at the lateral line system made sense for someone who had already shown an interest in teleost fishes and who had entered the discussion of the development and structure of the nervous system. Here was an accessible case of a special component of the nervous system which promised practical results from further study.

For his work, Harrison used various differently pigmented species of *Rana* and *Amblystoma* (now *Ambystoma*), as he had earlier. He sought to examine experimentally which cells make up the lateral line sense organs as well as when and how they become differentiated. Whereas earlier work had remained traditionally morphological in its reliance on observation and description and therefore limited, according to Harrison, his current approach could be properly experimental and analytical.[26]

Rather than plunge right into the experimental results, as he had in 1898, and rather than first introducing the current debates on the subject, as he had in 1901, Harrison began this paper with a section establishing the normal course of development to provide a solid basis for comparison. He then detailed his experiments, again using interchange of tissue from different organisms, and this time focusing especially on nerve development. At the conclusion, he considered the relevance of his study for the persistent debate between Hans Driesch and Wilhelm Roux concerning whether differentiation occurs primarily epigenetically or because of some sort of preformation. Though it seemed clear to Harrison that nerve connections are set at very early stages of development, well before the later embryonic stages with which he was experimenting and even before they became visible, nonetheless this did not warrant Roux's preformationist conclusions. It was not at all evident, given the information at hand, that the lateral line sense organs are built into the egg. Nor did Harrison have any reason to believe that some sort of special system resides in the nucleus to bring about the neural connections. He simply lacked sufficient solid experimental data to draw conclusions about this central embryological question.[27]

In another paper, initially reported in brief to the Association of American Anatomists in 1902 but expanded and published in 1904, Harrison looked at the relation of developing nerves and muscles. In particular, he followed up questions that Nussbaum had asked earlier: first, whether a nervous impulse is necessary to stimulate differentiation of muscles and, second, whether such impulses are required to continue muscle development. Researchers cannot expect answers without the use of carefully controlled experiments, Harrison maintained. By removing the spinal cord (which contains the source of nerves) before any peripheral nerve fibers appear, he could test the contribution of nerves. After answering the specific question, Harrison proposed also to address the more general question of "whether the normal processes of ontogeny are regulated by functional stimuli."[28]

The experiments showed that muscles could develop perfectly well, even in the absence of nerve development, and thus provided "crucial evidence" that muscle tissue must arise independently in the developing embryo, he concluded. Functional stimuli from the nerves simply are not necessary to cause the production of muscle tissue. It does not follow, however, that the muscles develop absolutely normally in the experimental case. Rather, as with experiments on the lateral line sense organs, some degeneration occurs. Yet the conclusion stands that normally muscles can arise and nerves can develop independently of each other and of each other's functioning.[29]

The *Journal of Experimental Zoology*

In the midst of his continuing research on nerve development and his teaching responsibilities at Hopkins, Harrison also took on another major responsibility in 1904. As editor of a new journal, the *Journal of Experimental Zoology*, he moved even farther to the forefront of the zoological world and closer to becoming, as *Fortune* later put it, "one of the most famous unknown men in America."[30] For Harrison remained shy and reserved, though effective, even in the many positions of power he came to hold during his long career.

For some time it had become increasingly clear that the United States needed its own biological publication. Prior to the 1890s, Americans had to send their papers to Germany and wait for rather a long time for the work to appear. Since some required publication in German, this also restricted the flow of information. With the establishment of an active community of research biologists in the United States by 1900, stimulated in part by the successes of the Johns Hopkins University program, the need became critical.[31]

Charles Otis Whitman had first responded to the crisis with a set of publications. He began the *Journal of Morphology* in 1887, with the financial support of Edward Phelps Allis, Jr., to publish substantial articles in the field which he felt needed the most attention and offered the most significant research results.[32] In 1890, Whitman inaugurated the *Biological Lectures* to put into print the public lectures on central biological issues presented at the MBL. That series lasted until 1899. In 1898, Whitman had also joined William Morton Wheeler to begin the *Zoological Bulletin* (changed to the *Biological Bulletin* two years later when the MBL took over editorship) to publish shorter pieces and abstracts. All of these served their purpose well, but by 1903 further needs had arisen, in part because financial troubles had caused what turned out to be a five-year suspension of the *Journal of Morphology*.

In particular, those two Hopkins men at Columbia, Wilson and Morgan, had decided that a journal for experimental and analytical work was in order. They consulted with Conklin, and the three began the drive to found a new publication. Throughout 1903, they met a number of times to decide who should serve on the editorial board, how to fund the journal, and who should take the responsibilities of managing editor. With the help of Henry Fairfield Osborn, the senior biologist at Columbia, Wilson and Morgan managed to secure a donation of $5,000 to start the journal. The arrangement was that they would organize a company to retain responsibilities for the publication. This group would remain independent of Columbia, though the funding was channeled through that institution, since the organizers felt it imperative to gain the cooperation of all the leading American biologists from a variety of universities.[33]

In November 1903, the plans and financial details were in place. They had decided to produce their *Journal of Experimental Zoölogy*, dedicated to research "understood to include experimental researches on growth and development, whether normal or pathological, on regeneration, cellular phenomena, and all related subjects, and also those relating to evolution, variation, heredity, adaptation, ecology, and the like."[34] Wilson then wrote to Harrison to offer him the editorship, saying:

> We know, of course, that it is a position involving a good deal of work and a very considerable sacrifice of time on your part; but we feel strongly that the time will be well spent in a cause which is of vital importance to American zoology, and I am sure that you will win the appreciation and gratitude of all the zoologists if you will consent to make the sacrifice. I feel pretty sure that when once the matter is under way, the business will become largely a matter of routine that will not involve as much labor as it may seem at the start, and of course,

> all the other editors will give you every possible aid that is in their power. . . . The position really ought to be a salaried one, but you will appreciate the fact that this is hardly possible at present, and it would have to be a labor of love—not to say glory![35]

Harrison accepted and remained managing editor until 1946, just after the one hundredth volume appeared. The first issue of the journal was published in 1904. Though business was not always routine and though Harrison sometimes let the paperwork pile up, the journal became—and remained—a success.

Further Nerve Studies

Nineteen hundred and four proved a busy year for Harrison. In addition to beginning the *Journal*, he also made a major trip to Europe. Early that year, he requested permission to take leave from Hopkins to attend an Anatomical Congress at Jena and to continue working in Germany for the duration of the spring and early summer months. He would, he assured the dean of the medical school, have completed his regular work for the semester by 15 March and therefore could take leave without neglecting his duties.[36] With permission granted, he left for Germany once again.

In April, the Anatomische Gesellschaft met in Jena to discuss a variety of current controversial topics, including nerve fiber development. Harrison joined well-known anatomists Albert von Kölliker and Oskar Schultze in a session that considered the role of Schwann cells in nerve development. Von Kölliker argued against and Schultze for their importance. Drawing on his own experimental study, Harrison maintained that the Schwann, or sheath, cells could not possibly be crucial for nerve development, because, when he had removed their source before the cells had formed, the nerve fibers had nonetheless developed. It clearly could not have been the case that the Schwann cells were required. If they play any role at all in normal development of nerves, he concluded, it must be only secondary.[37]

In July 1904, Harrison offered another discussion of his recent work on nerves, this time in Bonn. Acknowledging that His's neuron theory had not achieved full acceptance as yet, Harrison insisted that support would continue to come from embryological research. New methods would provide the necessary evidence, he felt. In fact, he said that since the Anatomical Congress he had had the chance to pursue further experiments and had found "experimental proof" in favor of the neuron and outgrowth and against the cell chain theories.[38] The key, he felt, lay in the role of the Schwann cells.

His studies on embryos without Schwann cells yet formed, and with the source of the Schwann cells removed, showed him that the development of peripheral nerves can only be understood in terms of His's neuron theory. The experimental cases produced naked nerve fibers without any Schwann cells, and those fibers made their way to establish peripheral connections. The connections appeared quite normal, which certainly suggested strongly that neither the Schwann cells nor a preexisting functioning had been necessary. Studies clearly showed, Harrison concluded, that ganglion cells send out extensions, which become nerve fibers without either material or direct help from other cells. These fibers then form a peripheral nerve network with other cells. The evidence also showed that Schwann cells play no role in the formation of nerve axons or the peripheral network. In his lecture, Harrison stressed and restressed the certainty of this conclusion: the role of Schwann cells was "completely impossible" and "proven meaningless."[39]

Or so he thought. The continuing debate, with additional studies proffered in support of reticular and bridge theories, suggested otherwise. His experimental work would never convince some critics such as Leipzig anatomist Hans Held, who rejected the very legitimacy of any interventionist experimental methods that necessarily disrupted normal processes.[40] Despite his brave declarations of certainty during his visit to Germany, it must have become abundantly clear to Harrison that it would take more to convince advocates of the alternative views.[41]

Upon his return to the United States in late summer 1904, Harrison settled in once again in the laboratory at Hopkins to launch additional attempts to make his work as convincing and definitive as it needed to be. In a paper read to the Association of American Anatomists in 1905, he began by explaining the need to move to different and more efficacious forms of experimentation. "Prior to the year 1904 all attempts to solve these problems were based on observations made upon successive stages of normal embryos. When one compares the careful analyses of their observations, as given by various authors, one cannot but be convinced of the futility of trying by this method to satisfy everyone that any particular view is correct. The only hope of settling these problems definitely lies, therefore, in experimentation."[42] In an earlier draft of the paper, he had called the older method the "embryological histological method." He had also stressed the "inadequacy" of the older method to "convince everyone of the truth."[43] By the time it went to press, talk of "truth" had dropped out, while emphasis on the value of experimenting with living, growing organisms had strengthened even further.

Harrison did intend to use experimentation to determine both the constitution of the nerve fiber (whether one cell or a cell chain) and its manner of connection between center and periphery (whether preguided or the product of outgrowth). His answers are already evident, given his earlier work, but his method of attack had changed.

Instead of stressing the support that his research results provided for the neuron theory and their disproof of alternative views, as he had before, here he struck a more conciliatory tone. He reported some of the experimental results he had discussed in Germany. He then added a set of experiments pursued back at Hopkins to test the claim of the protoplasmic bridge advocates that Schwann (or sheath) cells alone can produce motor nerves. By microsurgically removing the tissues known to make up motor nuclei (the ventral half of the medullary cord) but leaving the dorsal part of the medullary cord and the ganglion crest, which gives rise to Schwann cells, he sought to determine what the Schwann cells can accomplish by themselves. Because of the difficulties of the procedure, only a few specimens survived. In all cases, they contained only very small motor fibers and no motor neurons. Therefore, even though not yet "established beyond all question," the conclusion seemed secure that sheath cells cannot produce motor nerves by themselves, as the advocates of the protoplasmic bridge hypothesis maintained.

With his experiments on living, functioning frog embryos, Harrison could show what studies restricted to the use of prepared dead materials could not. In particular, he reported:

> The foregoing results can be interpreted in but one way. The nerve center (ganglion cells) is shown to be the one necessary factor in the formation of the peripheral nerve. When the former is removed from the body of the embryo the latter fails to develop. When it is transplanted to abnormal positions in the body of the embryo it then gives rise to nerves which may follow paths, where normally no nerves run, and likewise when the tissues surrounding the center are changed entirely, nerves proceeding from that center may develop as normally. The nerve is therefore a product of the ganglion cell. The histological findings indicate that it is an outflow of the substance of the ganglion cell and not a mere activation by contact of indifferent extra ganglionic substance.[44]

Yet, while disagreeing with his opponents, Harrison did not explicitly criticize them. Instead, he sought to show how their results could also be interpreted in accordance with the neuron and outgrowth theories. Here, too, he moved away from the emphasis on "truth" in his earlier draft to seek reconciliation and interpretation that made sense of all the results. Hensen had produced valuable observations, Harrison stressed, and despite Hensen's emphasis on the causal necessity of preexisting nerve function, he was not entirely wrong. For, "while the facts necessitate our deciding against the validity of Hensen's view, as far as the question of *primary* continuity is concerned," nonetheless the string of Schwann cells might play a secondary role. In fact, "this view is correct in so far as in many instances of nervous connection between center and end organs is established when the two are very close together, and the long nerve paths originate in such cases by the moving apart of center and innervated organ after the establishment of the connection." The individual nerve fiber remains the outgrowth of a single ganglion cell, according to Harrison. Yet he assigned the Schwann cells a new function, namely "an important role in the nutrition and protection of the fibers."[45] With this conciliatory paper, Harrison seems to have launched a campaign to win the day by positive persuasion rather than by violently overthrowing the opposition.

In 1907, Harrison added an additional experimental approach to his collection of assets.[46] In 1906, again in a paper initially presented to the Association of American Anatomists, he explored the evidence gained from transplanting limbs in amphibian larvae. In particular, he adopted the methods that anatomist-embryologist Hermann Braus had used to support the bridge theory. Braus had studied at Bonn, then at Jena, and finally had moved to Würzburg, where he developed a close friendship with embryologist Hans Spemann. Braus had worked with many of the men Harrison respected most, and he adopted experimental methods that Harrison valued most highly. Thus, of all the advocates of the reticular view and bridge theory, Braus was one most likely to attract Harrison's close attention and respect. Harrison praised Braus's work for his "wealth of interesting and important observations."[47]

By detailing precisely what happens during each developmental step, Harrison showed once again, as he had with Hensen, that Braus's results were consistent with the outgrowth theory as well as with the bridge theory, which Braus favored. He then strengthened the argument by piling on additional experimental cases of his own. He proposed to show that Braus (and others) offered evidence that was logically consistent with the sort of bridge theory they wanted. Yet the support relied heavily on analogy and "collateral facts," which "may be interpreted quite as readily, if not more so, in accordance with the outgrowth theory." Ultimately, he concluded that "there are too many loopholes left to permit of a rigid proof."[48]

Yet he nonetheless believed that he could show the outgrowth theory's superiority, even if not inevitability.

Braus had argued that he could transplant a limb bud (the part of the embryo known to give rise to the limb under normal conditions) before nerve fibers had appeared. He could transplant this bud to nearly any part of a normal host tadpole and have it grow. In all cases, nerve pathways would develop which he saw as just like those in normal cases. For Braus, it simply made no sense to hold that random outgrowth of individual nerve fibers from all different parts of the body could move into the abnormal area and make workable connections at all. The further claim that such an outgrowth could produce apparently normal connections seemed ridiculous to him, as well as to that majority of researchers who supported some versions of bridge and reticular theories.

Harrison acknowledged the force of Braus's arguments and pointed out that the existence of functional bridges would have special appeal for the physiologically inclined, such as Braus and Hensen. Yet limb buds already have experienced some neural development of both structure and function. Further, something in the degenerating neural tissue of the donor transplant bud might serve to attract the outgrowing nerve fiber and thus explain how outgrowth could occur and produce functional connections. In other words, Harrison saw Braus's interpretations as ingenious but not compelling. Similarly for Braus's purported evidence: Harrison could show how each piece could be turned to his advantage. His own research showed the way in which wandering nerve fibers moved into "aneurogenic" tissue, without any normal development, which Braus claimed could never happen since he believed that such material never develops nerves. In contrast, Harrison found that, as long as the nerve center remains, fibers will grow and find their way to apparently functional connections from center to periphery. Thus outgrowth occurs as the mode of fiber development.

Yet aside from mere growth, the precise finding of the proper functional pathways depends on two factors. What Braus's findings showed, according to Harrison, was the way in which the internal structure of the limb itself directs the way in which the nerve fibers grow within it. For example, "the manner of branching cannot possibly be predetermined in the ingrowing nerves themselves, because in the normal body these same nerves have an entirely different distribution." Ultimately, the exact source of the nerves that move into a given area and the structure and growth of the area itself will both contribute to determining just what each nerve's pathway is and how it becomes established.[49] There was room for outgrowth and for some structural direction of pathways.

For some time, Harrison had realized the need to construct some sort of experiments that would allow him to separate the living nerve from its normal surroundings in order better to test the powers of the nerve fiber itself. He could try to grow it outside the body altogether, grow it in dead tissue, or perhaps grow it in one of the large body cavities that have no normal structured tissue or neural connections. For several years he had attempted to make one of these approaches work.[50] Then, during the spring of 1907, Harrison reported, he began a new series of experiments that pursued the first option, growing nerve tissue outside the body. The first discussion of his results came on 22 May 1907, when he read a paper to the Society for Experimental Biology and Medicine in New York. There he explained that "the immediate object of the following experiments was to obtain a method by which the end of a growing nerve could be brought under direct observation while alive, in order that a correct conception might be had regarding what takes place as the fiber extends during embryonic development from the nerve center out to the periphery."[51]

Harrison isolated pieces of embryonic tissue which he knew from close study of normal developing embryos would normally give rise to nerve fibers. This included pieces of the medullary tube, for example, and branchial ectoderm from frog embryos of a stage just after the medullary folds had closed. Adapting Born's transplantation technique, he instead performed an explantation. After removing the small pieces of presumptive neural tissue, he placed them on a cover slip in a medium of lymph extracted from an adult frog. He then inverted the cover slip, with its lymph and embedded tissue, over a depression slide and sealed it with paraffin. The tissue survived under these conditions and provided just the sort of living and developing material separated from its normal surroundings which he had sought.

This work represented the first successful tissue culture ever, the first time that living tissue had been successfully cultivated outside the living body. For this innovation, with its myriad practical results for both medicine and biology, Harrison was nominated for a Nobel Prize in 1917, although no prize was given because of the war.[52] He did pursue the promising technique with further experiments on nerve development, and others later received Nobel Prizes and other awards for their tissue culture work. But Harrison was not really interested in the technique for its own sake. For him it was just a tool, useful only so long as it moved him toward definite results for the particular embryological questions he wished to address.[53]

Under these experimental conditions, the different kinds of cells differentiated in ways characteristic of their type, even though they did not join together in a normal way to make up more complex organs. Nerve fibers did precisely what Harrison would have predicted: they moved out beyond the ganglion cells and extended into the surrounding medium by outgrowth of the protoplasm. Though the tissue preparations remained very delicate and he had not succeeded in preserving any of the developing fibers, Harrison recognized that he had the makings of a major breakthrough in getting the sorts of experimental evidence he felt was crucial. He concluded his exciting paper in a characteristically modest manner: "While at present it seems certain that the mere outgrowth of the fibers is largely independent of external stimuli, it is, of course, probable that in the body of the embryo there are many influences which guide the moving end and bring about contact with the proper end structure. The method here employed may be of value in analyzing these factors."[54] In fact, he realized that if only he could improve the success rate and make observations and preservation more successful, he could use the method very effectively to criticize the interpretations of the bridge theorists.

The Move to Yale

The spring and early summer of 1907 had brought more to Harrison than this important work, however. He also decided at that time to move from the familiar Hopkins environment and to branch out on his own. At that time, Yale University was rather traditional, conservative, and even a bit backward in its views of biology. Comparative anatomist Sidney I. Smith had just retired after a series of health problems, and the other departmental leader, A. E. Verrill, had also recently retired. In addition, Yale had no university-wide biology or zoology program because undergraduate courses in the sciences all fell within the domain of the Sheffield Scientific School.

"Sheff" had given the first Ph.D. degrees in the United States in 1861 and had thereafter undergone reorganization and strengthening so that it had gained its academic independence from the rest of the university. By the 1890s, Yale had just begun to allow "crossing over," so that students in either Yale College or the Sheffield School could take courses from the other school. Yet "Sheff" retained control over science. It was with considerable surprise and annoyance, then, that the Sheffield School greeted the 1906 suggestion from Yale College that it would hire a faculty member of its own to teach biology. As biology became increasingly popular, more and more of the students were crossing over to the Sheffield School. Why should the college not keep its own? they must have asked. The faculty of the Academical Department, as Yale College was called, resolved in 1906 that it would hire a biologist into its school. The administration was "fully convinced that the great development of Biological Science in recent years, together with the continually increasing demand for opportunity to study science among our students makes it imperative for us to consider from the beginning just what courses are needed to meet this demand to serve as a preparation for medicine, for psychology, anthropology and geology and still more as an introduction to modern scientific thought."[55]

This decision sounded the death knell for "Sheff" as a separate school, but an immediate attempt to avoid hard feelings led to the creation of a new position. This would be Yale's first university professor, someone appointed in both Yale College and the Sheffield School. They would hire in comparative anatomy, to replace Smith, and the salary would be augmented by a special gift from the Bronson Fund. George Howard Parker, assistant professor at Harvard, reported in his autobiography that he had originally received an offer of the position, which he declined.[56] In June 1907, the Yale trustees voted to offer Harrison the job, which they did. He accepted and became the Bronson Professor of Comparative Anatomy of Yale University. He also became chairman of the new Biology Department when it began a few years later.

Moving from Baltimore after the many years he had spent there proved difficult. Harrison wrote to his friend Conklin shortly after arriving and beginning teaching in New Haven, "Of course I have been homesick since [leaving Baltimore], but my surroundings are so pleasant here that I am fast getting over that." Indeed, the Harrisons lived in a lovely house very near campus, the house that became the well-commemorated Mory's after Harrison moved out. The considerably higher salary, the prestige of the unique university position (which soon included appointments in the Yale Medical School and the Graduate College as well), and his role as innovator and scientific leader appealed to Harrison. The undergraduate teaching did not. In the same letter to Conklin, he continued: "I don't feel that my course in elementary biology is getting along very well and do feel as if I had been floundering a good deal. Still I am now quite sure that we can arrange our courses next year on a pretty satisfactory plan and that gives me encouragement."[57] Basically that meant arranging things so that he would not have to teach the introductory students.

At Yale, Harrison continued his research. Shortly after his arrival, he received a letter from Mall encouraging him. Mall reported the contents of a letter he had received from cytologist and anatomist Gustav Retzius, who had visited the Johns Hopkins University Medical School from Germany in its first year. Retzius was interested in nerve development and in the neuron theory. He applauded Harrison's recent efforts and reported that even the ardent advocate of the cell chain theory and director of the Naples Zoological Station, Anton Dohrn, had abandoned that theory in favor of outgrowth.[58] No doubt Mall's letter cheered Harrison.

The biologists at Yale were housed in one of the Sheffield School buildings, an old house, really, and Harrison found himself down the hall from the bacteriologist Leo Rettger. Rettger helped obtain aseptic techniques, which insured a higher success rate in Harrison's transplant experiments. Rettger also made suggestions about apparatus and procedures.[59] In addition, Warren Lewis, Harrison's former colleague at Hopkins, and Hans Spemann, Harrison's counterpart in Germany, also offered suggestions for improvements of technique, including the use of Zeiss binocular dissecting microscopes and special surgical apparatus. Spemann also used tiny precise glass needles for his operations.[60] Despite the unfamiliar and less well equipped surroundings, Harrison was well prepared to carry on his research.

A "Crucial Experiment"

In an invited lecture to the Harvey Society in New York in March 1908, Harrison explained the point of his experimental work on the development of nerve fibers. He outlined the basic assumptions of the competing outgrowth and the protoplasmic bridge theories and acknowledged that the trend recently had definitely been toward the latter. Yet, "brilliant and attractive as it seems," he pointed out that "very little real evidence has ever been brought forth to support it. In fact, its mainstay has been the imaginary difficulty of conceiving how the alternative view could be true." He explained that the two theories were not sufficiently logically distinct to allow a "clean-cut and crucial experiment" to decide definitely between them.[61]

Even after detailing his earlier series of experiments in favor of outgrowth, he admitted that the case had thereby only been made to incline farther in the direction of outgrowth. There still remained "one defect in the conditions of experimentation which stands in the way of rigorous proof."[62] Namely, the nerve fibers in all cases remained in contact with their normal surrounding tissue, a circumstance that might very well disturb the experimental results. It was necessary to find a way to isolate the developing tissue. That he had done with the tissue culture experiments. With the improvements, he had even kept tissues alive for up to five weeks. Beautiful nerve fibers grew out from individual ganglion cells, as any observer could see. Furthermore, they behaved in a manner characteristic of normal fibers under normal conditions. As a result, he felt that he had shown "beyond question that the nerve fiber begins as an outflow of hyaline protoplasm from cells situated within the central nervous system." Therefore, he concluded, "we have in the foregoing a positive proof" of the outgrowth hypothesis.[63]

Yet two years later, Harrison recognized that not all the opponents of the outgrowth theory had indeed rolled over and played dead. In addition to several shorter articles, he also launched one last "crucial" attack in a lengthy paper published in the journal he edited. Repeating the work during 1909 had provided still better results than the limited set at Hopkins or even those during the first year at Yale. They were, Harrison felt, "much more convincing," largely because the results were much clearer, the fibers had grown better and longer, and the illustrations and preparations had succeeded better.[64]

As before, Harrison reported his results and presented his conclusions about the mode of nerve fiber development in careful detail. He then extended his discussion further to explain what difference fiber outgrowth makes for the nervous system and for understanding development more generally. Since the ganglion cell clearly has the ability to extend outward into a nerve fiber, that ability must exist before the cell begins to function in any way. It must be, in Roux's sense, self-differentiating rather than primarily driven from outside the cell itself. Whatever normally guides the nerve paths must be laid down at a very early developmental stage, perhaps by a few fibers that might be termed "path finders," he concluded on the basis of these and earlier research results.[65] Some as yet unknown specific reaction must then effect the final connection between outgrowing fiber and its end point in the periphery of the organism. While much of the picture remained for the moment unresolved but ripe for exploration, Harrison felt that his chief contribution lay in having taken the outgrowth of fibers out of the realm of speculation and having placed it on definite ground with a crucial test. The rest would follow with further properly experimental study.

As Harrison's research succeeded and he continued to gain international attention for his work, he found his position at Yale frustrating. He did not like teaching large classes, especially large elementary classes. Yet the program continued to succeed and to attract more and more students, which made the large size necessary. Harrison's tendency toward perfectionism probably frustrated him in these large classes as well. When in 1909 he received word that he was being considered for a position at the Wistar Institute—a research institution with no teaching duties—he was extremely tempted and did not discourage the invitation. As he complained to his friend Conklin, who was then at Princeton, "In the laboratory we are swamped with students again and the outlook is discouraging. The expenditure of two hundred dollars seems to be for us almost as grave a question as the disposal of your two thousand. But this is a nasty thing for me to say and I am going to stop."[66]

By November, Harrison's tone had changed, so that, when the final offer came from the Wistar, he turned it down. Not, however, before extracting some significant promises from Yale. He wanted assurance that his position would be primarily concerned with research and graduate instruction, with little undergraduate teaching and then only when necessary; that he would receive a technical research assistant; that the department would be reorganized in accordance with a plan he advanced; and that his salary would be raised to meet the Wistar offer. The Yale president agreed, and Harrison resolved to stay.[67]

With his improved position, and with the promises of expansion and a new biological laboratory building, Harrison had achieved a solid position. Given his editorship of the *Journal* and his successful research with its invention of tissue culture as well, he was clearly one of the leading biologists in the United States. As a result of his international reputation built on his entry into the nerve development debates, he had joined his Hopkins colleagues in achieving leadership of international rank as well.

Like Wilson, Morgan, and Conklin, Harrison also attracted a cluster of graduate students to work with him. Also like his colleagues, he had a solid and established research program to which he could introduce those prospective biologists. As one such recent college graduate, Victor Twitty, put it, Harrison had a "gravitational pull for many students." Many did not know much about him before they arrived at Yale. After all, unlike his friends, Harrison had published no major textbooks—indeed no books at all.[68] Further, a number of his early publications had appeared in foreign journals that were not accessible everywhere. So some, like Twitty, ended up at Yale because that was where they received the best offer of an assistantship rather than because of an active desire to work with Harrison. Once there, however, they felt the pull and entered into Harrison's world of experimental embryology, probably working on problems of frog or salamander development using the latest available techniques.[69]

After a building frenzy—which was partially stimulated by the Wistar offer in 1910—finally ended with the opening of a wonderful modern building in 1914, Yale had even more to offer. After that time, Harrison's research program expanded. He came increasingly to ask questions about the patterns and causes of differentiation, which his students then took up as well. Limb development formed a favorite subject for a while, then studies of eyes and ears, the balancer organ, and the establishment of symmetry and axes during development. The 1920s brought the greatest increase in numbers of graduate students, as well as many visitors from abroad. The tide of foreign study in biology had begun to turn, and Europeans came to the United States after World War I to visit their professional colleagues.

Harrison accepted that evolution of species occurs, but he never joined the discussion about the mechanics of evolution. The subject was too speculative for him. Similarly, he never really accepted genetics and never hired a geneticist during his tenure at Yale. He reported to one of his students that, on a visit to his friend Morgan at Columbia before 1910, Morgan had told him something like "Look at all these flies. I haven't gotten a thing out of them."[70] At that time, Morgan remained sceptical about the value of genetics. Unlike Morgan, Harrison was never persuaded to change his mind.

Harrison adopted a "sink or swim" approach with his students which many found disconcerting but on which some thrived. As Twitty put it, "I was not the first to learn that the Chief, kindly as he was, did not exactly smother his students with care and supervision."[71] Harrison wanted the students to learn for themselves, and to make mistakes, as Brooks and Mall had let their students and younger colleagues make their own errors. Many saw at Yale the existence of a "Harrison school," which used transplantation, explanation, and other experimental techniques to address basic embryological questions.

Harrison remained much more formal and distant than his Hopkins friends. He was shy and reserved, so that it took a few brave students to get through to him and establish a rapport that then continued throughout his long career at Yale. They called him the "Chief" and began to eat lunch with him

Fig. 19. Harrison on a canoeing trip, possibly in Canada. From the Harrison Papers.

Fig. 20. Old friends at the Marine Biological Laboratory, probably in the 1950s. Harrison and Conklin met nearly every year for the trustees' meeting. From the Marine Biological Laboratory Archives.

upstairs in the attic of the new biological laboratory building.

A few students joined Harrison on walking trips around the country, for Harrison loved walking. He had walked the countryside at Hopkins and continued to hike whenever and wherever he could. One former student, J. P. Trinkaus, recalls that Harrison was known to dress for church on a Sunday morning, then decide it was too beautiful a day, change, and head off for a lovely walk in the woods.[72] He took occasional canoe trips with Conklin and other friends and recounted these as high points of his life. By 1915, then, Harrison had achieved a successful life in New Haven, with an established reputation as a researcher and as a scientific leader. His research program, his family, and his life had settled into a comfortable pattern that continued until he retired in 1938. He resisted efforts to lure him back to Baltimore and to other attractive jobs.

Throughout, Harrison remained a committed experimenter. More than any of the other three protagonists, he stressed the necessity of an analytical approach instead of mere description alone. Careful observation and description certainly had their proper place in embryology, but experimentation had proven its value for testing among existing alternative hypotheses and for providing definite answers to complex problems. Improved methods and tools made experimentation all the more effective. In a lecture before a medical group at the University of Michigan in 1911, Harrison exhorted the audience to use experiment to explore biology and medicine. Do not feel constrained to produce something practical, he urged, for there is a danger in overemphasizing the useful. Instead, seek scientific knowledge and "the value of truth for its own sake."[73]

Throughout his career, Harrison remained much more the careful "classic" than the opportunistic "romantic." Yet unlike Wilson, who carefully and wisely tried to pick his way between Scylla and Charybdis, Harrison set a strong and definite course that he tried to make good. His dogged effort to make nerve fiber outgrowth convincing and his continued exploration of transplantation research in frog embryonic tissue demonstrate his patience and his commitment to keep trying even in the face of frustration and temporary failures. One problem at a time, carefully tackled, could produce knowledge, and hence scientific progress.

Conclusion

THIS BOOK is the story of the morphological tradition and how it developed in the decades around 1900. That tradition was already strong in 1880, and while it underwent transformation, it did not die. Importantly, those changes are best described as epistemological rather than theoretical. What changed were the questions asked and what could be counted as permissible answers. Ultimately the tradition fragmented, in part because of the success of the more specialized lines of research carried on within it. Perhaps there is no need to settle the question of whether the tradition is now gone, or whether instead it lives on in its offspring. The story still has much to teach us about science.

By 1880, the morphological tradition was concerned with problems of form and patterns of change in form, and it relied on a mixture of descriptive, observational, and comparative methods. Morphologists had the goal of describing this form and its changes more specifically in individuals, as well as exploring the causes of the changes. Since evolution was taken by some morphologists as a given force for change, they assumed that ontogenetic change was caused by phylogenetic change. The primary goal for those morphologists, then, was to establish the pattern of evolutionary, phylogenic development. For other morphologists who relied on some understanding of design as the explanation of form, the goal remained fixed on individuals instead of phylogenies.

Morphology prior to the 1880s did not generally ask questions about the proximate or mechanical and local causes of change. Nor did it explore details of early stages of embryogenesis, which were thought not to be important. At first, most morphologists held, the egg just grows and produces more raw material for the later stages of differentiation to build upon.

The morphological tradition included such diverse work as that in comparative anatomy from early in the nineteenth century and study of germ layers in the latter half of the century. It involved researchers in Germany and also in France, England, the United States, and elsewhere. Morphology was not a localized research program, nor was it the work of a particular school centered on only one dominant leader. Morphology never became a field or discipline since it never achieved the institutional base that fields or disciplines are thought to require.[1] Instead, it involved a set of approaches to a set of problems. Those approaches and problems underwent modification themselves and yet still remained within the tradition. For the morphological tradition involved a broad set of basic commitments with considerable long-term stability across different contexts.

It takes a great deal to eliminate such a tradition. Yet transformation is possible, certainly, and even expected in response to ongoing historical change more generally. In fact, the tradition had already expanded and developed to include evolutionary explanations just after mid century. In the 1880s, it began to change relatively rapidly and in relatively more far-reaching ways.

One important aspect of the changing morphological tradition involved its transplantation to the United States. At midcentury a recent arrival from Switzerland, Louis Agassiz, was the outstanding morphologist in the United States. At Harvard and at Penikese Island, he trained most of the men and women who carried the morphological tradition onward to other institutions. William Keith Brooks took with him to the Johns Hopkins University and Charles Otis Whitman to the Marine Biological Laboratory and the University of Chicago much that they had learned from Agassiz. Yet each also brought his own emphasis and his own revisions to his morphological work and the way he taught his own students. Thus, we have seen that Brooks gave the particular program that our four friends entered at Hopkins a special flavor that emphasized marine research, for example, and embryology almost exclusively.

In addition, the Hopkins program also incorporated a mixture of physiology with its morphology. As a result, the students encountered the morphological tradition within a context that encouraged the use of other, more traditionally physiological approaches and problems as well. Their experience at Hopkins involved what we might think of as a special morphological adaptation. These young Americans were particularly well prepared to remain open-minded about the types of problems and approaches they would find appropriate for their morphological work. They were in a good position to adapt to changes, to adopt the new, and to forge new research programs of their own. The relative flexibility of the American academic traditions also allowed them to move easily into new directions of research, to obtain support rapidly, and thereby to achieve remarkable success. They were well prepared to contribute to the expansion of the morphological tradition.

At first, the 1880s brought expansion. Morphologists moved into new areas of research previously unexplored. They moved into a middle ground between the older morphology and physiology. In doing so, they made contact with reformers among the physiologists and learned from them as well. In particular, they adopted new methods and expanded their goals.

Physiologists considered problems of functioning of organisms and focused on processes in the living system. Since they could not gain access to information about internal processes simply by peering at the outside of an organism, they could either cut it open and watch or devise indirect, experimental tricks to gain further information. They did both. Though vivisection was not of immediate use to the morphologists, experimental intervention could be. It could help to illuminate the patterns of changing internal form during embryonic development, for example. So the morphologists embraced the wider range of methods and approaches in their research.

They also began to realize the importance of heredity as related to but distinct from development. Heredity was not simply something conservative that guaranteed that the resulting adult would be the right type of thing. Heredity also influenced embryonic forms as well. Similarly, development responded to the constraints imposed by heredity as well as environment at all stages. As a result, some morphologists turned to explore heredity. Others, including the four Americans, provided a new focus on embryology, on all stages of embryonic development. They asked, what are the patterns of change, starting with the very first unfertilized egg stage?

Also, morphologists began to ask what causes the changes from one developmental stage to the next. It no longer sufficed to respond simply, "phylogeny." Nor did "design" or any form of teleology provide an adequate account. Proximate, mechanical factors must be at work in guiding the change, it now seemed clear. A new set of problems to get at these proximate factors arose. This meant establishing the relative importance of factors internal or external to the organism and asking questions about how any one part of the organism affects others. For example, what are the respective roles of nucleus and cytoplasm for heredity and development? New problems required new methods of attack as well, so that microscopic techniques evolved, equipment changed, and experimental intervention was incorporated into morphology.

These changes brought expansion in the old morphological tradition, and they brought revision. The new work clearly remained part of that tradition, but, as with the ship of Theseus, we can reasonably ask whether the tradition remained the same after a substantial number of its parts had been changed.

A classic answer makes sense here. We have the same tradition so long as there is historical continuity of all parts and their functioning interaction as a whole. In other words, if a tradition is revised but remains fundamentally and functionally continuous at all times, then it is reasonable to regard it as the same tradition. This might be true even if all parts changed, but we do not have to address that concern here since all parts of morphology did not change. Much remained. The tradition remained focused on problems of understanding form and patterns and on changes in form. It remained centrally dependent on observations and descriptions of form and on comparisons.

Furthermore, the participants continued to think of themselves as morphologists and to identify themselves as such. They all became members of the American Morphological Society, and Edmund Beecher Wilson, Edwin Grant Conklin, and Thomas Hunt Morgan each served as president. (Wilson was first president in 1890 and again in 1895, Conklin in 1899, and Morgan in 1900.) In that capacity, each took a turn explaining what morphology was and placing it in the rapidly changing biological field more generally. Ross Granville Harrison also was president, in 1924, but by then the organization had changed to its current name, the American Society of Zoologists.[2] During the period considered here, to 1915, each clearly felt comfortable with the label morphologist. Each therefore participated in the morphological tradition.

Yet they also all helped to transform that tradition in ways that ultimately led to its replacement with alternative sets of commitments. In part because of their unique Hopkins experience, as they worked within the expanding morphological tradition in the 1890s, each incorporated his own revised set of epistemological standards. Each worked through in his own way what he saw as the proper methods and goals for his research. Reinforced by the pragmatic experimental approach that Henry Newell Martin had exhibited at Hopkins, each called for positive knowledge and definite results as the product of biological research. This entailed addressing within each research program a narrower set of questions, since not all could yield positive results—at least for now. Furthermore, they felt that the goals of each particular line of research should remain relatively narrowly focused in order to achieve results. Rather than large-scale speculative theories, each sought established results guided by what Morgan continually reminded his readers were temporary useful guiding or working hypotheses. The four thereby brought epistemic change that transformed the morphological tradition yet further.

Thus, by contributing to the expansion of the tradition into new areas of concern, the Americans helped to narrow it as well. They helped to expand the tradition to look at new problems concerning the earliest styles of development, including fertilization and heredity as well. They joined their European counterparts in pursuing new methods, equipment, and techniques. This therefore brought expansion of the problems and methods within the tradition.

Yet at the same time, the four Americans each stressed the importance of narrowing the goals and questions for each particular individual line of research. Wilson concentrated on the cell and then on the chromosomes; Conklin continued to explore cell lineage and its implications; Morgan moved through various highly focused research projects to genetics; and Harrison used transplantation techniques to get at questions of tissue development in a few organisms. Each research program therefore narrowed and became productive. Each could solve the carefully focused problems it posed and therefore achieved definitive results and positive knowledge. Thus we have both narrowing at this individual level and expansion of problems and approaches at the level of the tradition. The tradition had changed at several levels.

This fundamental change in tradition did not come about because something anomalous in a particular theory drove the four friends from Hopkins to a new theory. Nor did any particular failure in the older tradition bring a revolution or rejection of the past. Nor did anything within the logic of the scientific work itself require such an epistemic revision. The background at Hopkins, with its peculiar blend of morphology and physiology, certainly suggested and then probably reinforced the change. Yet the institution itself did not dictate the particular epistemic norms the researchers adopted. We cannot understand the change as a direct social or institutional construction, that is. The change was rational, certainly, and made sense in an increasingly practical American setting more generally, but that setting alone cannot explain the change either. In fact, no one of these factors alone explains it. Rather, the confluence of intellectual, institutional, social, and personal factors acted to transform the morphological tradition and to reinforce this transformation in epistemological norms. This episode in the history of biology shows that science can at root be guided more by practical community consensus about how to get good science done than by logic strictly understood.

Beyond the Morphological Tradition

This narrative carries the story only to 1915, though the story of the four careers obviously does not end there. By 1915, all of the principals had reached intellectual maturity and had founded research programs they would pursue for the rest of their long careers. By 1915, each had established a thriving research community of students and visitors, had achieved recognition as an international leader in biology, and had built an environment of solid institutional support for his research program. In addition, together they had established a leading publication in the *Journal of Experimental Zoology*. They had helped keep the Marine Biological Laboratory solid and successful despite financial and ideological crises. They had worked together for several biological societies and for other causes. Soon they also joined forces to work for the National Research Council and, in various ways, to strengthen the state of academic biology in the United States. They tried to hire one another at various times as well, and Wilson did eventually get Morgan to Columbia. By 1915, all were clearly at the forefront of biological research, and they worked together and individually to advance biology.[3]

By 1915 also, their research directions had diverged in ways that ultimately led to the fragmentation of the morphological tradition into various smaller intellectual lines of research and commitment. This is not to suggest that only these four were involved. Yet they did remain central players, specially placed by their Hopkins training to respond to innovation and to innovate themselves. This is not to suggest, either, that everything they did was revisionist or progressive. Yet each of their lines of work did prove remarkably successful by many standards.

What our protagonists did was to help with transforming the morphological tradition. They contributed centrally to transforming its epistemological core commitments by stressing the need for definite results. This reinforced specialization, as each researcher pursued a narrower line of research, with a more restricted collection of organisms and methods. Wilson remained with cytology and focused more and more closely on the earliest developmental stages and on the relative roles of nucleus and cytoplasm, or development and heredity. Conklin explored evolutionary questions by looking at a small range of developmental phenomena in a few organisms. Morgan sequentially explored each of a number of more and more specific problems about development and heredity before moving to chromosomal inheritance and then to genetics. Harrison pursued what some saw as a "gold mine" with his transplantation techniques applied to a relatively few cases of frog development.

Each of these lines of research was rich and productive; each proved capable of supporting a lifetime of research. Yet, compared to the large sets of problems earlier morphologists had attacked, the narrowing of focus and method was noteworthy. It is not the subjects alone which narrowed, however, but also the kinds of conclusions that each drew from his work. Unlike many of their European counterparts of the 1880s and 1890s, these four Americans by the early 1900s resisted drawing conclusions that went much beyond the data at hand. The search for definite results and positive knowledge required sticking close to the facts. This does not mean that they did not explore larger theoretical questions or make general pronouncements about the nature of science or of the world at large. But those remained separate, theoretical declarations. Wilson, Conklin, Morgan, and Harrison sought to avoid confusing what they knew and what they wished they knew.

For all of their innovations, these four remained within the morphological tradition. Yet their students did not have the same set of commitments or the same background training at Hopkins to shape their work. Also, the morphological tradition they encountered at Hopkins had already substantially changed by the time their students entered biology. By 1915, then, it was not so clear to the next generation that they should think of themselves as morphologists. Instead, they were cytologists, evolutionists, embryologists, or geneticists, for example. Each of these lines of research had become sufficiently strong and had gained sufficient support to be considered a field within biology by 1915 or soon thereafter. Our researchers had become so well established that at least some felt they had each established a school. Each had its own identity, and the identity of the larger tradition within which it had arisen had faded into the background and even beyond. Furthermore, other lines of research had led to other successful programs and schools as well. Ecology, for example, had several strong leaders, and numerous other specialties so diverse as behavioral biology and agricultural genetics were also becoming established.

Perhaps in part, expansion—in problems, approaches, and personnel—had made morphology too big to hold together as one tradition. Perhaps in part, the alluring glitter of one or another well-defined more specialized and productive program had caused commitment to the older tradition to fade. In part, the nature of funding for science and the changing institutional settings reinforced the move to specialization. Specialization, in turn, brought competition and fragmentation. The tradition was no longer the same. It did not disappear in precisely 1915. Nor was it overthrown forcefully at any particular point after that. Rather, the morphological road broadened and divided in a way that made it into a major highway. That highway in turn carried biology and biologists in new directions. New lines of research such as cell biology, evolutionary biology, genetics, and developmental biology have arisen, and revised programs, schools, and traditions have emerged and gained their own identities. These were historically connected with the old but also substantially different and more like offspring in a new generation. Their story remains to be told.

So What?

This is fundamentally a story of scientific change, so we should be able to learn something about the nature of that change by stepping beyond the individual case at hand. Can this story help to sort through the myriad alternative theories about scientific change which philosophers and sociologists have been rushing to put forth over the past few decades, for example? Does it help to demonstrate which are the useful categories of analysis?

Yes on both counts.

First, we see that scientific change can be slow, gradual, and nonetheless effective. Thus, any theories of science which stress revolution as necessary for major change do not fit the facts. There need be nothing that looks like a revolution in order to effect real and important change, that is. Biology in 1915 looked very different and was carried out in very different ways from that of 1880 without there having been any obvious or self-conscious revolution. Instead, a persistent tradition was significantly transformed. The process is essentially evolutionary, with continual gradual modifications within the species (or the particular localized research programs) within the larger phylogenetic line (or the tradition).

In addition, the change need not center on any belief statements or propositions in the usual sense. There need be no core truths or even claims about the world to which everyone adheres. Instead, what is shared by the participants in a change may be, for example, a set of epistemic commitments. Rather than being metaphysical, then, the central commitments may concern such things as what it means to know, how to achieve knowledge, how to warrant facts, and how to produce new facts. It may even be more important to the group to share convictions about what sorts of things are legitimate than to worry about which particular claims are the best. This preoccupation with what should count as good scientific work certainly characterized part of the contributions by the four from Hopkins.

If this picture of the changing nature of American biology is correct, then much recent discussion of science has centered on categories of analysis that turn out to be peripheral. Instead of focusing so intently at all times on scientific theories, by which is generally meant a set of propositional claims about the natural world, we must in some cases at least look to epistemic claims, by which is meant a set of claims about what might count as knowledge about the natural world. Instead of yet more theories about change in theories, we instead need a different historical approach focused in a different way. We need to study the science closely, especially since the epistemic norms may not be articulated at all, or the researcher may simply be wrong about the epistemology adopted. In addition, since epistemology is heavily influenced by community conventions, it is important to study the scientific work within its institutional and historical context as well as to look at individual biographical details to provide clues. Work in the history and philosophy of science will best advance these goals by adopting an eclectic approach, demanding neither social construction nor logical analysis alone.

If standards of what counts as knowledge in science can lie at the center of significant scientific change, then this might appear to lead to a pernicious relativism. It might also seem that there are no standards by which to judge whether science ever improves or progresses. Yet this is not so, as others such as David Hull and Robert Richards have so eloquently demonstrated in their recent studies.[4] For though there may be no objective absolute best way to know that will hold for all scientists for all times and circumstances, there may nonetheless be a measure of fitness that allows evaluation of a particular set of scientific norms within its context at any given time. Some science is simply better than other science in this sense. Laudan's interpretation, in which science is better or more progressive when it solves more or more important problems or solves them more effectively, seems on the right track.[5] Yet we need further work on exactly what counts as a problem and what counts as having solved one.

By placing the focus here on what the researchers saw as the proper goals of their research, we get a clearer sense of their problems and of when and how they thought they had achieved knowledge about them. In particular, the problems of embryonic development and heredity were properly attacked by addressing working hypotheses with careful observation, comparison, and experiment that yielded definitive and reproducible facts and therefore positive knowledge. With such an epistemology, morphology was progressive, good science and, ironically, became sufficiently successful that it moved beyond morphology into a realm increasingly specialized and fragmented.

Notes

Introduction

1. Coleman, *Biology in the Nineteenth Century* (1977); Allen, *Life Science in the Twentieth Century* (1978); Cravens, *The Triumph of Evolution* (1978); and others on changes in biology around 1900.
2. Moore, "Wilson" (1987), p. 785.
3. Ibid.
4. Nicholas, "Harrison" (1961), pp. 148–149.

CHAPTER 1 The Hopkins Ideal

1. Morgan, "Wilson" (1941), p. 316. This is the longest biography and includes a full bibliography of Wilson's work. Others include Morgan, "Wilson" (1939); Morgan, "Wilson" (1942); Muller, "Wilson" (1943); Allen, "Wilson" (1970–80). Just what this "adoption" meant is not entirely clear, though Morgan refers also to Wilson's "foster parents" in a way that makes the arrangement sound rather more formal or serious than simply family closeness.
2. Morgan, "Wilson" (1941), p. 315.
3. Ibid., pp. 316, 317.
4. Ibid., p. 318. Also see Chittenden, *Sheffield Scientific School* (1928); Havemeyer, *Sheff Days and Ways* (1958).
5. Muller, "Wilson" (1943), p. 7.
6. Wilson, quoted by Morgan, "Wilson" (1941), p. 318.
7. Morgan, "Wilson" (1941), p. 319.
8. Harvey, "Conklin" (1953), p. 703. Other biographies include Harvey, "Conklin" (1958); Butler, "Conklin" (1952); Richards, "Conklin" (1935); Allen, "Conklin" (1970–80).
9. Richards, "Conklin" (1935), p. 187. On Conklin's religious life, see his autobiographical "Conklin" (1953).

10. Harvey, "Conklin" (1953), p. 704; Conklin, "Conklin" (1953), pp. 52–53; Butler, "Conklin" (1952), p. 5.

11. Harvey, "Conklin" (1958), p. 60. Conklin, "Conklin" (1953), p. 51, reported that he received $175 for one hundred days.

12. Conklin, "Conklin" (1953), pp. 51–52, says that they studied molluscs especially, and that he served as assistant in the university museum during his senior year and learned especially about the classification of molluscs. On the role of museums more generally, see Kohlstedt, "Museums on Campus" (1988).

13. Conklin, "Conklin" (1953), p. 54.

14. Harvey, "Conklin" (1958), p. 61.

15. Allen, *Morgan* (1978), p. 10. This is the best and fullest study of Morgan's life and scientific career. Also see the classic older biographies on which Allen drew, such as Sturtevant, "Morgan" (1959), with a full bibliography; Allen, "Morgan" (1970–80); Shine and Wrobel, *Morgan* (1976).

16. Allen, *Morgan* (1978), pp. 11–12; Sturtevant, "Morgan" (1959), p. 284.

17. Dexter, "The Annisquam Sea-Side Laboratory" (1980).

18. Allen, *Morgan*, (1978), p. 24.

19. Stephenson, in *Filson Club Historical Quarterly* (1946), p. 99. I thank Garland Allen for bringing this to my attention.

20. Letter from Thomas Hunt Morgan to A. H. Sturtevant, about 1943, Sturtevant Papers; quoted in Allen, *Morgan* (1978), p. 25.

21. Nicholas, "Harrison" (1961). Other biographies include: Nicholas, "Harrison" (1960); Abercrombie, "Harrison" (1961); Rudnick, Haymaker, and Schiller, eds., "Harrison" (1970); Oppenheimer, "Harrison" (1970–80).

22. Nicholas, "Harrison" (1961), p. 133.

23. Ibid.

24. Weygandt, *On the Edge of Evening* (1946), p. 42.

25. Many letters in the Harrison Papers at Yale and the Conklin Papers at Princeton show the continued importance for both men of "tramping," that is, hiking and camping.

26. The students are listed in files of correspondence, Gilman Papers. Also on Martin and his students, see Fye, "Martin" (1985).

27. "Doctoral Dissertations, 1876–1926."

28. "What Is Life?" (1925).

29. French, *A History* (1946), pp. 1–40; Hawkins, *Pioneer* (1960), pp. 3–20; Thom, *Johns Hopkins* (1929), p. 103.

30. Bond, *When the Hopkins* (1927), p. 9.

31. Franklin et al., *Gilman* (1910); Flexner, *Gilman* (1946).

32. Herrick, "The Johns Hopkins University" (1880); French, *A History* (1946), p. 2; Gilman, "Reminiscences" (1906), pp. 4–5, 10, 48.

33. Gilman, "Reminiscences" (1906), p. 48.

34. French, *A History* (1946), pp. 34–40.

35. Hawkins, *Pioneer* (1960), p. 309.

36. French, *A History* (1946), p. 45; Hawkins, *Pioneer* (1960), p. 307; Royce, in *Celebration of the Twenty-fifth Anniversary* (1902), pp. 116–118.

37. French, *A History* (1946), p. 79.
38. Hawkins, *Pioneer* (1960), pp. 114–116; French, *A History* (1946), pp. 46–48; Class of 1890, *Hopkins Medley* (1890), p. 126.
39. Gilman, inaugural address (1876), pp. 19, 43.
40. Pauly, "The Appearance of Academic Biology" (1984).
41. Fye, *The Development of American Physiology* (1987); Rosenberg, "Martin" (1970–80), p. 142.
42. Hawkins, *Pioneer* (1960), p. 48; Gilman, inaugural address (1876), p. 43.
43. Rosenberg, "Martin" (1970–80), p. 142; Hawkins, *Pioneer* (1960), p. 48.
44. Owens, "Pure and Sound Government" (1985); Rosenberg, "Martin" (1970–80), p. 142.
45. Martin, "Study and Teaching" (1876), pp. 193, 194.
46. Ibid., pp. 199, 201; Martin, "Modern Physiological Laboratories" (1884), p. 230; *Annual Report of the President of the Johns Hopkins University*, 13th (1888), p. 25; 15th (1890), pp. 34–36; *Johns Hopkins University Circulars* 10 (1890): 16–17.
47. French, *A History* (1946), p. 82; *Annual Report of the President of the Johns Hopkins University*, 3rd (1878), p. 33.
48. Hawkins, *Pioneer* (1960), p. 262; *Johns Hopkins University Circulars* 3 (1884): 93; *Annual Report of the President of the Johns Hopkins University*, 3rd (1878), pp. 33–34.
49. Hawkins, *A History* (1960), p. 262.
50. Gilman, inaugural address (1876), p. 53; Bond, *When the Hopkins* (1927), p. 35; Hawkins, *Pioneer* (1960), pp. 259–267.
51. French, *A History* (1946), p. 93.
52. Benson, "Brooks" (1979).
53. Conklin, "Brooks" (1910), pp. 31, 32, 38–39.
54. French, *A History* (1946), pp. 40, 41.
55. Martin, "Modern Physiological Laboratories" (1884), p. 228. See Benson, "H. Newell Martin, W. K. Brooks, and the Reformation of American Biology" (1987) for further discussion of the ideal. In addition, see other papers in "The Role of Johns Hopkins University" (1987).
56. Gilman, inaugural address (1876), p. 43.
57. Gilman, "The Original Faculty" (1906), p. 51.
58. French, *A History* (1946), p. 82; Huxley, "Address on University Education" (1877), p. 99.
59. Gilman, "Reminiscences" (1906), p. 20. Gilman recollected that Huxley's lecture lacked the "glow" of Huxley's usual work, perhaps because he had been required to allow newspaper reporters to capture the speech ahead of time on paper, something that Huxley evidently never liked to do.
60. Huxley, "Address on University Education" (1877), p. 119.
61. Huxley, "Lecture on the Study of Biology" (1877), pp. 132, 138. On the nature of natural history, see Barber, *The Heyday of Natural History* (1980); Benson, "From Museum Research to Laboratory Research" (1988).

62. Martin, "Study and Teaching" (1876), p. 196.
63. Pauly, "The Appearance of Academic Biology" (1984), p. 379.
64. Reading lists appear in *Johns Hopkins University Circulars*. For example, December 1890, p. 37.
65. The fact that the two subjects are always listed separately and offer separate official reports each year supports the suggestion. The rhetoric about cooperation faded after the first decade.
66. Rosenberg, "Martin" (1970–80), pp. 142–143.
67. MacMillan, "On the Emergence of a Sham Biology" (1893), p. 184; follow-up note; and response by Brooks.
68. Swanson, "A History of Biology" (1951), p. 227.
69. As is evident from the *Johns Hopkins University Circulars*, which list classes, and from numerous notes documenting that botany had not yet developed.
70. Charles Otis Whitman to President Harper, 3 May 1899, Whitman Collection; and other Whitman correspondence.

CHAPTER 2 The Hopkins Experience in Biology

1. Class of 1890, *Hopkins Medley* (1890), p. 57. For further discussion see Maienschein, "H. N. Martin and W. K. Brooks" (1987).
2. Henry Newell Martin to Daniel Coit Gilman, 21 June 1876, Gilman Collection, describes the basic equipment that he considered crucial for a decent physiological laboratory at the new university.
3. Edmund Beecher Wilson to Gilman, 11 July 1880 (copy, 28 May 1883), Gilman Collection.
4. Martin, "Study and Teaching" (1876), p. 193. On Martin's research, see Breathnach, "Martin" (1969).
5. Martin, "Modern Physiological Laboratories" (1884). On Michael Foster, see Geison, *Michael Foster and the Cambridge School of Physiology* (1978). Geison's excellent study illustrates the physiological tradition and thus the context of Martin's work.
6. Hawkins, *Pioneer* (1960), pp. 143–144; Martin, "Study and Teaching" (1876), pp. 181–191.
7. For discussion of physiology in the United States, see Fye, *The Development of American Physiology* (1987), especially chapter 4; Kohler, *From Medical Chemistry to Biochemistry* (1982), especially chapter 11; Pauly, *Controlling Life* (1987); and Geison, ed., *Physiology in the American Context* (1987), especially essays by Fye, Maienschein, Pauly, and Borell.
8. Martin, "The Influence upon the Pulse Rate" (1882), p. 23.
9. Martin, "The Normal Respiratory Movements of the Frog" (1895), p. 141.
10. Martin, "Study and Teaching" (1876), pp. 194–195.
11. Martin, "Modern Physiological Laboratories" (1884), p. 233.
12. Hawkins, *Pioneer* (1960), p. 295.
13. Feibleman, *An Introduction to Peirce's Philosophy* (1946), p. 21.

14. Hawkins, *Pioneer* (1960), p. 293.

15. Numerous documents, including letters, notes, and reports by Martin, attest to his continual involvement.

16. Martin, address at the opening of the laboratory; and Martin to Gilman, 7 May 1881, Gilman Collection, makes the same bias clear, though less emphatically.

17. Pauly, "The Appearance of Academic Biology" (1984).

18. Howell, "Martin" (1908), p. 56.

19. Hawkins, *Pioneer* (1960), p. 143. The administration tried to lower Martin's salary, but he refused to accept that arrangement.

20. Bond, *When the Hopkins* (1927), p. 34.

21. Sturtevant, "Morgan" (1959), p. 285.

22. Nicholas, "Harrison" (1961).

23. Harrison, "On the Origin and Development of the Nervous System" (1935), p. 155.

24. Sturtevant, "Morgan" (1959), p. 285.

25. Packard and Cope, "Editors' Table" (1884). I thank Sheila Weiss for this reference.

26. Hawkins referred to "Brooks, who was an observer and philosopher rather than an experimenter." Garland Allen has often suggested that the Hopkins students such as Morgan rebelled against the speculative philosophy of Brooks and turned to experimentation instead.

27. Andrews, "Biographical Sketch" (1908), p. 33.

28. Benson, "Brooks" (1979), chapter 5, especially p. 302.

29. Brooks, *The Foundations of Zoology* (1915), p. 9.

30. William Keith Brooks to Gilman, 10 September 1887, Gilman Collection.

31. Conklin, "Brooks" (1910), pp. 68–69.

32. Morgan, "Wilson" (1941), p. 319.

33. Ibid.

34. Brooks, "Speculative Philosophy" pt. 1 (1882), pp. 200, 204; pt. 2 (1883), p. 365.

35. Ibid., pt. 2 (1883), pp. 377, 380.

36. Brooks, *The Law of Heredity* (1883), p. 15.

37. Ibid., p. 320.

38. Brooks, "On the Development of Salpa" (1876), p. 339.

39. Brooks, "Observations upon the Early Stages" (1880), p. 73.

40. For discussion of Agassiz's life, see especially Lurie, *Agassiz* (1960). For a family perspective, see Elizabeth Cary Agassiz, ed., *Agassiz* (1885).

41. Louis Agassiz, *The Structure of Animal Life* (1865).

42. Wallace Craig, memo, 27 August 1910, Whitman Collection.

43. Alexander Agassiz Papers. On Alexander Agassiz, see especially G. R. Agassiz, *Letters and Recollections of Alexander Agassiz* (1913); and Murray, "Alexander Agassiz" (1911); Goodale, "Alexander Agassiz" (1912).

44. Brooks to Uhler, 18 June 1876, Gilman Collection.

45. Benson, "Brooks" (1979); Benson, "H. Newell Martin, W. K. Brooks, and the Reformation of American Biology" (1987), especially pp. 767–769.

Also see Werdinger, "Embryology at Woods Hole" (1980), especially chapter 1 on William King [*sic*] Brooks.

46. Benson, "Brooks" (1979), considers Brooks's publications and establishes this point convincingly, p. 294.

47. Conklin, "Address at the Jubilee of the Department of Biology at Case Western Reserve University" (1938), Conklin Papers. Thanks to Keith Benson for bringing this unpublished paper to my attention.

48. Benson, "Brooks" (1979), p. 160.

49. Class of 1890, *Hopkins Medley* (1890), p. 57.

50. Brooks, "The Zoölogical Work of the Johns Hopkins University" (1886), p. 37.

51. Brooks, especially "Chesapeake Zoological Laboratory" (1884). Each year recorded who went, where, accomplishments, etc. See biographies for details on individual attendance.

52. Brooks, "The Zoölogical Work of the Johns Hopkins University" (1886).

53. Andrews, "Johns Hopkins in Jamaica" (1946), (8 pp.), p. 6.

54. Wilson and Conklin contributed most significantly to cell lineage work, as will be discussed in chapter 5.

55. Brooks to Gilman, 7 September 1880, pp. 1–3, and 11 June 1886, Gilman Collection.

56. Andrews, "Johns Hopkins in Jamaica" (1946). This documents the purpose and details of the trip.

57. Harrison, response to Andrews memo, 27 December 1943, Harrison Papers.

58. J. E. Humphrey was associate professor of botany and died in Jamaica. He reportedly went there against his will after Brooks persuaded him, which undoubtedly made Brooks's anguish all the worse. Student Clarke became ill on the boat home and died in the Boston hospital. Brooks to Gilman, various letters, 1893, Gilman Collection.

59. Benson, "Brooks" (1979), especially chapter 3; also Benson, "H. Newell Martin, W. K. Brooks, and the Reformation of American Biology" (1987), pp. 766–768. See Conklin, "Brooks" (1910), for a full bibliography including the oyster work.

60. Brooks to Gilman, 11 and 4 July 1880, Gilman Collection.

61. Benson, "Brooks" (1979), p. 92.

62. Brooks, *Handbook of Invertebrate Zoology* (1882). On techniques, see Benson, "Brooks" (1979), pp. 159–166.

63. See, for example, *Scientific Results, Chesapeake Zoological Laboratory, Studies from the Biological Laboratory.*

64. Brooks withdrew in large part because of his wife's health problems, while Martin withdrew because of his own alcoholism and related troubles.

65. Conklin, "Brooks" (1910), p. 46; McCullough, "Brooks' Role" (1969), p. 413.

66. Conklin, "Brooks" (1910), pp. 42–43, 45–46; Howell, "Brooks" (1908), p. 12.

67. Howell, "Brooks" (1908), p. 13.

68. Andrews, "Biographical Sketch" (1908), pp. 35–37.
69. Brooks to Gilman, 4 July 1880, Gilman Collection.
70. Conklin, "Brooks" (1910), p. 46.
71. McCullough, "Brooks' Role" (1969), p. 424; Allen, *Life Science in the Twentieth Century* (1978); Allen, "Naturalists and Experimentalists" (1979).
72. We see, for example, that Brooks's two kinds of publications either are philosophically oriented or include the name of the particular species or group in the title. In contrast, the students came increasingly to follow the physiological tendency to focus on generalized problems, such as the significance of cells or cleavage or metamerism. That shift reflects a larger shift within the morphological tradition, influenced by forces outside Hopkins as well as within.
73. For a different and excellent detailed view of Wilson, see Baxter, "Wilson and the Problem of Development" (1974).
74. Russell, *Form and Function* (1916), pp. 274–277, considers these theories; Wilson, "Early Stages of some Polychaetous Annelids" (1880).
75. Wilson, "A Problem of Morphology" (1880), p. 66; Wilson, "The Development of Phoronis" (1880), p. 82.
76. Wilson, "The Origin and Significance of the Metamorphosis of Actinotrocha" (1881). Baxter, "Wilson and the Problem of Development" (1974), p. 31. In a postcard to Harrison in 1936, Wilson wrote that he was "sure that" the *Actinotrocha* work was his dissertation but that he seemed "to remember having offered also another m.s.—God only knows what." (Wilson to Ross Granville Harrison, 28 November 1936, Harrison Papers.) Muller, "Wilson" (1943), p. 8, says *Renilla* was, but all other evidence sides with Baxter and with Wilson's postcard.
77. Brooks, report of the Chesapeake Zoological Laboratory, 12 August 1888, and report of 1887, Gilman Collection.
78. Brooks, report of the Chesapeake Zoological Laboratory, 1880, Gilman Collection.
79. Brooks to Gilman, 4 June 1881, Gilman Collection.
80. Wilson, with Osborn and J. M. Wilson, "Variation in the yolk-cleavage of Renilla" (1882).
81. Wilson, "Observations upon the Structure and Development of Renilla and Leptogorgia" (1882), p. 247.
82. Brooks to Gilman, 2 June 1882, Gilman Collection.
83. Wilson, "The Development of Renilla" (1883); Wilson, "The Development of Renilla" (1884), p. 735.
84. Conklin, "The Embryology of Crepidula" (1897). In a number of places, Conklin recounts that he had almost despaired of getting his dissertation published, since Morgan's had used up the available Hopkins funds. Whitman immediately offered to print it in the *Journal of Morphology*, saying of its considerable cost, "What is money for?"
85. Conklin with Brooks, "Structure and Development of the Gonophores" (1890).
86. Harvey, "Conklin" (1958), p. 61.
87. Morgan, "Origin of the Test Cells" (1889), p. 63; Morgan, "On the

Amphibian Blastopore" (1889), pp. 360, 371–373.

88. Spencer F. Baird to Gilman, 19 September 1882, Gilman Collection, and other correspondence in that collection discuss the arrangements.

89. Morgan, "Origin of the Test Cells" (1889); Morgan, "The Origin of the Test Cells of Ascidians" (1890); Thomas Hunt Morgan to Gilman, 20 March 1891, Gilman Collection, reports on Morgan's tenure as a Bruce Fellow from July through September at the MBL, where he worked on test cells, metamorphosis, and other projects.

90. Martin to Gilman, 14 May 1888, Gilman Papers.

91. Morgan, "The Relationships of the Sea-spiders" (1891); Morgan, "Embryology and Phylogeny of the Pycnogonids" (1891), pp. 28–29.

92. Allen, *Morgan* (1978), pp. 70, 90.

93. Sturtevant, "Morgan" (1959), p. 287, says that Morgan visited Naples in 1890, but a list compiled by Christiane Groeben from the Stazione Zoologica archival records does not cite his attendance there, so he must have visited only briefly, if at all.

94. Morgan, "The Growth and Metamorphosis of Tornaria" (1891), p. 407.

95. Ibid., p. 447.

96. Brooks to Gilman, 1 December 1890 and 21 June 1891, Gilman Collection.

97. A. H. Sturtevant, "Reminiscences of T. H. Morgan," transcribed from a talk given 16 August 1967 at the MBL, MBL Archives. I thank the current MBL director, Harlyn Halvorson, for bringing this to my attention.

98. Morgan, "The Dance of the Lady Crab" (1889); Morgan, "Embryology of the Sea Bass" (1891); Morgan, "A New Larval Form from Jamaica" (1891). Morgan reportedly also collected frog materials that were later used by his students at Bryn Mawr and later his wife, Lilian Sampson. On Sampson, see Keenan, "Lilian Vaughan Morgan" (1983).

99. Morgan to Gilman, 1891 [?], Gilman Collection.

100. Harrison to Gilman, 14 December 1891, Gilman Collection.

101. Harrison to Ethan Allen Andrews, 27 December 1946, Harrison Papers.

102. Brooks to Gilman, 4 September 1893, and Harrison to Gilman, 6 August 189[3], Gilman Collection.

CHAPTER 3 Morphology Abroad

1. Morgan, "Wilson" (1941), p. 319.

2. On Balfour's importance at Cambridge and the reactions to his untimely death, see Geison, *Michael Foster and the Cambridge School of Physiology* (1978), pp. 124–130. Wilson's letters to Gilman show his participation.

3. Edmund Beecher Wilson to Daniel Coit Gilman, 9 March 1883, Gilman Collection.

4. Wunderlich, *Leuckart* (1978), presents a list of students who did doctoral work under Leuckart, pp. 41–49.

5. Morgan, "Wilson" (1941), p. 319.
6. Wilson to Gilman, 9 March 1883, Gilman Collection.
7. Laudan, *Progress and Its Problems* (1977), pp. 81, 95–100.
8. Although I do not agree with all of Kuhn's *Structure of Scientific Revolutions* (1962), his remains a useful recognition that scientists operate in groups because of their shared and often unarticulated assumptions.
9. See especially Darden and Maull, "Interfield Theories" (1977).
10. Geison, *Michael Foster and the Cambridge School of Physiology* (1978), offers the best example. Also see Gerson, "Scientific Work and Social Worlds" (1983), who offers an alternative, and very useful, way of defining research clusters.
11. Mayr, "The Species Concept" (1949), p. 373.
12. Russell, *Form and Function* (1916), p. 70.
13. Mayr, "Cause and Effect in Biology" (1961).
14. Nyhart, "Morphology and the German University, 1860–1900" (1986). Nyhart, "The Disciplinary Breakdown of German Morphology" (1987), p. 365, sees morphology as a "field of scientific research" but not as an institutionally established discipline. She concentrates on that study of structure which was Darwinian and employed the methodologies of comparative anatomy and embryology. As a result, she leaves out work that was not strictly Darwinian (or evolutionary) or depended on different methods—work that made up part of the "morphological tradition" under consideration here, but not a central part of her "field."
15. Quoted in Russell, *Form and Function* (1916), pp. 260–261. Gegenbaur, *Grundzüge der Vergleichenden Anatomie* (1859), p. vi; Gegenbaur, "The Condition and Significance of Morphology" (1876), pp. 39–54.
16. Russell, *Form and Function* (1916), chapter 7; Lenoir, *The Strategy of Life* (1982).
17. Meckel, *System der Vergleichenden Anatomie* (1821), pp. xiii–xv. Lenoir, *The Strategy of Life* (1982), pp. 56–61, discusses Meckel. Also see Broman, "Transformation of Academic Medicine in Germany" (1987), especially pp. 237–246. As Broman sees it (p. 241), when Johann Friedrich Meckel began the *Deutsches Archiv für die Physiologie*, "the physiologists had become a self-conscious community."
18. Russell, *Form and Function* (1916), p. 93.
19. Haeckel, *Generelle Morphologie der Organismen* (1866); Haeckel, *Natürliche Schöpfungsgeschicht* (1868). For more on Haeckel's life, see Uschmann, ed., *Haeckel* (1983); Allen, *Life Science in the Twentieth Century* (1978), chapter 1, stresses Haeckel's importance, for example.
20. Baxter, "Wilson's 'Destruction' of the Germ-Layer Theory" (1977), especially pp. 362–366; Russell, *Form and Function* (1916), chapters 14, 16; Oppenheimer, "The Non-Specificity of the Germ-Layers" (1967), pp. 256–294; Haeckel, especially "Gastraea-Theorie" (1875).
21. Russell, *Form and Function* (1916), p. 268.
22. Allen, *Life Science in the Twentieth Century* (1978), pp. 2–6, for example, follows the usual practice and suggests that Haeckel was the only representative of morphology. Nyhart, "Morphology and the German Uni-

versity" (1986), and Russell, *Form and Function* (1916), show otherwise by introducing the host of characters who also played leading roles.

23. Nyhart, "The Disciplinary Breakdown of German Morphology" (1987), p. 386, provides an example in which Haeckel's approach was explicitly rejected by a job search committee at Heidelberg.

24. Oppenheimer, ed., *Autobiography of Dr. Karl Ernst von Baer* (1986).

25. Russell, *Form and Function* (1916), pp. 115–116; von Baer, *Über Entwickelungsgeschichte der Thiere* (1828), p. 153.

26. Ernst Haeckel to Richard Hertwig, 19 November 1901, in Uschmann, ed., *Haeckel* (1983), p. 281.

27. Lankester, "Notes on the Embryology and Classification of the Animal Kingdom" (1877).

28. Coleman, *Biology in the Nineteenth Century* (1977), p. 23; Schleiden, *Beiträge zur Phytogenesis* (1838), pp. 137–176; Schwann, *Mikroskopische Untersuchungen* (1839). Also see Hughes, *A History of Cytology* (1959), for discussion of the earlier work; Maienschein, "Cell Theory and Development" (1990).

29. As the *Oxford English Dictionary* and other dictionaries reveal.

30. Coleman, *Biology in the Nineteenth Century* (1977), p. 33; Virchow, *Die Cellularpathologie* (1858).

31. Remak, *Untersuchungen über die Entwickelung der Wirbelthiere* (1850–55).

32. van Beneden, "La segmentation chez les ascidians" (1884); Whitman, "The Embryology of Clepsine" (1878).

33. Mayr, *The Growth of Biological Thought* (1982), p. 665; Bracegirdle, *A History of Microtechnique* (1978). Bracegirdle (pp. 129–130, 77, 81) explains that His improved the microtome in 1866 by mounting the object to be cut on a microscope stand, thereby permitting regular and continuous cutting. Though not produced commercially, the instrument heralded the production of new models in the following decade. By 1882, a fully useful microtome was available which offered a method of producing uniform, connected serial sections. In 1882 also appeared the highly successful idea of floating sections in warm water to flatten them. On cell theory, see especially Maienschein, "Cell Theory and Development" (1990); and Coleman's outstanding, "Cell, Nucleus, and Inheritance" (1965).

34. As Whitman recognized: "The Inadequacy of the Cell Theory" (1893). See Hughes, *A History of Cytology* (1959), for useful discussion of the contributions of each individual.

35. Mayr, "Cause and Effect in Biology" (1961).

36. Bütschli, "Studien über die ersten Entwickelungsvorgänge" (1876); quotation translated by Berger, "Bütschli" (1970–80), pp. 626–627.

37. His, *Unsere Körperform* (1874), letters 1 and 7 especially.

38. Ibid., pp. 142–144.

39. Ibid., p. 19. I have borrowed Conklin's translation of this passage from his "Cleavage and Differentiation" (1898), pp. 18–19.

40. His, "On the Principles of Animal Morphology" (1888), pp. 173–

174. This is His's most theoretical paper in many ways and reflects his reaction to responses to his earlier *Unsere Körperform.*

41. His, "On the Principles of Animal Morphology" (1888), pp. 174–175.

42. Ibid., p. 175.

43. His, *Unsere Körperform* (1874), pp. 131–144, 145–155. He wrote that "scientific theories of generation can be none other than theories of transmitted movement," p. 150.

44. Ibid., pp. 147, 148, 152.

45. Maienschein, "Why Do Research at the Seashore?" (1988).

46. Kofoid, *The Biological Stations of Europe* (1910), especially pp. 28–30 on individual laboratory equipment.

47. The best single recent discussion of the *stazione's* history is Groeben, "Dohrn" (1985), (references to greenhouses, p. 15). Others include other papers in the same special volume. Also Müller, *Die Geschichte der Zoologischen Station Neapel* (1876). For a detailed discussion of the facilities and equipment around 1910, see Kofoid, *Biological Stations* (1910).

48. Groeben, "Dohrn" (1985), p. 14.

49. Groeben, "The Naples Zoological Station and Woods Hole" (1984), p. 61; Maienschein, "First Impressions: American Biologists at Naples" (1985); Pauly, "American Biologists in Wilhelmian Germany" (1984).

50. Morse, "Whitman" (1912); Lillie, "Whitman" (1911), especially p. xxiv.

51. Whitman, "The Advantages of Study" (1883), pp. 93–94.

52. Charles Otis Whitman to Helen (Mrs. William E.) Frost, 23 February 1882, Whitman Collection.

53. In 1877, Emily Nunn sought admission to the Hopkins program, evidently very persistently. She did attend the teachers' class in 1877–78 and the Chesapeake Zoological Laboratory session in 1879, but the Hopkins officials resolved not to admit her or other women to the graduate program in biology. See "Admission of Women" file, Gilman Papers.

54. Nunn, "The Naples Zoölogical Station" (1883); Nunn (Whitman), "The Zoölogical Station at Naples" (1886).

55. Morgan, "Wilson" (1941), p. 323.

56. Letters between Wilson and Anton Dohrn from 1883, Stazione Archives, document their discussion. Wilson wrote on 29 January that he wished to visit Naples. Dohrn responded on 2 February that the rules forbade it since the Americans had not subscribed to any table. Wilson, 9 February, replied that he would try to arrange for an American table. I thank Christiane Groeben for her generosity in bringing these and many other valuable materials to my attention.

57. Wilson to Gilman, 9 March 1883, Student Records, Record Group #13.010.

58. Though Gilman did offer, as is recorded in letters from Wilson to Gilman, Student Records, Record Group #13.010. Also see other letters, including Wilson to Dohrn, 12 March 1883, Stazione Zoologica Archives.

59. Morgan, "Wilson" (1941), p. 320, quoting Wilson.

60. Wilson to Gilman, 13 April 1883, Student Records, Record Group #13.010.

61. Wilson, "The Mesenterial Filaments of the *Alcynoia*" (1884).

62. Dohrn showed the depth of this musical friendship when he later wrote to H. F. Osborn at Columbia and lamented that he had not seen Wilson for a while. He sent a teasing message for his friend: "Should you see sometimes an old friend of ours, called Edmund B. Wilson, will you tell him, that we all consider him a most unreliable fellow, who seems not to care a bit for 'auld acquaintance'? I have a portrait of him also, in my study: I have covered it with a veil, not wishing to be remembered of such an egoistical, faithless individual, who sacrifices old affections for new ones! To punish him, tell him, that in March I was in Rome, to hear the Joachim-Quartetto play all the 16 [Quartets] of Beethoven in the Carracci-Fresco saloon of the Palazzo Farnese!! I was there together with the Mendelssohns. That will be enough for him. And we hope, to bring the Quartetto next or after-next year to Naples, and they shall play in the library of the Zool. Station,—but Wilson will not be invited, being an infidel to old affections. Give him that message!" Dohrn to Wilson, 15 June 1905, in the collection of Wilson's granddaughter Linda Timmons.

63. Wilson to Dohrn, 20 September 1883, Stazione Archives.

CHAPTER 4 Transforming Traditions at Home

1. Morgan, "Wilson" (1941), p. 320, and as reflected in letters from Wilson to Gilman, Gilman Collection.

2. Huxley and Martin, *A Course of Elementary Instruction* (1883).

3. Sedgwick and Wilson, *General Biology* (1886), p. 2.

4. Morgan, review of Sedgwick and Wilson, *General Biology,* 2d ed. (1895), pp. 740–741, outlines the changes from the first edition and offers a critical review of Sedgwick and Wilson's approach.

5. Sedgwick and Wilson, *General Biology* (1886), chapters 1–3.

6. Ibid., 2d ed. (1895), pp. 80–81.

7. Hawkins, *Pioneer* (1960), pp. 265–266. On President Thomas, see Finch, *Carey Thomas of Bryn Mawr* (1947). On education more generally, see Curtis and Nash, *Philanthropy in the Shaping of American Higher Education* (1965).

8. Baxter, "Wilson and the Problem of Development" (1974).

9. Baxter has established the first of these points in two excellent articles: Baxter, "Wilson as a Preformationist" (1976); Baxter, "Wilson's 'Destruction' of the Germ-Layer Theory" (1977).

10. Whitman, "The Germ Layers of *Clepsine*" (1886). His publications since the dissertation in 1878 had largely focused on microscopic methods and on various aspects of life history and development of a range of organisms. See Lillie, "Whitman" (1911), pp. lxxiv–lxxvii, for a complete bibliography.

11. Whitman, "The Embryo of *Clepsine*" (1878), p. 263.
12. Lankester, *Notes on Embryology and Classification for the Use of Students* (1877), pp. 14–15.
13. Whitman, "A Contribution to the History of the Germ-layers in Clepsine" (1887), p. 107.
14. Ibid., pp. 138–140.
15. Ibid., p. 169.
16. On Whitman's stay in Milwaukee until 1898, when he moved to head the Biology Department at Clark University, see Dornfeld, "The Allis Lake Laboratory" (1956).
17. Wilson, "The Embryology of the Earthworm" (1889), p. 395.
18. Wilson, "The Origin of the Mesoblast-Bands" (1890), p. 207.
19. Ibid., p. 208.
20. Lillie, *Marine Biological Laboratory* (1944), p. 124.
21. Wilson, "Some Problems of Annelid Morphology" (1890), p. 54.
22. Ibid., p. 78.
23. Dexter, "From Penikese to the Marine Biological Laboratory" (1974); Maienschein, "Agassiz, Hyatt, Whitman, and the Birth of the Marine Biological Laboratory" (1985); Lillie, *Marine Biological Laboratory* (1944).
24. Elizabeth Cary Agassiz, ed., *Agassiz* (1885), chapter 25; Conklin, "The Beginning of Biology at Woods Hole" (1927). Also see Morse, "Agassiz and the School at Penikese" (1923).
25. Louis Agassiz to William Wykoff, 14 June 1873, Louis Agassiz Papers.
26. "Penikese Island" (1873), p. 378.
27. Louis Agassiz said: "That is the charm of teaching from Nature herself. No one can warp her to suit his own views. She brings us back to absolute trust as often as we wander," in Elizabeth Cary Agassiz, ed., *Agassiz* (1885), p. 775. The plaque from Penikese reading "Study Nature, not Books," is housed in the MBL Library. See also Lillie, *Marine Biological Laboratory* (1944), pp. 20–21. For discussion of Agassiz's opening address, see also Wright and Wright, "Agassiz's Address at the Opening of Agassiz's Academy" (1950).
28. Craig, memo, 27 August 1910, Whitman Collection.
29. "Penikese Island" (1873), p. 378.
30. Craig, memo, 27 August 1910, Whitman Collection.
31. Dexter, "From Penikese to the Marine Biological Laboratory" (1974).
32. Lillie, *Marine Biological Laboratory* (1944), p. 22; Whitman, address at the opening of MBL, 17 July 1888, MBL *Annual Report*, 1888, p. 25.
33. Kofoid, *Biological Stations* (1910), p. 14.
34. Dexter, "From Penikese to the Marine Biological Laboratory" (1974), p. 159; Wilder, "What We Owe to Agassiz" (1907), p. 12.
35. Galtsoff, *The Story of the Bureau of Commercial Fisheries* (1962), p. 27.
36. Ibid., p. 20.
37. Conklin, "The United States Bureau of Fisheries," in Lillie, *Marine Biological Laboratory* (1944), pp. 24–26. Many of the early MBL attendees were women, and the Fish Commission evidently had no women investigators. This alone would explain nearly half the population difference and

perhaps some of the lack of enthusiasm.

38. Clapp, "Some Recollections of the First Summer at Woods Hole" (1927), p. 3; MBL *Trustees' Minutes*, 6 June 1888, p. 35.

39. MBL *Trustees' Minutes*, 13 March 1888, p. 1. Also see Maienschein, *One Hundred Years Exploring Life* (1989), for further discussion of early MBL years.

40. Samuel F. Clarke to William Sedgwick, 18 December 1887, MBL *Trustees' Minutes*, 1888.

41. MBL *Trustees' Minutes*, 7 April 1888, pp. 11, 13; 28 April 1888, p. 23; 12 May 1888, p. 29.

42. Whitman, MBL *Annual Report*, 1890, p. 22; Whitman, "Some of the Functions and Features of a Biological Station" (1898).

43. Whitman, MBL *Annual Report*, 1890, p. 5; 1895, pp. 17–31.

44. Whitman, "Some of the Functions and Features of a Biological Station" (1898), p. 39.

45. Lillie, *Marine Biological Laboratory* (1944), p. 36.

46. Clapp, "Some Recollections" (1927), p. 3. Confirmed by Whitman, MBL *Annual Report*, 1895, pp. 21–22.

47. Clapp, "Some Recollections" (1927), p. 10.

48. Whitman, "The Kinetic Phenomena of the Egg During Maturation and Fecundation (Oökinesis)" (1887), p. 227.

49. Maienschein, "Preformation or New Formation—or Neither or Both?" (1986).

50. Whitman, "Prefatory Note" (1895), pp. vi–vii.

51. Whitman, "Inadequacy of the Cell-Theory of Development" (1894), p. 111.

52. Ibid., pp. 123–124.

53. Maienschein, ed., *Defining Biology* (1987), especially the introduction, discusses the lectures presented and the central topics of interest.

54. Whitman, MBL *Annual Report*, 1892, pp. 34–35.

55. Maienschein, ed., *Defining Biology* (1987), lists the lectures given at the MBL through the 1890s, pp. 51–56. Also on Whitman's look at the MBL, see Whitman, "Marine Biological Laboratory" (1898). All this spirit of cooperation and shared interests could become a bit much, however. Jacques Loeb wrote to Wilson in 1902 that perhaps it was time to look to the west coast. "Of course we all feel that Woods Hole is our ideal place, but we must not forget, that there is a good deal of sentimentality about Woods Hole and the idea of 'cooperation' which must not make us blind in regard to our real problems and opportunities." Loeb to Wilson, 12 April 1902, in the collection of Wilson's granddaughter Linda Timmons.

56. In 1976, Donald P. Costello introduced me to the time-honored procedure for gathering *Nereis*, once a popular pastime at the MBL.

57. Donald Costello, interview at University of North Carolina, 1967, in MBL Archives.

58. Wilson, "Some Problems of Annelid Morphology" (1890), p. 54.

59. For a discussion of cell lineage work, see Maienschein, "Cell Lineage, Ancestral Reminiscence, and the Biogenetic Law" (1978); Werdinger, "Em-

bryology at Woods Hole" (1980), especially chapter 4. See also Lillie, *Marine Biological Laboratory* (1944), p. 125: "Cell lineage bore upon one of the main embryological problems of the day, 'isotropy' or the ovum, as it was called—the view that all parts of the ovum are of equal value whether with reference to the main axes of the embryo or to the prospective significance of its parts. This was the radical epigenetic view of development held strongly for a time by Driesch and Jacques Loeb, among others, but anathema to most embryologists at Woods Hole. It was impossible for any student of cell lineage who observed the regularity and precision of the first cleavage planes with reference to the preformed poles of the egg, and the determination of direction, rate, and place of subsequent cleavages in relation to the first, together with the invariable prospective significance of cells so derived, to be misled by any such oversimplification."

60. Lillie noted that Wilson had expressed his intention to do so. And Baxter has convincingly documented the move: Baxter, "Wilson's 'Destruction' of the Germ-Layer Theory" (1977).

61. Wilson, "Cell-Lineage of *Nereis*" (1892), p. 377, 387.

62. Ibid., p. 453.

63. Wilson, "The Heliotropism of Hydra" (1891), pp. 420, 417–418.

64. Whitman, "Animal Behavior" (1899).

65. Lillie, *Marine Biological Laboratory* (1944), p. 124.

66. Conklin, "Early Days at Woods Hole" (1968), p. 114.

67. Richards, "Conklin" (1935), p. 192.

68. Harvey, "Conklin" (1958), p. 64.

69. Bonner, "What Is Money For?" (1984).

70. Conklin, "Early Days at Woods Hole" (1968), p. 116.

71. Conklin, "Preliminary Note on the Embryology of Crepidula fornicata and of Urosalpinx cinerea" (1891); Conklin, "The Cleavage of the Ovum in Crepidula fornicata" (1892).

72. Conklin, "The Marine Biological Laboratory" (1900), p. 340.

73. Costello, personal interview (1976), reported that Harrison had given him a set of slides and notes for cell lineage work on *Polycoerus*. Harrison had received materials initially from Morgan, had continued the work briefly, then moved on to other projects.

74. Morgan, *The Development of the Frog's Egg* (1897).

75. *Biological Lectures delivered at the Marine Biological Laboratory, 1898* (1899). Only in the last decade or so have researchers returned in significant numbers to cell lineage work, this time primarily on nematodes, which have a relatively small number of cells.

76. Wilson, "The Structure of Protoplasm" (1899).

77. Wilson, "Cell Lineage and Ancestral Reminiscence" (1899), p. 24.

78. Kuhn, *The Structure of Scientific Revolutions* (1970).

79. Allen, *Life Science in the Twentieth Century* (1978), especially pp. xi–xxiii. Allen's interpretations there and elsewhere were discussed by Maienschein, Rainger, and Benson in "American Morphology at the Turn of the Century" (1981). Allen and Frederick B. Churchill provided responses.

80. Sapp, *Beyond the Gene* (1987).

81. Laudan, *Progress and Its Problems* (1977), especially pp. 70–120; Churchill, especially "From Heredity Theory to *Vererbung*" (1987) and "Regeneration" (1989); Farley, *Gametes and Spores* (1982).

82. Several papers in the series of *Biological Lectures delivered at the Marine Biological Laboratory* in the 1890s reflect this concern and the growing awareness that not all cleavages were as regular and determinate as others.

CHAPTER 5 Responding to Innovations Abroad

1. For further discussion of Osborn's career at the American Museum of Natural History, see especially Rainger, *An Agenda for Antiquity* (forthcoming).

2. Charles Otis Whitman to Henry Fairfield Osborn, n.d., but probably late 1890 or early 1891, Osborn Collection. I thank Ronald Rainger for bringing this letter and other related materials to my attention.

3. Crampton, *The Department of Zoology of Columbia University* (1942).

4. Baltzer, *Life and Work of a Great Biologist* (1967). "Chronology" outlines his career; see also pp. 10–13.

5. Boveri, "Zellenstudien I" (1887); "Zellenstudien II" (1888); "Zellenstudien III" (1890).

6. Morgan, "Wilson" (1941), p. 321.

7. Baltzer, *Life and Work of a Great Biologist* (1967), p. 58.

8. Wilson, *The Cell in Development and Inheritance* (1896).

9. Maienschein, "Heredity/Development in the United States" (1987).

10. For more detailed discussion of Oskar Hertwig's contributions, see Churchill, "Hertwig, Weismann, and the Meaning of Reduction Division" (1970). More generally, see Baxter and Farley, "Mendel and Meiosis" (1979); Farley, *Gametes and Spores* (1982), especially chapters 6, 8.

11. The best discussions of this and other cytological contributions of the late nineteenth century appear in Coleman, "Cell, Nucleus, and Inheritance" (1965); and Farley, *Gametes and Spores* (1982), especially chapters 6 and 8. Also see Hughes, *A History of Cytology* (1959); Baker, "The Cell-Theory" (1948–55).

12. Roux, *Über die Bedeutung der Kerntheilungsfiguren* (1883), especially p. 6; Weismann, *Die Kontinuität des Keimplasmas* (1885), pp. 161–249; Weismann, *Über die Zahl der Richtungskörper (1887), pp. 333*–384; Weismann, *Das Keimplasm* (1893). For further discussion of Weismann, see Churchill, "August Weismann and a Break from Tradition" (1968); Churchill, "Hertwig, Weismann, and Reduction Division" (1970); Churchill, "From Heredity Theory to *Vererbung*" (1987). Also see Churchill, "Weismann's Continuity of the Germ-Plasm" (1985).

13. Hertwig and Hertwig, "Über den Befruchtungs und Teilungsvorgang des tierischen Eies" (1887). Also see Oskar Hertwig, "Welche Einfluss übt die Schwerkraft auf die Theilung der Zellen?" (1884–85); Oskar Hertwig, "Das Problem der Befruchtung und der Isotropie des Eies" (1885), pp. 276–318.

14. Boveri, "Ein geschlechtlich erzeugter Organismus" (1889).
15. Baltzer, *Life and Work of a Great Biologist* (1967), pt. 2, discusses Boveri's research program in more detail.
16. Boveri, "Ein geschlechtlich erzeugter Organismus" (1889), p. 80; cited in Baltzer, *Life and Work of a Great Biologist* (1967), p. 81.
17. Boveri, "Über die Niere des Amphioxus" (1890).
18. Wilson, *The Cell in Development and Inheritance* (1896).
19. Edmund Beecher Wilson to Anton Dohrn, 10 December 1891, Stazione Zoologica Archives. On 15 July 1891, Wilson had first written from Woods Hole to tell Dohrn that he had accepted the Columbia position, that he would visit Europe during the next year, and that he most hoped to spend some time in Naples.
20. Pflüger, "Einige Beobachtungen" (1881); Pflüger, "Hat die Concentration des Samens einen Einfluss" (1882); Pflüger, "Über die das Geschlecht bestimmenden Ursachen" (1882); and others in *Pflügers Archiv.*
21. Pflüger, "Die Bastardirung bei den Batrachien" (1882), p. 51.
22. Pflüger, "Über den Einfluss der Schwerkraft" (1883); Pflüger, "Über die Einwirkung der Schwerkraft" (1884).
23. Pflüger, "Über den Einfluss der Schwerkraft" (1883), p. 68.
24. References in the papers of each to each other, to correspondence, and to exchange of materials demonstrated the close connections. Also, they were colleagues in Breslau.
25. Born, "Über die Entstehung der Geschlechtsunterschiede" (1881).
26. On Spemann, see Hamburger, *The Heritage of Experimental Embryology* (1988).
27. Born, "Über den Einfluss der Schwere" (1884), p. 2; Born, "Biologische Untersuchungen I" (1885); Born, "Biologische Untersuchungen II" (1886).
28. Roux responded to Born's point of view in "Über die Zeit der Bestimmung der Hauptrichtungen des Froschembryo" (1883), especially pp. 6, 14, 23.
29. Roux, "Beiträge zur Entwickelungsmechanik des Embryo" (1885). Here Roux lays out his convictions that study of development must be causal and mechanical and asks in particular (p. 422) whether the process is interactive or self-directing.
30. Roux, "Über die Bestimmung der Hauptrichtungen" (1885), pp. 5, 24–25.
31. Pflüger, "Über die Einwirkung der Schwerkraft" (1884), p. 608.
32. Weismann, *Die Kontinuität des Keimplasmas* (1885). Roux, "Beiträge zur Entwickelungsmechanik des Embryo" (1885), p. 8. Attacks did come. Oskar Hertwig became Weismann's most outspoken opponent. The nucleus may undergo changes, but those do not direct differentiation, Hertwig asserted. Nor does the future organism exist preformed or even predetermined in the egg. Predeterminist theories are pernicious and explain nothing. Weismann's theory "merely transfers to an invisible germ region the solution of a problem that we are trying to solve, at least partially, by investigation of visible characters; and in the invisible region it is impossible

to apply the methods of science. So, by its very nature, it is barren to investigation, as there is no means by which investigation may be put to the proof." Oskar Hertwig, *The Biological Problem of To-day* (1896), p. 136. For further discussion see Maienschein, "Preformation or New Formation—or Neither or Both?" (1986). Also see Gould, *Ontogeny and Phylogeny* (1977), pp. 186–202.

33. Weismann, *Die Kontinuität des Keimplasmas* (1885), partially reprinted in Moore, ed., *Readings in Heredity and Development* (1972), pp. 56–76.

34. Roux, "Beiträge zur Entwickelungsmechanik des Embryo, No. 5" (1888); trans. in Willier and Oppenheimer, eds., *Foundations of Experimental Embryology* (1964), p. 4. Page citations here refer to the more accessible translation.

35. Ibid., p. 8.

36. Ibid., pp. 25–26.

37. Ibid., p. 28.

38. See Churchill, "Wilhelm Roux and a Program for Embryology" (1966).

39. Driesch, "Entwicklungsmechanische Studien. I" (1891–92).

40. Driesch, especially in *Die mathematisch-mechanische Betrachtung* (1890); Driesch, "Entwicklungsmechanische Studien. VI" (1893). There he outlined what he saw as the various alternative approaches to zoology and the limitations and strengths of each.

41. Driesch, "Entwicklungsmechanische Studien. I" (1891); trans. in Willier and Oppenheimer, eds., *Foundations of Experimental Embryology* (1964) p. 49. But such "idle speculation" takes a stronger form and results in a distinct controversy. Also see Churchill's excellent discussion of the Roux-Driesch experiments in "From Machine-Theory to Entelechy" (1969); Churchill, "Chabry, Roux, and the Experimental Method in Nineteenth Century Embryology" (1973); Mocek, *Wilhelm Roux–Hans Driesch* (1974).

42. Wilson to Hans Driesch, 15 June 1892, Stazione Zoologica Archives, reports that he had worked on separating up to the eight-cell stage in *Serpula*, a genus of annelids.

43. Ibid.

44. Wilson to Driesch, 25 September 1892, Stazione Zoologica Archives.

45. Wilson, "On Multiple and Partial Development in Amphioxus" (1892), pp. 735, 739–740.

46. Morgan, "Wilson" (1941), p. 322.

47. Wilson, "*Amphioxus*, and the Mosaic Theory of Development" (1893). The first part of the paper reviews in detail the results of his and other studies of isolated blastomeres; the later parts address the various theories put forth and discuss which evidence weighs in favor of which theories, and why. This was presumably responding to Driesch, "Entwicklungsmechanische Studien. III, IV, V, VI."

48. Wilson to Driesch, 5 June 1893, Zoologica Stazione Archives.

49. As Driesch put forth especially in "Entwicklungsmechanische Studien. II" (1892), pp. 50–55. Wilson had developed this point of view in "*Amphioxus*, and the Mosaic Theory of Development" (1893), pp. 306, 315: "Con-

sidered as a purely formal explanation, the Roux-Weismann hypothesis is perfectly logical and complete. Its weakness lies in its highly artificial character; for both of its two fundamental postulates—viz: qualitative nuclear division, and accessory latent idioplasm—are purely imaginary. They are complicated assumptions in regard to phenomena of which we are really quite ignorant, and they lie at present beyond the reach not only of verification, but also of disproof. The 'explanation' is, therefore, unreal." Yet "we are thus enabled, in a measure, to reconcile the apparently conflicting results of Roux on the one hand, and those of Hertwig and Driesch on the other. It is true that no middle ground is possible in the question of qualitative *versus* quantitative division; but it is otherwise with the external phenomena of cleavage. I have endeavored to show that the phenomena of regeneration are not incompatible with a modified form of the mosaic theory, in which the hypothesis of qualitative division is repudiated. Thus modified, the mosaic theory is of the utmost importance, and is destined, I believe, to form the basis of all exact and thorough investigations on animal ontogeny."

50. Wilson, "The Mosaic Theory of Development" (1894), p. 3.

51. Ibid., p. 5.

52. Ibid., pp. 5, 7.

53. Ibid., p. 8.

54. For the best study of Loeb's life and career, see Pauly, *Controlling Life* (1987). Chapter 3 discusses his move to the United States. Also see Anne Leonard Loeb's (his wife's) biographical notes on him, Jacques Loeb Papers.

55. Pauly, *Controlling Life* (1987), p. 62.

56. Ibid., pp. 62–63.

57. Pauly, *Controlling Life* (1987), suggests that Loeb resented Morgan's directorship of the department and that the two did not always get along well. Nonetheless, they were in contact virtually every summer at the MBL and participated together often enough in enough projects that they can be considered effectively as friends. On Loeb, also see Maienschein, "Physiology, Biology, and the Advent of Physiological Morphology" (1987).

58. This is the intriguing thesis of Pauly's *Controlling Life* (1987).

59. Loeb, *Untersuchungen zur physiologischen Morphologie der Thiere. I. Über Heteromorphose* (1891) and *II. Organbildung und Wachstum* (1892). This theme runs throughout and is articulated most explicitly in II, pp. 6–7; it followed that "physiological morphology was not only a possible, but a necessary part of animal physiology" as well as of animal morphology.

60. Loeb, "Investigations in Physiological Morphology. III" (1892), pp. 257, 261, 262. On his resistance to *Naturphilosophie*, see Loeb, "On Some Facts and Principles of Physiological Morphology" (1894), p. 56: "But it would be a mistake and a falling back into the German Naturphilosophie to attempt at present an explanation of how the unknown chemical nature of the germ determines all the different organs and characters that belong to the species." Also Pauly, *Controlling Life* (1987), documents Loeb's con-

cern to reject the approach and metaphysics of *Naturphilosophie* and to establish an "engineering standpoint."

61. On Loeb's place in American physiology, see Kohler, *From Medical Chemistry to Biochemistry* (1982), pp. 109–114; Pauly, *Controlling Life* (1987). Also see Maienschein, "Physiology, Biology, and the Advent of Physiological Morphology" (1987), pp. 177–193.

62. Manning, *Black Apollo of Science. The Life of Ernest Everett Just* (1983), pp. 254–256.

63. Morgan, "Balanoglossus and Tornaria in New England" (1892); Morgan, "The Development of Balanoglossus" (1894). Also, on Morgan at Bryn Mawr more generally, see Oppenheimer, "Morgan as an Embryologist" (1983).

64. Morgan, "Spiral Modification of Metamerism" (1892). He returned to the subject and, so he said, gave a "full consideration of the facts"—though he offered no new theories—in Morgan, "A Study of Metamerism" (1895).

65. Boveri, "Ein geschlechtlich erzeugter Organismus" trans. by Morgan as "An Organism Produced Sexually" (1893), p. 232.

66. Morgan's introduction to Boveri's, "An Organism Produced Sexually" (1893), p. 222.

67. As was discussed later in a general MBL lecture: Clapp, "Relation of the Axis of the Embryo to the First Cleavage Plane" (1899).

68. Morgan, "Experimental Studies on the Teleost Eggs" (1893), p. 806.

69. Ibid., pp. 807, 810. For further discussion, see Atz, "*Fundulus heterochitus* in the Laboratory" (1986).

70. Morgan, "Experimental Studies on the Teleost Eggs" (1893), p. 814.

71. Chamberlin, "The Method of Multiple Working Hypotheses" (1890). The paper was much discussed and was revised and reprinted elsewhere. Also see Greene, *Geology in the Nineteenth Century* (1982), pp. 260–267, for discussion of Chamberlin's method and its applications in geology.

72. Maienschein, "Arguments for Experimentation in Biology" (1987).

73. Chamberlin, "The Method of Multiple Working Hypotheses" (1890), p. 96. Gilbert favored a similar approach in "The Origin of Hypotheses" (1891).

74. Morgan, "Experimental Studies on Echinoderm Eggs" (1894), p. 143.

75. Ibid., p. 145.

76. Presumably here Morgan was responding to Driesch, "Entwicklungsmechanische Studien. VIII." It is interesting to compare Morgan's drawings with Wilson's, Conklin's, or Harrison's. He simply does not provide the detail or show concern with verisimilitude that the others did. One does not get the feeling in looking at these illustrations that their creator was providing a record for posterity. They really are illustrations rather than standards against which to compare other results, which Wilson and Conklin provided in their cell lineage work.

77. Morgan, "Experimental Studies on Echinoderms" (1894), p. 149.

78. Ibid., p. 152.

79. Allen, *Morgan* (1978), p. 76, says that Morgan "had little direct contact with experimental embryology prior to 1894." He then refers to the one 1893 paper on teleosts as if it were the only experimental work Morgan did. However, Morgan clearly had read widely and had heard much about the German experimental approach. He had reproduced many of the major experiments himself and had begun to devise his own experimental program during 1892 and 1893.

80. Harrison, unpublished biography of Nussbaum (unpaginated), Harrison Papers, second page.

81. A few of Harrison's early notebooks from Germany are on deposit in the Harrison Papers. There he recorded that Nussbaum progressed slowly through the academic ranks at Bonn until he finally received a chair as *ordentlicher* professor in biology and histology in 1907. Harrison's long-time research assistant Sally Wilens has provided an excellent guide to the extensive collection that makes up the Harrison Papers.

82. Nicholas, "Harrison" (1960), p. 137.

83. William Keith Brooks to Daniel Coit Gilman, 3 May 1893, Gilman Collection.

84. Ross Granville Harrison to Gilman, 6 August 1894, Gilman Collection.

85. Brooks to Gilman, 4 September [1894], Gilman Collection.

86. Harrison, "Über die Entwicklung der nicht knorpelig vorgebildet Skelettheile in den Flossen der Teleostier" (1893), p. 272.

87. Harrison, "The Development of the Fins of Teleosts" (1894).

88. Harrison, "Die Entwicklung der unpaaren und paarigen Flossen der Teleostier" (1895).

89. Richard Edes Harrison, 20 February 1990, personal correspondence with the author.

90. The astute reader will have noticed the lack of discussion of Conklin in this chapter. Unlike the others, he did not go to Europe until much later and was much more involved with the Philadelphia scientific world.

CHAPTER 6 Edmund Beecher Wilson

1. Wilson, "Embryological Criterion of Homology" (1896), pp. 103–104.

2. Ibid., p. 114.

3. Ibid., pp. 104, 124, 114.

4. Ibid., p. 123.

5. Wilson and Mathews, "Maturation, Fertilization, and Polarity in the Echinoderm Egg" (1895). They were responding to Fol, "Le Quadrille des Centres" (1891), pp. 393–420.

6. Wilson, "Archoplasm, Centrosome, and Chromatin" (1895), p. 470.

7. Ibid., pp. 471–472.

8. Ibid.

9. Wilson and Leaming, *An Atlas of the Fertilization and Karyokinesis of the Ovum* (1895).

10. Wilson, *The Cell in Development and Inheritance* (1896).
11. Ibid., p. 296.
12. Ibid., pp. 306–307, 315.
13. Ibid., p. 321.
14. Ibid., p. 323.
15. Ibid., p. viii.
16. Ibid., pp. 327, 328.
17. Ibid., p. 330.
18. Conklin, review of Wilson, *The Cell in Development and Inheritance* (1897), p. 321.
19. Edmund Beecher Wilson to Edwin Grant Conklin, 3 August 1897, Conklin Papers.
20. Wilson, "On Cleavage and Mosaic-Work" (1896), p. 21.
21. Driesch and Morgan, "Zur Analysis der ersten Entwickelungsstudien des Ctenophoreneis" (1895).
22. Wilson, "Cell-Lineage and Ancestral Reminiscence" (1899), p. 21.
23. Conklin to Wilson, 6 June 1898, in the collection of Wilson's granddaughter Linda Timmons.
24. Wilson, "Considerations on Cell-Lineage and Ancestral Reminiscence" (1898), p. 23.
25. Wilson, "The Structure of Protoplasm" (1899), pp. 10, 19.
26. Ibid., pp. 19, 20.
27. Wilson, "Some Aspects of Recent Biological Research" (1900), pp. 21–22.
28. Ibid.
29. Wilson, "Aims and Methods of Study of Natural History" (1901).
30. Ibid.
31. For further discussion of this point, see Maienschein, "Arguments for Experimentation in Biology" (1987).
32. For details on the department, see Crampton, *The Department of Zoology of Columbia University* (1942).
33. There was considerable discussion at the time about who deserved credit for the technique, and the Columbia group tended to accord priority to Morgan. Yet we have seen earlier that Morgan actually got the idea of placing eggs in altered salt solutions from Loeb but extended the technique to different developmental questions. It seems fair to give both credit.
34. Wilson, "The Chemical Fertilization of the Sea Urchin (eggs)" (1900), provides a note announcing Wilson's results, which he discussed in more detail later.
35. Wilson, "Experimental Studies in Cytology. I" (1901), pp. 575–576.
36. Wilson, "Experimental Studies in Cytology. II and III" (1901), pp. 373–395.
37. For further discussion, see Maienschein, "Heredity/Development in the United States" (1987).
38. For example, Morgan, "Some Problems of Regeneration" (1899); and Morgan, "Regeneration: Old and New Interpretations" (1899) and (1900).

The latter lecture is reprinted in Maienschein, ed., *Defining Biology* (1987), pp. 295–320.

39. Wilson, "Notes on the Reversal of Asymmetry in the Regeneration of the Chelae in Alpheus Heterochelis" (1903), p. 210.

40. Wilson, "Notes on Merogony and Regeneration in Renilla" (1903), p. 225.

41. Wilson, "Experiments on Cleavage and Localization in the Nemertine-egg" (1903), p. 448.

42. Wilson, *The Cell in Development and Inheritance* (1900), pp. 423–425, cited in Wilson, "Experiments on Cleavage and Localization in the Nemertine-egg" (1903), pp. 457–458.

43. Wilson, "Experimental Studies on Germinal Localization" (1904), pp. 17, 58, 64, 265–266. He also discussed the subject in a more popular article: Wilson, "Mosaic Development in the Annelid Egg" (1904).

44. Wilson, "Mendel's Principles of Heredity and the Maturation of the Germ Cells" (1902).

45. Montgomery, "A Study of the Chromosomes of the Germ Cells of Metazoa" (1902); and other papers of 1901 and 1902 especially.

46. Montgomery, "A Study of the Chromosomes of the Germ Cells of Metazoa" (1902); Sutton, "The Chromosome in Heredity" (1902–3).

47. For example, Sutton, "The Chromosome in Heredity" (1902–3).

48. Wilson, discussion article, "Mr. Cook on Evolution, Cytology, and Mendel's Laws" (1903). Cook had expressed disagreement with Wilson's sympathetic look at chromosomes and Mendelism in the *Science* article of the previous year.

49. Wilson, "The Problem of Development" (1905). For further discussion of this move from a different perspective, see Gilbert, "The Embryological Origins of the Gene Theory" (1978).

50. Wilson, "The Problem of Development" (1905), p. 292.

51. For a fuller discussion of this trend, see Maienschein, "What Determines Sex?" (1984).

52. McClung, "The Accessory Chromosome—Sex Determinant?" (1902). Actually McClung was confused about which sex has the accessory, which rather called his conclusions into doubt. As a result, historians who worry about assigning priorities for discoveries often worry about whether to give McClung credit. For present purposes, that does not really matter. The point is that Wilson read McClung's work, and that work reinforced Wilson's own emerging interest in chromosomes.

53. Wilson, "The Chromosomes in Relation to the Determination of Sex in Insects" (1905), p. 500.

54. Stevens, *Studies in Spermatogenesis* (1905), pp. 53, 55, 56. For further discussion of this intriguing scientist and her contributions, see Brush, "Stevens and the Discovery of Sex Determination by Chromosomes" (1978); Ogilvie and Choquette, "Stevens (1861–1912)" (1981); Maienschein, "Stevens" (forthcoming). Much of the discussion of Stevens's work by other than these careful scholars has centered on the question of whether Wilson "stole" credit for her work. Once again, such historians focus on questions

of priority in making a particular contribution later judged important rather than on trying to understand the logic and dynamics of scientific change.

55. Wilson, "Some Recent Studies on Heredity" (1906–7), p. 210.

56. Wilson, "The Biological Significance of Sex" (1907), pp. 378, 379.

57. Wilson, "Some Aspects of Progress in Modern Zoology" (1915), pp. 3–4.

58. Ibid., p. 6.

59. Harrison, "Response on Behalf of the Medallist" (1935), p. 566.

CHAPTER 7 Edwin Grant Conklin

1. Edwin Grant Conklin recorded his concerns in a draft of a letter to President Bashford of Ohio Wesleyan, February or March 1891, Conklin Papers.

2. "National Traits in Science" (1883).

3. Conklin took meticulous reading and lecture notes for all his courses at Hopkins and probably relied heavily on those as he taught physics and other courses along with biology at Ohio Wesleyan and after.

4. Bard, "Conklin, 1863–1952" (1964), p. 55.

5. Whitman, "Ookinesis" (1887), and work following up on that paper.

6. Conklin, "The Fertilization of the Ovum" (1894), pp. 15, 16.

7. Ibid., p. 31.

8. Ibid., p. 34.

9. Maienschein, ed., *Defining Biology* (1987), pp. 3–56.

10. Conklin to David Starr Jordan, 23 February 1897, Conklin Papers.

11. Conklin, "Conklin" (1953), p. 56.

12. Conklin, "Discussion of the Factors of Organic Evolution" (1896), p. 78.

13. Ibid., p. 87.

14. "Old-Fashioned" (1939), p. 40.

15. Conklin, "Weismann's Germinal Selection" (1896), pp. 854, 856. Conklin later wrote a biographical sketch of Weismann (1915), which also considers Weismann's ideas.

16. Conklin, "Cleavage and Differentiation" (1898), pp. 17–18.

17. Ibid., p. 43.

18. Conklin, review of Wilson, *The Cell in Development and Inheritance* (1897).

19. Bonner, "What Is Money For?" (1984), p. 83. Also see Richards, "Conklin" (1935), p. 192.

20. Various letters in the Conklin Papers came from editors urging Conklin to finish some paper or other, especially when no absolute deadline had been imposed.

21. Conklin, "The Embryology of *Crepidula*" (1897), p. 4.

22. Ibid., p. 59, 79, 80, 108.

23. Ibid., p. 204.

24. Galileo, "Letter to Madame Christina" (1957, original 1615), p. 186.

25. Conklin, "Evolution and Revelation" (1897), pp. 6, 8. For further discussion of related issues, see Atkinson, "Conklin on Evolution" (1985).

26. Charles Darwin, *On the Origin of Species* (1859). This is the last sentence of the first edition; the bracketed phrase was added later.

27. Harvey, "Conklin" (1958), p. 67. Also see Conklin, "Conklin" (1953), p. 59.

28. Conklin, "The Evidence and Factors of Organic Evolution" (1898).

29. Conklin, "Advances in Methods of Teaching Zoology" (1899), p. 82.

30. Conklin, "The Marine Biological Laboratory" (1900), pp. 334, 341.

31. Conklin, "Protoplasmic Movement as a Factor of Differentiation" (1899), p. 69.

32. Ibid., p. 90.

33. Conklin, "The Fertilization of the Egg and Early Differentiation of the Embryo" (1900), p. 20.

34. Conklin, "Phenomena and Mechanism of Inheritance" (1899), p. 536. Conklin must have used this example repeatedly for classes as well, since many of his students quote this as evidence of his solid scientific commitments and his failure to be taken in by fads.

35. Conklin, "Karyokinesis and Cytokinesis" (1902), p. 6.

36. Ibid., pp. 13, 28–30.

37. Ibid., pp. 74, 107.

38. Ibid., p. 115.

39. Whitman and William Morton Wheeler had especially argued for this in-between position in their lectures to the MBL during the 1890s.

40. Conklin, "The Organization and Cell-Lineage of the Ascidian Egg" (1905), p. 73.

41. Thomas Hunt Morgan to Conklin, 8 June 1905, Conklin Papers. Conklin's response acknowledged that Morgan had, in fact, given the initial suggestions to use these ascidians, and he expressed regret that he had failed to give due credit.

42. Conklin, "Organization and Cell-Lineage of Ascidians" (1905), p. 87.

43. Ibid., p. 101.

44. Conklin, "Mosaic Development of Ascidian Eggs" (1905), p. 209.

45. Morgan to Conklin, 8 June 1905, Conklin Papers.

46. Conklin to Morgan, 10 June 1905, Conklin Papers.

47. Conklin, "Organ-Forming Substances in the Eggs of Ascidians" (1905).

48. Conklin, "Does Half of an Ascidian Egg give rise to a Whole Larva?" (1906), p. 751. Here he concluded that the different cells and substances were so specialized from the beginning "that they can give rise to no other types of structures than those which they form under normal conditions."

49. Conklin, "Organ-Forming Substances in the Eggs of Ascidians" (1905), p. 224. Conklin followed up on this suggestion in more detail in "The Mutation Theory from the Standpoint of Cytology" (1905).

50. Conklin, "The Embryology of *Fulgur*" (1907), p. 353.

51. Conklin, review of Morgan, *Regeneration* (1902), p. 621.
52. Conklin, "The Cause of Inverse Symmetry" (1903), p. 587.
53. Conklin, "Experiments on the Origin of Cleavage Centrosomes" (1904).
54. Conklin, "Does Half of an Ascidian Egg give rise to a Whole Larva?" (1906), pp. 736, 751.
55. Conklin, review of Morgan, *Experimental Zoology* (1908), p. 140.
56. Conklin, "The Application of Experiment to the Study of the Organization and Early Differentiation of the Egg" (1909).
57. Richards, "Conklin" (1935), p. 203.
58. Wenrich, "Biology at the University of Pennsylvania" (1951), pp. 157–159.
59. Ross Granville Harrison to Conklin, 16 November 1907, Conklin Papers.
60. Richards, "Conklin" (1935), pp. 202–203.
61. Conklin, "The Mechanism of Heredity" (1908), pp. 98, 92.
62. Richards, "Conklin" (1935), p. 203, partial list.
63. Many of these souvenirs remain in the Conklin Papers.

CHAPTER 8 Thomas Hunt Morgan

1. Morgan, *The Development of the Frog's Egg* (1897), pp. v–vi explains the genesis of the project.
2. Allen, *Morgan* (1978), pp. 53–54. The careful reader will note many ways in which this interpretation of Morgan's work departs from that of my friend Garland Allen. In part this results from the emphasis here on the period up to 1915. I have concentrated on the many research papers during that time and only secondarily on the textbooks, while Allen's full-scale biography draws more heavily on the books. In addition, my approach has been more empirical: I have tried to detail the progress of science by looking carefully at the writings of four friends from Hopkins. Rather than pointing to differences of interpretation at every step, I leave it to the reader to note them and to draw appropriate conclusions from the evidence at hand. Manier, "The Experimental Method in Biology" (1969), more closely agrees with my interpretation, though he focuses more on Morgan's textbooks than on the gradual emergence of his views as revealed by his sequence of many papers. I have not observed Morgan's emphasis on the convergence of several lines of research, which, Manier discusses, p. 202.
3. Morgan went to Zurich at the end of his stay in Naples, in the summer of 1895, with Hans Driesch and embryologist Curt Herbst. He returned during the summer of 1896.
4. Morgan, *The Development of the Frog's Egg* (1897), p. v.
5. This was a recurrent theme in Morgan's work, as it was in Wilson's and Conklin's.
6. Sturtevant, "Morgan" (1959), p. 289.
7. Morgan, "The Orientation of the Frog's Egg" (1894), p. 383.
8. Ibid., p. 387.

9. Ibid.
10. Morgan, "Half-Embryos and Whole-Embryos" (1895), p. 627.
11. Morgan, *The Development of the Frog's Egg* (1897), p. 136.
12. Ibid., p. 135.
13. Morgan, "The Formation of the Embryo of the Frog" (1894); Morgan, "The Formation of the Fish Embryo" (1895).
14. Morgan, "Impressions of the Naples Zoölogical Station" (1896), pp. 17, 18. Morgan wrote this sketch of the laboratory in response to a request from Henry Fairfield Osborn and as part of the campaign to gain American support for the institution. For a more detailed account of the political reasons for Morgan's involvement, see the letters between Morgan and Anton Dohrn, 1895, 1896, Stazione Zoologica Archives. Also see Allen, *Morgan* (1978), pp. 60–63. Actually, Morgan was isolated from the latest techniques and equipment, but at the MBL and Bryn Mawr, he worked amidst a first-rate group of people and with excellent students. In fact, he collaborated with several of those students and reportedly found the experience rewarding.
15. Correspondence between Morgan and Dohrn documents the trouble that Agassiz was causing for the *stazione* and Morgan's part in trying to salvage the situation and restore the Americans' reputation and position there. Allen, *Morgan* (1978), p. 58, gives the dates when Morgan most likely returned to Europe which accord with my findings.
16. See the correspondence between Morgan and Hans Driesch, American Philosophical Society, microfilmed letters that Garland Allen collected and donated.
17. Sturtevant reports in "Reminiscences of T. H. Morgan," p. 5, that Morgan also pursued other related research, some of which he did not publish until later. He was interested, for example, in why the ascidian *Ciona* is generally but not always self-sterile. "Morgan had a nice hypothesis: maybe the acidity of the water is responsible. Let's see what pH changes will do. But being Morgan, he didn't set up measured amounts or concentrations. What he did was to take a dish in which eggs and sperm were present and squeeze a lemon over it. And it worked." See also Maienschein, "Morgan as Invertebrate Embryologist" (1989).
18. Morgan, "Studies of the 'Partial' Larvae of Sphaerechinus" (1895), pp. 122–124. See also Morgan's other papers in the same volume, two of which were submitted with the above in February 1895, with five other similar studies following in the same journal within the year.
19. Driesch and Morgan, "Zur Analysis der ersten Enwickelungsstadien des Ctenophoreneies. I. and II." (1895). Gilbert, "In Friendly Disagreement" (1987), p. 799, also discusses this work and calls it Morgan's "most crucial experiments."
20. Morgan, "The Number of Cells in Larvae from Isolated Blastomeres of Amphioxus" (1896), p. 292.
21. Ibid.
22. Weismann, *Das Keimplasm (1893). Chapter 2 considers regeneration.*
23. Morgan, "Regeneration in Allolobophora foetida" (1897), p. 582.

24. Morgan, "Regeneration in the Hydromedusa, Gonionemus Vertens" (1899), p. 951.

25. Morgan, "Regeneration in Allolobophora foetida" (1897), p. 583.

26. Morgan, "Experimental Studies of the Regeneration of Planaria Maculata" (1898), p. 389. For further detailed discussion of Morgan's regeneration, see Maienschein, "Morgan's Regeneration, Epigenesis, and (W)holism" (forthcoming).

27. Allen, *Morgan* (1978), pp. 86–87, cites a letter from Morgan to Driesch which documents the enthusiasm he felt for the older work. He also wrote that he felt "one thinks a good deal while one reads."

28. Morgan, "Some Problems of Regeneration" (1899), p. 198.

29. Ibid., pp. 205–206.

30. Morgan, "Regeneration: Old and New Interpretations" (1900), p. 195.

31. Ibid., p. 207.

32. Morgan, "Regeneration in Teleosts" (1900), p. 131. Morgan actually spelled *morphallaxis* in various ways at different times. I use the version from Morgan, *Regeneration* (1901), pp. 270–271.

33. Morgan, *Regeneration* (1901), p. 275; Morgan, "The Physiology of Regeneration" (1906), p. 500.

34. Morgan, *Regeneration* (1901), p. 269.

35. Morgan to Driesch, 14 March, 1904, in Morgan-Driesch correspondence, cited in Allen, *Morgan* (1978), p. 69.

36. On Morgan's wife, see Keenan, "Lilian Vaughan Morgan" (1983).

37. Boring, "Morgan" (1946).

38. Morgan, "The Dynamic Factor in Regeneration" (1909).

39. Wilson, "Welcoming Address" (1932), cited in Gilbert, "In Friendly Disagreement" (1907), p. 805.

40. Morgan, "Developmental Mechanics" (1898), p. 158. Also see Morgan, *Experimental Zoology* (1907), p. 4.

41. Morgan, *Experimental Zoology* (1907), pp. 6–7.

42. Ibid., p. 9.

43. Morgan, "Developmental Mechanics" (1898).

44. Morgan, "The Relation between Normal and Abnormal Development of the Embryo of the Frog" (1902), reference to heterotropism on p. 306.

45. Morgan, "Cytological Studies of Centrifuged Eggs" (1910).

46. Morgan, "The Relation between Normal and Abnormal Development of the Embryo of the Frog" (1902), p. 239.

47. Morgan, *Evolution and Adaptation* (1903), pp. ix–x.

48. Morgan, "Darwinism in the Light of Modern Criticism" (1903), p. 476. A few years later, Morgan gave a much more positive look at Darwin's contribution to biology, though he still had his reservations about the efficacy of natural selection as the primary mechanism: Morgan, "For Darwin" (1909).

49. Morgan, "Darwinism in the Light of Modern Criticism" (1903), p. 477.

50. Morgan, "Recent Theories in Regard to the Determination of Sex"

(1903), p. 116. See Maienschein, "What Determines Sex?" (1984), pp. 467–468, 476–480.

51. Morgan, "Self-fertilization induced by Artificial Means" (1904).

52. Morgan, "An Alternative Interpretation of the Origin of Gynandromorphous Insects" (1905), p. 634.

53. Ibid.

54. Morgan, "Sex-determining Factors in Animals" (1907), p. 384.

55. Morgan, "What are 'Factors' in Mendelian Explanations?" (1909), p. 366.

56. Morgan, *Evolution and Adaptation* (1903), pp. 278–287; and other papers thereafter.

57. Morgan, "Chromosomes and Heredity" (1910).

58. Morgan, "What are 'Factors' in Mendelian Explanations?" (1909), p. 377. This represents exactly the same view expressed in *Experimental Zoology* (1907), p. 80.

59. Morgan, "Chromosomes and Heredity" (1910), pp. 454, 468–469.

60. Ibid., p. 477.

61. Ibid., pp. 495–496. For an important alternative interpretation of Morgan's move to genetics, see Roll-Hanson, "Drosophila Genetics: A Reductionist Research Program" (1978). Also see Lederman, "Research Note: Genes or Chromosomes: The Conversion of Thomas Hunt Morgan" (1989). In a related discussion, Vicedo, "Morgan" (1989), suggests that we must see Morgan not as a straightforward empiricist, as some have claimed, but as a realist with respect to genes.

62. Morgan, *Heredity and Sex* (1913), p. iv.

63. Harrison, "Response on Behalf of the Medallist" (1935), p. 566.

CHAPTER 9 Ross Granville Harrison

1. Bryn Mawr course catalog for 1894–95, pp. 104–106.

2. Harrison's son Richard Edes Harrison told this story in personal correspondence with the author, 20 February 1990.

3. At Naples, he occupied the Harvard table from February through March or April and had the use of microtomes and other equipment made available for visiting American researchers, as recorded in letters to Anton Dohrn, Stazione Zoologica Archives. I thank archivist Christiane Groeben for this material. Richard Edes Harrison, personal correspondence, 20 February 1990, says that his parents spent the winter in Naples, with plenty of time in the "Aquario."

4. Sabin, *Franklin Paine Mall* (1934), p. 134.

5. Chesney, *The Johns Hopkins Hospital and the Johns Hopkins University School of Medicine* (1943), p. 225. Sabin, *Franklin Paine Mall* (1934), pp. 138–140, also discusses Mall's teaching methods.

6. Chesney, *The Johns Hopkins Hospital and the Johns Hopkins University School of Medicine* (1943), p. 225.

7. Nicholas, "Harrison" (1960), p. 407.

8. Born, "Über Verwachsungsversuche mit Amphibienlarven" (1896–7), set out the method and some experimental results. Harrison, "The Growth and Regeneration of the Tail of the Frog Larva" (1898), discusses Harrison's first results. For discussion of Harrison's embryological work more generally, see Oppenheimer, "Harrison's Contributions to Experimental Embryology" (1967), pp. 92–116.

9. Harrison, "The Growth and Regeneration of the Tail of the Frog Larva" (1898), pp. 430–431.

10. Gustav Born to Ross Granville Harrison, 23 November 1898, Harrison Papers. Another letter of 5 May 1899 followed up on suggestions about other applications and improvements of the technique and even asked Harrison to share some of his organisms to try crossbreeding the next spring.

11. A number of his former students and associates have told me of his dislike for writing, which caused Harrison to turn down many invitations and to publish less than he easily could have done.

12. Harrison, "The Growth and Regeneration of the Tail of the Frog Larva" (1898), p. 437.

13. Sabin, *Franklin Paine Mall* (1934), p. 93, explains that Mall had become interested in the nervous system and its development while in Germany. In 1892, he entered discussion of the neuron and alternative theories of nervous structure and function, in which he was clearly stimulated by Wilhelm His but not persuaded to accept precisely His's point of view. With the increasing debate about the nature and development of nerve fibers in the 1890s, it is likely that the Hopkins group surrounding Mall often discussed the latest results and theories.

14. Maienschein, "Morgan's Regeneration, Epigenesis, and (W)holism" (forthcoming).

15. Harrison, "The Growth and Regeneration of the Tail of the Frog Larva" (1898), pp. 445–448.

16. Ibid., pp. 449, 464.

17. Loeb, *Untersuchungen zur physiologischen Morphologie der Thiere* (1891). Also presented in outline at the MBL: Loeb, "On Some Facts and Principles of Physiological Morphology" (1894).

18. Harrison "The Growth and Regeneration of the Tail of the Frog Larva" (1898), p. 468.

19. Nicholas, "Harrison" (1961), p. 138, quoting Richard Edes Harrison.

20. Harrison, "Über die Histogenese des peripheren Nervensystems bei Salmo salar" (1901), p. 355, where he wrote: "Es ist der Zweck dieses Studiums, die Umbildung der einzelnen Zellarten Schritt für Schritt und möglichst zusammenhängend zu beschreiben und ausserdem Vergleiche mit den Befunden bei anderen Wirbelthieren anzustellen. Um diesen Plan auszuführen, wird es aber unvermeidlich sein, auf manche Einzelheiten einzugehen, die schon von früheren Autoren bei Embryonen anderer Arten beschreiben worder sind."

21. His, "Die Neuroblasten u. deren Enstehung um embryonalen Mark" (1889), p. 291–300 for summary. For further detailed discussion of the

debates among the following theories, see Billings, "Concepts of Nerve Fiber Development, 1839–1930" (1971); Maienschein, "Harrison's Crucial Experiment as a Foundation for Modern American Experimental Embryology" (1978).

22. Golgi, "Recherches sur l'histologie des centres nerveux" (1883) and (1884).

23. Ramon y Cajal, *Textura del Sistema Nervioso del Hombre y de los Vertebrados* (1899).

24. Schultze, "Über die multizelluläre Enstehung der peripheren sensiblen Nervenfaser" (1905). Theodore Schwann had been the first to suggest this interpretation.

25. Ramon y Cajal, *Recollections of My Life* (1937), p. 537.

26. Harrison, "Experimentelle Untersuchungen über die Entwicklung der Sinnesorgane" (1903), p. 36.

27. Ibid., pp. 142–143.

28. Harrison, "An Experimental Study of the Relation of the Nervous System" (1904), p. 200.

29. Ibid., pp. 216–218.

30. "New Yale" (1934).

31. For discussion of the state of biology around 1900, see Rainger, et al., eds., *The American Development of Biology* (1988).

32. See Dornfeld, "The Allis Lake Laboratory" (1956), pp. 124, 133, for discussion of publications.

33. "Retrospect" (1945). Edmund Beecher Wilson, in a letter to Harrison, 24 March 1933, Harrison Papers, provides further details.

34. "Facsimile of First Announcement" (1945), pp. vii–ix.

35. Wilson to Harrison, 7 December 1903, reprinted in "Retrospect" (1945), pp. xv–xvi.

36. Harrison to W. H. Howell, dean of the medical faculty, 15 February 1904, Harrison Papers.

37. Harrison, "Nachtrag zu der Diskussion zu den Vorträgen von Schultze und v. Koelliker" (1904). As Wilens put it in the introduction to Harrison, *Organization and Development of the Embryo* (1969), p. xiii, the photographs of the meeting showed that it was a major event. The group included anatomists, histologists, and pathologists with their divergent views and different methods.

38. Harrison, "Neue Versuche und Beobachtungen über die Entwicklung der peripheren Nerven der Wirbeltiere" (1904).

39. Ibid., p. 7.

40. Held, *Die Entwicklung des Nervengewebes bei den Wirbeltieren* (1909), puts forth his views about proper methodologies as well as criticisms of the neuron theory.

41. See Maienschein, "Experimental Biology in Transition: Harrison's Embryology 1895–1910" (1983) for further discussion.

42. Harrison, "Further Experiments on the Development of Peripheral Nerves" (1906), p. 121.

43. Harrison, "Further Experiments, etc.," n.d., Harrison Papers.

44. Harrison, "Further Experiments on the Development of Peripheral Nerves" (1906), p. 129.

45. Ibid., pp. 130, 131.

46. For further discussion, see Oppenheimer, "Historical Relationships between Tissue Culture and Transplantation Experiments" (1971).

47. Harrison, "Experiments in Transplanting Limbs" (1907), p. 241. Harrison's notes on Braus's 1905 article are in the Harrison Papers and reveal the depth of his respect but also his frustration with Braus's results. On Braus, see Eggeling, "Braus" (1926); Braus, "Experimentelle Beiträge zur Frage nach der Entwickelung peripheren Nerven" (1905). Also see Nyhart, "Morphology and the German University" (1986).

48. Harrison, "Experiments in Transplanting Limbs" (1907), p. 241.

49. Ibid., pp. 276–278.

50. Harrison discusses this effort in later notes attached to his early draft for "Further Experiments, etc.," n.d., Harrison Papers. He reported there that "for several years I have been carrying on experiments with this end in view."

51. Harrison, "Observations on the Living Developing Nerve Fiber" (1907).

52. Nicholas, "Harrison" (1960), pp. 147–148.

53. Witkowski, "Harrison and the Experimental Analysis of Nerve Growth" (1985).

54. Harrison, "Observations on the Living Developing Nerve Fiber" (1907), p. 118.

55. Chittenden, *Sheffield Scientific School. 1846–1922* (1928), p. 434; Chittenden and Furniss, *The Graduate School of Yale* (1965), pp. 1–39.

56. Parker, *The World Expands* (1946), pp. 145–146.

57. Harrison to Edwin Grant Conklin, 16 November 1907, Conklin Papers.

58. F. P. Mall to Harrison, 8 August 1907, Harrison Papers.

59. Harrison, "The Outgrowth of the Nerve Fiber" (1910), p. 799.

60. Harrison, "Embryonic Transplantation and Development of the Nervous System" (1908), p. 388.

61. Ibid., p. 397.

62. Ibid., p. 400.

63. Ibid., p. 409.

64. Harrison, "The Outgrowth of the Nerve Fiber" (1910), p. 790.

65. Ibid., pp. 833–834, 838.

66. Harrison to Conklin, 20 October 1909, Conklin Papers.

67. Harrison to President Hadley, 6 January 1910, Hadley Papers.

68. Harrison's only book was edited by his longtime research assistant Sally Wilens: *Organization and Development of the Embryo* (1969).

69. Twitty, *Of Scientists and Salamanders* (1966), especially chapter 1.

70. J. P. Trinkaus, personal interview by the author, MBL, 1987. Actually, "Trink" put it in somewhat more colorful language.

71. Twitty, *Of Scientists and Salamanders* (1966), pp. 16–17.

72. Trinkaus interview, 1987.
73. Harrison, "Experimental Biology and Medicine" (1912), p. 63.

Conclusion

1. Nyhart, "Morphology and the German University" (1986).
2. Benson and Quinn, "The American Society of Zoologists" (1989).
3. For discussion on the term *biology,* see Joseph A. Caron, "Biology in the Life Sciences" (1988).
4. Hull, *Science as a Process* (1988); Richards, *Darwin and the Emergence of Evolutionary Theories of Mind and Behavior* (1987).
5. Laudan, *Progress and Its Problems* (1977).

Sources Cited

Archival Materials

I used the following collections, from which all material used is cited with permission:

Alexander Agassiz Papers. Museum of Comparative Zoology Archives, Harvard University.

Louis Agassiz Papers. Museum of Comparative Zoology Archives, Harvard University.

Edwin Grant Conklin Papers. Manuscript Collection, Princeton University Library.

Anton Dohrn Papers. Stazione Zoologica Archives, Naples.

Daniel Coit Gilman Collection (as President). Special Collections, Milton S. Eisenhower Library, Johns Hopkins University.

Arthur Twining Hadley Papers (Presidential Papers). Manuscripts and Archives, Yale University Library.

Ross Granville Harrison Papers. Manuscripts and Archives, Yale University Library.

Jacques Loeb Papers. Manuscript Division, Library of Congress.

Marine Biological Laboratory Archives.

Marine Biological Laboratory Director's Archives. Director's Office, MBL.

Morgan-Driesch Correspondence. Given to me by Garland Allen and also now available at the American Philosophical Society, Philadelphia.

H. F. Osborn Collection. Department of Vertebrate Paleontology, American Museum of Natural History.

Student Records. Ferdinand Hamburger, Jr. Archives and Milton S. Eisenhower Library, Johns Hopkins University.

A. H. Sturtevant Papers. Institute Archives, Robert A. Millikan Memorial Library, California Institute of Technology.

M. Carey Thomas Papers. Bryn Mawr College Archives.

Charles Otis Whitman Collection. Department of Special Collections, Joseph Regenstein Library, University of Chicago.

Other materials from special collections:

Bryn Mawr College Catalogs.
Johns Hopkins University. *Annual Reports of the President.*
Marine Biological Laboratory. *Annual Reports* and *Trustees' Minutes.*
Linda Timmons. Personal letters and photographs of E. B. Wilson.

Other Sources

Abercrombie, M. "Ross Granville Harrison." *Royal Society of London, Biographical Memoirs* 7 (1961): 111–126.

Agassiz, Elizabeth Cary, ed. *Louis Agassiz: His Life and Correspondence.* Boston: Houghton, Mifflin, and Co., 1885.

Agassiz, G. R. *Letters and Recollections of Alexander Agassiz.* Boston: Houghton Mifflin Co., 1913.

Agassiz, Louis. *The Structure of Animal Life.* New York: Scribner, Armstrong, and Co., 1865.

Allen, Garland E. "Edwin Grant Conklin." *Dictionary of Scientific Biography,* 3:389–391. New York: Charles Scribner's Sons, 1970–80.

Allen, Garland E. "Thomas Hunt Morgan." *Dictionary of Scientific Biography,* 9:515–526. New York: Charles Scribner's Sons, 1970–80.

Allen, Garland E. "Edmund Beecher Wilson." *Dictionary of Scientific Biography,* 14:423–436. New York: Charles Scribner's Sons, 1970–80.

Allen, Garland E. *Life Science in the Twentieth Century.* Cambridge: Cambridge University Press, 1978.

Allen, Garland E. *Thomas Hunt Morgan.* Princeton: Princeton University Press, 1978.

Allen, Garland E. "Naturalists and Experimentalists: The Genotype and the Phenotype." *Studies in the History of Biology* 3 (1979): 179–209.

Andrews, Ethan Allen. "Biographical Sketch." *Johns Hopkins University Circulars,* 1908, pp. 22–37.

Atkinson, J. W. "E. G. Conklin on Evolution: The Popular Writings of an Embryologist." *Journal of the History of Biology* 18 (1985): 31–50.

Atz, James W. "*Fundulus heterochitus* in the Laboratory: A History." *American Zoologist* 26 (1986): 111–120.

Baer, Karl Ernst von. *Über Entwickelungsgeschichte der Thiere.* Königsberg: Gebruder Borntrager, 1828.

Baker, John R. "The Cell-Theory: A Restatement, History, and Critique." *Quarterly Review of Microscopical Science* 89 (1948): 103–125; 90 (1949): 87–108; 93 (1952): 157–190; 94 (1953): 407–440; 96 (1955): 449–481.

Baltzer, Fritz. *Life and Work of a Great Biologist 1862–1915.* Trans. Dorothea

Rudnick. Berkeley and Los Angeles: University of California Press, 1967.

Barber, Lynn. *The Heyday of Natural History.* Garden City, N.Y.: Doubleday and Co., 1980.

Bard, Philip. "Edwin Grant Conklin, 1863–1952." *Proceedings of the American Philosophical Society* 108 (1964): 55–56.

Baxter, Alice Levine. "Edmund Beecher Wilson and the Problem of Development: From the Germ Layer Theory to the Chromosome Theory of Inheritance." Ph.D. dissertation, Yale University, 1974.

Baxter, Alice Levine. "Edmund B. Wilson as a Preformationist: Some Reasons for His Acceptance of the Chromosome Theory." *Journal of the History of Biology* 9 (1976): 29–57.

Baxter, Alice Levine. "E. B. Wilson's 'Destruction' of the Germ-Layer Theory." *Isis* 68 (1977): 363–374.

Baxter, Alice, and John Farley. "Mendel and Meiosis." *Journal of the History of Biology* 12 (1979): 137–173.

Beneden, Edouard van. "La segmentation chez les ascidians et ses rapports avec l'organisation de la larve." *Archives de Biologie* 5 (1884): 111–126.

Benson, Keith R. "William Keith Brooks (1848–1908): A Case Study in Morphology and the Development of American Biology." Ph.D. dissertation, Oregon State University, 1979.

Benson, Keith R. "H. Newell Martin, W. K. Brooks, and the Reformation of American Biology." *American Zoologist* 27 (1987): 759–771.

Benson, Keith R. "From Museum Research to Laboratory Research: The Transformation of Natural History into Academic Biology." In Ronald Rainger, Keith Benson, and Jane Maienschein, eds., *The American Development of Biology,* pp. 49–83. Philadelphia: University of Pennsylvania Press, 1988.

Benson, Keith R., and Brother C. Edward Quinn. "The American Society of Zoologists, 1889–1989. A Century of Interpreting the Biological Sciences." Pamphlet distributed at the centennial meeting of American Society of Zoologists, Boston, 1989.

Berger, James D. "Otto Bütschli." *Dictionary of Scientific Biography,* 2:625–628. New York: Charles Scribner's Sons, 1970–80.

Billings, Susan M. "Concepts of Nerve Fiber Development, 1839–1930." *Journal of the History of Biology* 4 (1971): 275–305.

Bond, Allen Kerr. *When the Hopkins Came to Baltimore.* Baltimore: Pegasus, 1927.

Bonner, J., with notes by Whitfield J. Bell, Jr. "What Is Money For?: An Interview with Edwin Grant Conklin, 1952." *Proceedings of the American Philosophical Society* 128 (1984): 79–84.

Boring, Alice. "Thomas Hunt Morgan, Bryn Mawr." *Alumnae Bulletin,* 1946, p. 18.

Born, Gustav. "Über die Entstehung der Geschlechtsunterschiede nach experimentellen Untersuchungen." *Jahresberichte der Schlesischen Gesellschaft für vaterländische Kultur. Medezinische Sektion* 59 (1881): 2–22.

Born, Gustav. "Über den Einfluss der Schwere auf das Froschei." *Bres-*

lauer ärztlichen Zeitschrift, 1884, 1–14. Repaginated offprint.

Born, Gustav. "Biologische Untersuchungen I. Über den Einfluss der Schwere auf das Froschei." *Archiv für mikroskopische Anatomie* 24 (1885): 475–545.

Born, Gustav. "Biologische Untersuchungen II. Weitere Beiträge zur Bastardirung zwischen den einheimischen Anuren." *Archiv für mikroskopische Anatomie* 27 (1886): 192–271.

Born, Gustav. "Über Verwachsungsversuche mit Amphibienlarven." *Archiv für Entwickelungsmechanik der Organismen* 4 (1896–97): 349–466, 517–623.

Boveri, Theodor. "Zellenstudien I: Die Bildung der Richtungskörper bei *Ascaris megalocephala* und *Ascaris lumbricoides*." *Jenaische Zeitschrift für Naturwissenschaft* 21 (1887): 423–515.

Boveri, Theodor. "Zellenstudien II: Die Befruchtung und Teilung des Eies von *Ascaris megalocephala*." *Jenaische Zeitschrift für Naturwissenschaft* 22 (1888): 685–882.

Boveri, Theodor. "Ein geschlechtlich erzeugter Organismus ohne mütterliche Eigenschaften." *Sitzungsberichte der Gesellschaft für Morphologie und Physiologie zu München* 5 (1889): 73–80.

Boveri, Theodor. "Über die Niere des Amphioxus." *Sitzungsberichte der Gesellshaft für Morphologie und Physiologie zu München* 6 (1890): 65–77.

Boveri, Theodor. "Zellenstudien III: Über das Verhalten der chromatischen Kernsubstanz bei der Bildung der Richtungskörper und bei der Befruchtung." *Jenaische Zeitschrift für Naturwissenschaft* 24 (1890): 314–401.

Boveri, Theodor. "Ein geschlechtlich erzeugter Organismus." *Sitzungsberichte der Gesellschaft für Morphologie und Physiologie zu München* 5 (1889). Trans. Thomas Hunt Morgan as "An Organism Produced Sexually Without Characteristics of the Mother." *American Naturalist* 27 (1893): 222–232.

Bracegirdle, Brian. *A History of Microtechnique.* Ithaca, N. Y.: Cornell University Press, 1978.

Braus, Hermann. "Experimentelle Beiträge zur Frage nach der Entwickelung peripheren Nerven." *Anatomischer Anzeiger* 26 (1905): 433–479.

Breathnach, C. S. "Henry Newell Martin (1848–1892): A Pioneer Physiologist." *Medical History* 13 (1969): 271–279.

Broman, Thomas. "Transformation of Academic Medicine in Germany, 1780–1820," pp. 237–246. Ph.D. dissertation, Princeton University, 1987.

Brooks, William Keith. "On the Development of Salpa." *Bulletin: Museum of Comparative Zoology* 3 (1876): 291–348.

Brooks, William Keith. "Observations upon the Early Stages in the Development of the Fresh-Water Pulmonates." Johns Hopkins University, *Studies from the Biological Laboratory* 1 (1879): 73–104.

Brooks, William Keith. *Handbook of Invertebrate Zoology.* Boston: S. E. Cassino, 1882.

Brooks, William Keith. "Speculative Philosophy," pt. 1. *Popular Science Monthly* 22 (1882): 195–204.

Brooks, William Keith. *The Law of Heredity.* Baltimore: John Murray and Co., 1883.

Brooks, William Keith. "Speculative Philosophy," pt. 2. *Popular Science Monthly* 22 (1883): 364–380.

Brooks, William Keith. Articles in *Johns Hopkins University Circulars,* especially "Chesapeake Zoological Laboratory" 3 (1884): 91–94.

Brooks, William Keith. "The Zoölogical Work of the Johns Hopkins University, 1878–1886," *Johns Hopkins University Circulars* 6 (1886): 37–39.

Brooks, William Keith. *The Foundations of Zoology.* New York: Columbia University Press, 1915 (1st ed. 1898).

Brush, Stephen G. "Nettie M. Stevens and the Discovery of Sex Determination by Chromosomes." *Isis* 69 (1978): 163–172.

Butler, Elmer Grimshaw. "Edwin Grant Conklin." *American Philosophical Society Yearbook,* 1952, pp. 5–12.

Bütschli, Otto. "Studien über die ersten Entwickelungsvorgänge der Eizelle, die Zelltheilung u. die Konjugation der Infusorien." *Abhandlungen der Senkenbergische naturforschende Gesellschaft* 10 (1876): 1–250.

Caron, Joseph A. "'Biology' in the Life Sciences: A Historiographical Contribution." *History of Science* 26 (1988): 223–268.

Chamberlin, Thomas C. "The Method of Multiple Working Hypotheses." *Science* 15 (1890): 92–96; reprinted, *Science* 148 (1965): 754–759.

Chesney, Alan M. *The Johns Hopkins Hospital and the Johns Hopkins University School of Medicine.* Baltimore: The Johns Hopkins Press, 1943.

Chittenden, Russell H. *History of the Sheffield Scientific School of Yale University 1846–1922.* New Haven: Yale University Press, 1928.

Chittenden, Russell Henry, and Edgar S. Furniss. *The Graduate School of Yale.* New Haven: Carl Purlington Rollins Printing Office, 1965.

Churchill, Frederick B. "Wilhelm Roux and a Program for Embryology." Ph.D. dissertation, Harvard University, 1966.

Churchill, Frederick B. "August Weismann and a Break from Tradition." *Journal of the History of Biology* 1 (1968): 91–112.

Churchill, Frederick B. "From Machine-Theory to Entelechy: Two Studies in Developmental Teleology." *Journal of the History of Biology* 2 (1969): 165–185.

Churchill, Frederick B. "Hertwig, Weismann, and the Meaning of Reduction Division circa 1890." *Isis* 61 (1970): 429–457.

Churchill, Frederick B. "Chabry, Roux, and the Experimental Method in Nineteenth-Century Embryology." In Ronald Giere and Richard Westfall, eds., *Foundations of Scientific Method: the Nineteenth Century,* pp. 161–205. Bloomington: Indiana University Press, 1973.

Churchill, Frederick B. "Weismann's Continuity of the Germ-Plasm in Historical Perspective." In Klaus Sander, ed., *August Weismann (1834–1914) und die theoretische Biologie des 19. Jahrhunderts.* Special issue of *Freiburger Universitätsblätter* 87/88 (1985): 107–130.

Churchill, Frederick B. "From Heredity Theory to *Vererbung.* The Transmission Problem, 1850–1915" *Isis* 78 (1987): 337–364.

Churchill, Frederick B. "Regeneration: 1885–1901." Paper presented at the History of Science Society meeting, 1989.

Clapp, Cornelia. "Relation of the Axis of the Embryo to the First Cleavage Plane." *Biological Lectures delivered at the Marine Biological Laboratory, 1898,* 1899, pp. 139–151.

Clapp, Cornelia. "Some Recollections of the First Summer at Woods Hole, 1888." *Collecting Net* 2, no. 4 (1927): 3, 10.

Class of 1890. *Hopkins Medley.* Baltimore: Guggenheim, Weil and Co., 1890.

Coleman, William. "Cell, Nucleus, and Inheritance: An Historical Study." *Proceedings of the American Philosophical Society* 109 (1965): 124–158.

Coleman, William. *Biology in the Nineteenth Century.* Cambridge: Cambridge University Press, 1977.

Coleman, William, ed. *The Interpretation of Animal Form.* New York: Johnson Reprint Corp., 1967.

Conklin, Edwin Grant. "Preliminary Note on the Embryology of Crepidula fornicata and of Urosalpinx cinerea." *John Hopkins University Circulars* no. 88 (1891): 89–90.

Conklin, Edwin Grant. "The Cleavage of the Ovum in Crepidula fornicata." *Zoologischer Anzeiger* 15 (1892): 185–188.

Conklin, Edwin Grant. "The Fertilization of the Ovum." *Biological Lectures delivered at the Marine Biological Laboratory, 1893,* 1894, pp. 15–35.

Conklin, Edwin Grant. "Discussion of the Factors of Organic Evolution from the Embryological Standpoint." *Proceedings of the American Philosophical Society* 35 (1896): 78–88.

Conklin, Edwin Grant. "Weismann's Germinal Selection." *Science* 3 (1896): 853–857.

Conklin, Edwin Grant. "The Embryology of Crepidula." *Journal of Morphology* 13 (1897): 1–226.

Conklin, Edwin Grant. "Evolution and Revelation." Address to Methodist Episcopal Church Congress, Pittsburgh, 1897. Conklin Papers and Marine Biological Laboratory.

Conklin, Edwin Grant. Review of Edmund Beecher Wilson, *The Cell in Development and Inheritance. Psychological Review* 4 (1897): 318–322.

Conklin, Edwin Grant. "Cleavage and Differentiation." *Biological Lectures delivered at the Marine Biological Laboratory, 1896–1897,* 1898, pp. 17–43.

Conklin, Edwin Grant. "The Evidence and Factors of Organic Evolution." In *University Extension Lectures,* pp. 3–20. Philadelphia: American Society for the Extension of University Teaching, 1898.

Conklin, Edwin Grant. "Protoplasmic Movement as a Factor of Differentiation." *Biological Lectures delivered at the Marine Biological Laboratory, 1898,* 1899, pp. 69–92.

Conklin, Edwin Grant. "Advances in Methods of Teaching Zoology." *Science* 9 (1899): 81–84.

Conklin, Edwin Grant. "Phenomena and Mechanism of Inheritance." *Philadelphia Medical Journal* 2 (1899): 534–536.

Conklin, Edwin Grant. "The Fertilization of the Egg and Early Differentiation of the Embryo." *University of Pennsylvania Medical Bulletin,* 1900, pp. 14–22.

Conklin, Edwin Grant. "The Marine Biological Laboratory." *Science* 12 (1900): 333–344.

Conklin, Edwin Grant. "Karyokinesis and Cytokinesis in the Maturation, Fertilization and Cleavage of *Crepidula* and other Gasteropoda." *Journal of the Academy of Natural Sciences of Philadelphia* 12 (1902): 5–121.

Conklin, Edwin Grant. Review of Thomas Hunt Morgan, *Regeneration. Science* 15 (1902): 620–623.

Conklin, Edwin Grant. "The Cause of Inverse Symmetry." *Anatomischer Anzeiger* 23 (1903): 577–588.

Conklin, Edwin Grant. "Experiments on the Origin of Cleavage Centrosomes." *Biological Bulletin* 7 (1904): 221–226.

Conklin, Edwin Grant. "Mosaic Development of Ascidian Eggs." *Journal of Experimental Zoology* 2 (1905): 145–223.

Conklin, Edwin Grant. "The Mutation Theory from the Standpoint of Cytology." *Science* 21 (1905): 525–529.

Conklin, Edwin Grant. "Organ-Forming Substances in the Eggs of Ascidians." *Biological Bulletin* 8 (1905): 205–230.

Conklin, Edwin Grant. "The Organization and Cell-Lineage of the Ascidian Egg." *Journal of the Academy of Natural Sciences of Philadelphia* 13 (1905): 1–119.

Conklin, Edwin Grant. "Does Half of an Ascidian Egg give rise to a Whole Larva?" *Archiv für Entwickelungsmechanik der Organismen* 21 (1906): 727–753.

Conklin, Edwin Grant. "The Embryology of *Fulgur*: A Study of the Influence of Yolk on Development." *Proceedings of the Academy of Natural Sciences of Philadelphia* 59 (1907): 320–359.

Conklin, Edwin Grant. "The Mechanism of Heredity." *Science* 27 (1908): 89–99.

Conklin, Edwin Grant. Review of Thomas Hunt Morgan, *Experimental Zoology. Science* 27 (1908): 139–140.

Conklin, Edwin Grant. "The Application of Experiment to the Study of the Organization and Early Differentiation of the Egg." *Anatomical Record* 3 (1909): 149–154.

Conklin, Edwin Grant. "William Keith Brooks 1848–1908." *National Academy of Sciences Biographical Memoirs* 8 (1913): 25–88.

Conklin, Edwin Grant. "August Weismann." *Proceedings of the American Philosophical Society* 54, no. 216 (1915): iii–xii; also *Science* 41 (1915): 917–923.

Conklin, Edwin Grant. "The Beginning of Biology at Woods Hole. Laboratory at Penikese Forerunner of MBL." *Collecting Net* 2, no. 2 (1927): 1, 3, 6; no. 3: 7.

Conklin, Edwin Grant. "Edwin Grant Conklin." In Louis Finkelstein, ed., *Thirteen Americans: Their Spiritual Biographies*, pp. 47–76. New York: Harper and Brothers, 1953.

Conklin, Edwin Grant. "Early Days at Woods Hole." *American Scientist* 56 (1968): 112–120.

Conklin, Edwin Grant, and William Keith Brooks. "Structure and Development of the Gonophores of a Certain Siphonophore." *Johns Hopkins University Circulars* no. 88 (1890): 87–90.

Crampton, Henry E. "The Department of Zoology of Columbia University 1892–1942." Pamphlet. Morningside Heights, N.Y., 1942; also *Bios* 21 (1950): 219–246.

Cravens, Hamilton. *The Triumph of Evolution.* Philadelphia: University of Pennsylvania Press, 1978.

Curtis, Merle, and Roderick Nash. *Philanthropy in the Shaping of American Higher Education.* New Brunswick, N.J.: Rutgers University Press, 1965.

Darden, Lindley, and Nancy Maull. "Interfield Theories." *Philosophy of Science* 44 (1977): 43–64.

Dexter, Ralph. "From Penikese to the Marine Biological Laboratory at Woods Hole—The Role of Agassiz's Students." *Essex Institute Historical Collections* 110 (1974): 151–161.

Dexter, Ralph. "The Annisquam Sea-Side Laboratory of Alpheus Hyatt, Predecessor of the Marine Biological Laboratory at Woods Hole, 1880–1886." In Mary Sears and Daniel Merriam, eds., *Oceanography: The Past,* pp. 94–100. New York: Springer-Verlag, 1980.

"Doctoral Dissertations, 1876–1926." *Johns Hopkins University Circulars,* 1926.

Dornfeld, Ernest J. "The Allis Lake Laboratory." *Marquette Medical Review* 21 (1956): 115–144.

Driesch, Hans. *Die mathematisch-mechanische Betrachtung morphologischer Probleme der Biologie.* Jena: Gustav Fischer, 1891.

Driesch, Hans. "Entwicklungsmechanische Studien. I. Der Werth der beiden ersten Furchungszellen in der Echinodermentwicklung. Experimentelle Erzeugen von Theil- und Doppelbildung." *Zeitschrift für wissenschaftliche Zoologie* 53 (1891–92): 160–178. Trans. and abr. as "The Potency of the First Two Cleavage Cells in Echinoderm Development. Experimental Production of Partial and Double Formations." In Benjamin Willier and Jane M. Oppenheimer, eds., *Foundations of Experimental Embryology,* pp. 38–50. New York: Hafner Press, 1964. Page citations refer to the more accessible translation.

Driesch, Hans. "Entwicklungsmechanische Studien. II. Über die Beziehungen des Lichtes zur ersten Etappe der thierischen Formbildung." *Zeitschrift für wissenschaftliche Zoologie* 53 (1891–92): 178–184.

Driesch, Hans. "Entwicklungsmechanische Studien. VI. Über einige allgemeine Fragen der theoretischen Morphologie." *Zeitschrift für wissenschaftliche Zoologie* 55 (1893): 34–62.

Driesch, Hans, and T. H. Morgan. "Zur Analysis der ersten Entwickelungsstadien des Ctenophoreneies. I. Von der Entwickelung einzelner Ctenophorenblastomeren." *Archiv für Entwickelungsmechanik der Organismen* 2 (1895): 204–215.

Driesch, Hans, and T. H. Morgan. "Zur Analysis der ersten Entwickelungsstadien des Ctenophoreneis. II. Von der Entwickelung ungefurchten Eier mit Protoplasmadefekten." *Archiv für Entwickelungsmechanik der Organismen* 2 (1895): 216–224.

Eggeling, H. V. "Hermann Braus." *Anatomischer Anzeiger* 62 (1926): 255–291.

"Facsimile of First Announcement." *Journal of Experimental Zoology* 100 (1945): vii–ix.

Farley, John. *Gametes and Spores. Ideas about Sexual Reproduction 1750–1914*. Baltimore: Johns Hopkins University Press, 1982.

Feibleman, James. *An Introduction to Peirce's Philosophy*. New York: Harper and Bros., 1946.

Finch, Edith. *Carey Thomas of Bryn Mawr*. New York: Harper, 1947.

Flexner, Abraham. *Daniel Coit Gilman: Creator of the American Type of University*. New York: Harcourt, Brace, and Co., 1946.

Fol, Hermann. "Le Quadrille des Centres. Un Episode Nouveau dans l'Histoire de la Fécondation." *Archives des Sciences Physiques et Naturelles* 15 (1891): 393–420.

Franklin, Fabian, et al. *The Life of Daniel Coit Gilman*. New York: Dodd, Mead, and Co., 1910.

French, John C. *A History of the University Founded by Johns Hopkins*. Baltimore: The Johns Hopkins Press, 1946.

Fye, W. Bruce. "H. Newell Martin—A Remarkable Career Destroyed by Neurasthenia and Alcoholism." *Journal of the History of Medicine and Allied Sciences* 40 (1985): 133–166.

Fye, W. Bruce. *The Development of American Physiology*. Baltimore: Johns Hopkins University Press, 1987.

Galileo. "Letter to Madame Christina of Lorraine Grand Duchess of Thacone." In Stillman Drake, ed., *Discoveries and Opinions of Galileo*, pp. 175–216. New York: Doubleday, 1957.

Galtsoff, Paul S. *The Story of the Bureau of Commercial Fisheries Biological Laboratory. Woods Hole Massachusetts*. Washington, D.C.: Government Printing Office, 1962.

Gegenbaur, Carl. *Grundzüge der Vergleichenden Anatomie*. Leipzig: Wilhelm Engelmann, 1859.

Gegenbaur, Carl. "The Condition and Significance of Morphology," 1876. Reprinted in William Coleman, ed., *The Interpretation of Animal Form*, pp. 39–54. New York: Johnson Reprint Corp., 1967.

Geison, Gerald L. *Michael Foster and the Cambridge School of Physiology*. Princeton: Princeton University Press, 1978.

Geison, Gerald L., ed. *Physiology in the American Context 1850–1940*. Bethesda, Md.: American Physiological Society, 1987.

Gerson, Elihu M. "Scientific Work and Social Worlds." *Knowledge: Creation, Diffusion, Utilization* 4 (1983): 357–377.

Gilbert, Grove Karl. "The Origin of Hypotheses, Illustrated by the Discussion of a Topographic Problem." *Science* 3 (1891): 1–13.

Gilbert, Scott F. "The Embryological Origins of the Gene Theory," *Journal of the History of Biology* 11 (1978): 307–351.

Gilbert, Scott F. "In Friendly Disagreement: Wilson, Morgan, and the Embryological Origins of the Gene Theory." *American Zoologist* 27 (1987): 797–806.

Gilman, Daniel Coit. Inaugural address. In *Johns Hopkins University Inaugural Addresses*, pp. 14–64. Baltimore: John Murray and Co., 1876.

Gilman, Daniel Coit. "The Original Faculty." *The Launching of a University and Other Papers,* pp. 47–56. New York: Dodd, Mead, and Co., 1906.

Gilman, Daniel Coit. "Reminiscences of the Thirty Years in Baltimore, 1875–1905." *The Launching of a University and Other Papers,* pp. 3–24. New York: Dodd, Mead, and Co., 1906.

Golgi, Camillo. "Recherches sur l'histologie des centres nerveux." *Archives italiennes de Biologie* 3 (1883): 285–317; 4 (1884): 92–123.

Goodale, George Lincoln. "Alexander Agassiz." *National Academy of Sciences Biographical Memoirs* 7 (1912): 289–305.

Gould, Stephen Jay. *Ontogeny and Phylogeny.* Cambridge, Mass.: Harvard University Press, 1977.

Greene, Mott T. *Geology in the Nineteenth Century.* Ithaca, N.Y.: Cornell University Press, 1982.

Groeben, Christiane. "The Naples Zoological Station and Woods Hole." *Oceanus* 27 (1984): 60–69.

Groeben, Christiane. "Anton Dohrn—The Statesman of Darwinism." *Biological Bulletin* 168 Suppl. (1985): 4–25.

Haeckel, Ernst. *Generelle Morphologie der Organismen.* 2 vols. Berlin: Georg Reimer, 1866.

Haeckel, Ernst. *Natürliche Schöpfungsgeschichte.* Berlin: Georg Reimer, 1868.

Haeckel, Ernst. "Gastraea-Theorie." *Jenaische Zeitschrift für Naturwissenschaft* 9 (1875): 402–508.

Hamburger, Viktor. *The Heritage of Experimental Embryology.* New York: Oxford University Press, 1988.

Harrison, Ross Granville. "Über die Entwicklung der nicht knorpelig vorgebildet Skelettheile in den Flossen der Teleostier." *Archiv für Entwickelungsmechanik der Organismen* 42 (1893): 248–278.

Harrison, Ross Granville. "The Development of the Fins of Teleosts." *Johns Hopkins University Circulars* 111 (1894): 59–61.

Harrison, Ross Granville. "Die Entwicklung der unpaaren und paarigen Flossen der Teleostier." *Archiv für mikroskopische Anatomie* 46 (1895): 500–578.

Harrison, Ross Granville. "The Growth and Regeneration of the Tail of the Frog Larva." *Archiv für Entwickelungsmechanik der Organismen* 7 (1898): 430–485.

Harrison, Ross Granville. "Über die Histogenese des peripheren Nervensystems bei Salmo salar." *Archiv für mikroscopische Anatomie* 57 (1901): 354–444.

Harrison, Ross Granville. "Experimentelle Untersuchungen über die Entwicklung der Sinnesorgane der Seitenlinie bei den Amphibien." *Archiv für mikroscopische Anatomie* 63 (1903): 35–149.

Harrison, Ross Granville. "An Experimental Study of the Relation of the Nervous System to the Developing Musculature in the Embryo of the Frog." *American Journal of Anatomy* 3 (1904): 197–220.

Harrison, Ross Granville. "Nachtrag zu der Diskussion zu den Vorträgen von Schultze und v. Koelliker." *Verhandlungen der Anatomische Gesellschaft. 80th Versammlung, Jena,* 1904, p. 52.

Harrison, Ross Granville. "Neue Versuche und Beobachtungen über die Entwicklung der peripheren Nerven der Wirbeltiere," *Sitzungsberichte der Niederrheinische Gesellschaft für Natur und Heilkunde. Bonn,* 1904, pp. 1–7.

Harrison, Ross Granville. "Further Experiments on the Development of Peripheral Nerves." *American Journal of Anatomy* 5 (1906): 121–131.

Harrison, Ross Granville. "Experiments in Transplanting Limbs and their Bearing upon the Problems of the Development of Nerves." *Journal of Experimental Zoology* 4 (1907): 239–281.

Harrison, Ross Granville. "Observations on the Living Developing Nerve Fiber." *Anatomical Record* 1 (1907): 116–118.

Harrison, Ross Granville. "Embryonic Transplantation and Development of the Nervous System." *Anatomical Record* 2 (1908): 385–410.

Harrison, Ross Granville. "The Outgrowth of the Nerve Fiber as a Mode of Protoplasmic Movement." *Journal of Experimental Zoology* 9 (1910): 787–846.

Harrison, Ross Granville. "Experimental Biology and Medicine." *Physician and Surgeon* 34 (1912): 49–65.

Harrison, Ross Granville. "On the Origin and Development of the Nervous System." *Proceedings of the Royal Society of London,* Series B, 118 (1935): 155–196.

Harrison, Ross Granville. "Response on Behalf of the Medallist." *Science* 84 (1936): 565–567.

Harrison, Ross Granville. *Organization and Development of the Embryo.* Ed. Sally Wilens. New Haven: Yale University Press, 1969.

Harvey, E. Newton. "Edwin Grant Conklin: 1863–1952." *Science* 117 (1953): 703–705.

Harvey, E. Newton. "Edwin Grant Conklin." *National Academy of Sciences Biographical Memoirs* 31 (1958): 54–91.

Havemeyer, Loomis. *Sheff Days and Ways.* New Haven, Conn., 1958.

Hawkins, Hugh. *Pioneer: A History of the Johns Hopkins University.* Ithaca, N.Y.: Cornell University Press, 1960.

Held, Hans. *Die Entwicklung des Nervengewebes bei den Wirbeltieren.* Leipzig: Verlag Johann Ambrosius Barth, 1909.

Herrick, Sophie. "Thc Johns Hopkins University." *Scribner's Monthly* 19 (1880): 199–208.

Hertwig, Oskar. "Welche Einfluss übt die Schwerkraft auf die Theilung der Zellen?" *Jenaische Zeitschrift für Naturwissenschaft* 18 (1884–85): 175–205.

Hertwig, Oskar. "Das Problem der Befruchtung und der Isotropie des Eies, einer Theorie der Vererbung." *Jenaische Zeitschrift für Naturwissenschaft* 18 (1885): 276–318.

Hertwig, Oskar. *The Biological Problem of To-day, Preformation or Epigenesis? A Basis of the Organic Theory.* Trans. P. Chalmers Mitchell. London: William Heinemann, 1896.

Hertwig, Oskar and Richard Hertwig. "Über den Befruchtungs und Teilungsvorgang des tierischen Eies unter dem Einfluss äusserer

Agentien." *Jenaische Zeitschrift für Naturwissenschaft* 20 (1887): 477–510.

His, Wilhelm. *Unsere Körperform und das physiologische Problem ihrer Entstehung.* Leipzig: F. C. W. Vogel, 1874.

His, Wilhelm. "Die Neuroblasten u. deren Enstehung um embryonalen Mark." *Archiv für Anatomie und Physiologie,* Anatomischer Abtheilungen Suppl. (1889): 249–300.

His, Wilhelm. "On the Principles of Animal Morphology" (letter to John Murray, 1888). Reprinted in William Coleman, ed., *The Interpretation of Animal Form,* pp. 167–178. New York: Johnson Reprint Corp., 1967.

Howell, William H. "William Keith Brooks." *Johns Hopkins University Circulars,* 1908, pp. 53–58; also with reminiscences by former students, *Journal of Experimental Zoology* 9 (1910): 1–52.

Hughes, Arthur. *A History of Cytology.* London: Abelard-Schuman, 1959.

Hull, David L. *Science as a Process.* Chicago: University of Chicago Press, 1988.

Huxley, Thomas Henry. "Address on University Education." *American Lectures,* pp. 99–127. New York: D. Appleton and Co., 1877.

Huxley, Thomas Henry. "Lecture on the Study of Biology." *American Lectures,* pp. 131–164. New York: D. Appleton and Co., 1877.

Huxley, Thomas Henry, assisted by H. N. Martin. *A Course of Elementary Instruction in Practical Biology.* London: Macmillan, 1883.

Johns Hopkins University Inaugural Addresses. Baltimore: John Murray and Co., 1876.

Keenan, Katherine. "Lilian Vaughan Morgan (1870–1952): Her Life and Work." *American Zoologist* 23 (1983): 867–876.

Kofoid, Charles Atwood. *The Biological Stations of Europe,* pp. 7–32. Washington, D.C.: Government Printing Office, 1910.

Kohler, Robert E. *From Medical Chemistry to Biochemistry.* Cambridge: Cambridge University Press, 1982.

Kohlstedt, Sally Gregory. "Museums on Campus: A Tradition of Inquiry and Teaching." In Ronald Rainger, Keith Benson, and Jane Maienschein, eds., *The American Development of Biology,* pp. 15–47. Philadelphia: University of Pennsylvania Press, 1988.

Kuhn, Thomas S. *The Structure of Scientific Revolutions.* Chicago: University of Chicago Press, 1962.

Lankester, E. Ray. *Notes on Embryology and Classification for the Use of Students.* London: J. and A. Churchill, 1877.

Lankester, E. Ray. "Notes on the Embryology and Classification of the Animal Kingdom." *Quarterly Journal of Microscopical Science* 17 (1877): 399–454.

Laudan, Larry. *Progress and Its Problems. Toward a Theory of Scientific Growth.* Berkeley and Los Angeles: University of California Press, 1977.

Lederman, Muriel. "Research Note: Genes or Chromosomes: The Conversion of Thomas Hunt Morgan." *Journal of the History of Biology* 22 (1898): 449–496.

Lenoir, Timothy. *The Strategy of Life.* Dordrecht: D. Reidel, 1982.

Lillie, Frank R. "Charles Otis Whitman." *Journal of Morphology* 22 (1911): xv–lxxvii.

Lillie, Frank R. *The Woods Hole Marine Biological Laboratory.* Chicago: University of Chicago Press, 1944.

Loeb, Jacques. *Untersuchungen zur physiologischen Morphologie der Thiere. I. Über Heteromorphose.* Würzburg: Georg Hertz, 1891.

Loeb, Jacques. "Investigations in Physiological Morphology. III. Experiments on Cleavage." *Journal of Morphology* 7 (1892): 253–262.

Loeb, Jacques. *Untersuchungen zur physiologischen Morphologie der Thiere. II. Organbildung und Wachstum.* Würzburg: Georg Hertz, 1892.

Loeb, Jacques. "On Some Facts and Principles of Physiological Morphology." *Biological Lectures delivered at the Marine Biological Laboratory, 1893*, 1894, pp. 37–61.

Lurie, Edward. *Louis Agassiz: A Life in Science.* Chicago: University of Chicago Press, 1960.

MacMillan, Conway. "On the Emergence of a Sham Biology in America." *Science* o.s. 21 (1893): 184; followup note, n.s. 3 (1896): 634; response by William Keith Brooks, n.s. 3 (1896): 708.

Maienschein, Jane. "Cell Lineage, Ancestral Reminiscence, and the Biogenetic Law." *Journal of the History of Biology* 11 (1978): 129–158.

Maienschein, Jane. "Ross Harrison's Crucial Experiment as a Foundation for Modern American Experimental Embryology." Ph.D. dissertation, Indiana University, 1978.

Maienschein, Jane. "Experimental Biology in Transition: Harrison's Embryology 1895–1910." *Studies in the History of Biology* 6 (1983): 107–127.

Maienschein, Jane. "What Determines Sex? A Study of Converging Approaches, 1880–1916." *Isis* 75 (1984): 457–480.

Maienschein, Jane. "Agassiz, Hyatt, Whitman, and the Birth of the Marine Biological Laboratory." *Biological Bulletin* 168 Suppl. (1985): 26–34.

Maienschein, Jane. "First Impressions: American Biologists at Naples." *Biological Bulletin* 168 Suppl. (1985): 187–191.

Maienschein, Jane. "Preformation or New Formation—or Neither or Both?" In T. J. Horder, J. A. Witkowski, and C. C. Wylie, eds., *A History of Embryology*, pp. 73–108. Cambridge: Cambridge University Press, 1986.

Maienschein, Jane. "Arguments for Experimentation in Biology." *PSA 1986* 2 (1987): 180–195.

Maienschein, Jane. "Heredity/Development in the United States, circa 1900." *History and Philosophy of the Life Sciences* 9 (1987): 79–93.

Maienschein, Jane. "H. N. Martin and W. K. Brooks: Exemplars for American Biology?" *American Zoologist* 27 (1987): 773–783.

Maienschein, Jane. "Physiology, Biology, and the Advent of Physiological Morphology." In Gerald L. Geison, ed., *Physiology in the American Context. 1850–1940*, pp. 177–193. Bethesda, Md.: American Physiological Society, 1987.

Maienschein, Jane. "Why Do Research at the Seashore?" *American Zoologist* 28 (1988): 15–25.

Maienschein, Jane. "T. H. Morgan as Invertebrate Embryologist." *International Journal of Reproduction and Development* 15 (1989): 1–6.

Maienschein, Jane. *One Hundred Years Exploring Life, 1888–1988.* Boston: Jones and Bartlett, 1989.

Maienschein, Jane. "Cell Theory and Development." In G. N. Cantor, J. R. R. Christie, M. J. S. Hodge, and R. C. Olby, eds., *Companion to the History of Modern Science.* London: Routledge, 1990.

Maienschein, Jane. "Nettie Marie Stevens." *Dictionary of Scientific Biography. Supplement II.* New York: Charles Scribner's Sons, forthcoming.

Maienschein, Jane. "T. H. Morgan's Regeneration, Epigenesis, and (W)holism." In Charles Dinsmore, ed., *History of Regeneration Research.* Cambridge: Cambridge University Press, forthcoming.

Maienschein, Jane, ed. *Defining Biology. Lectures from the 1890s.* Cambridge, Mass.: Harvard University Press, 1987.

Maienschein, Jane, Ronald Rainger, and Keith Benson. Introduction to "American Morphology at the Turn of the Century." Special section of *Journal of the History of Biology* 14 (1981): 83–158.

Manier, Edward. "The Experimental Method in Biology." *Synthese* 20 (1969): 185–205.

Manning, Kenneth R. *Black Apollo of Science. The Life of Ernest Everett Just.* New York: Oxford University Press, 1983.

Martin, Henry Newell. "The Study and Teaching of Biology." *Memoirs from the Biological Laboratory of the Johns Hopkins University* 3 (1876): 192–204. Reprinted in William Coleman, ed., *The Interpretation of Animal Form,* pp. 181–191. New York: Johnson Reprint Corp., 1967.

Martin, Henry Newell. "The Influence upon the Pulse Rate of Variations of Arterial Pressure, of Venous Pressure, and of Temperature." *Memoirs from the Biological Laboratory of the Johns Hopkins University* 3 (1895): 12–24. Paper originally published in 1882.

Martin, Henry Newell. "Modern Physiological Laboratories—What They Are and Why They Are." *Johns Hopkins University Circulars* 3 (1884): 227–241; reprinted, *Science* 3 (1884): 73–76, 100–103.

Martin, Henry Newell. "The Normal Respiratory Movements of the Frog, and the Influence upon its Respiratory Centre of Stimulation of the Optic Lobes." *Memoirs from the Biological Laboratory of the Johns Hopkins University* 3 (1895): 117–147.

Mayr, Ernst. "The Species Concept." *Evolution* 3 (1949): 371–373.

Mayr, Ernst. "Cause and Effect in Biology." *Science* 134 (1961): 1501–1506.

Mayr, Ernst. *The Growth of Biological Thought.* Cambridge, Mass.: Harvard University Press, 1982.

McClung, Clarence E. "The Accessory Chromosome—Sex Determinant?" *Biological Bulletin* 3 (1902): 43–84.

McCullough, Dennis. "W. K. Brooks' Role in the History of American Biology." *Journal of the History of Biology* 2 (1969): 411–438.

Meckel, Johann Friedrich. *System der Vergleichenden Anatomie.* Halle: Rengerschen, 1821.

Mocek, Reinhard. *Wilhelm Roux–Hans Driesch. Zur Entwickelungsphysiologie der Tiere ("Entwicklungsmechanik").* Jena: Gustav Fischer, 1974.

Montgomery, Thomas H., Jr. "A Study of the Chromosomes of the Germ

Cells of Metazoa." *Transactions of the American Philosophical Society* 20 (1902): 154–236.

Moore, John A. "Edmund Beecher Wilson, Class of '81." *American Zoologist* 27 (1987): 785–796.

Moore, John, ed. *Readings in Heredity and Development.* New York: Oxford University Press, 1972.

Morgan, Thomas Hunt. "The Dance of the Lady Crab." *Popular Science Monthly* 34 (1889): 482–484.

Morgan, Thomas Hunt. "On the Amphibian Blastopore." *Johns Hopkins University Studies from the Biological Laboratory* 4 (1889): 355–377.

Morgan, Thomas Hunt. "Origin of the Test Cells of Ascidians." *Johns Hopkins University Circulars* 8 (1889): 63.

Morgan, Thomas Hunt. "The Origin of the Test Cells of Ascidians." *Journal of Morphology* 4 (1890): 195–204.

Morgan, Thomas Hunt. "A Contribution to the Embryology and Phylogeny of the Pycnogonids." *Johns Hopkins University Studies from the Biological Laboratory* 5 (1891): 1–76.

Morgan, Thomas Hunt. "Embryology of the Sea Bass." *American Naturalist* 25 (1891): 162–166.

Morgan, Thomas Hunt. "The Growth and Metamorphosis of Tornaria." *Journal of Morphology* 5 (1891): 407–458.

Morgan, Thomas Hunt. "A New Larval Form from Jamaica." *American Naturalist* 25 (1891): 1137–1139.

Morgan, Thomas Hunt. "The Relationships of the Sea-spiders." *Biological Lectures delivered at the Marine Biological Laboratory, 1890,* 1891, pp. 142–167.

Morgan, Thomas Hunt. "Balanoglossus and Tornaria in New England." *Zoologischer Anzeiger* 15 (1892): 456–457.

Morgan, Thomas Hunt. "Spiral Modification of Metamerism." *Journal of Morphology* 7 (1892): 245–251.

Morgan, Thomas Hunt. "Experimental Studies on the Teleost Eggs." *Anatomischer Anzeiger* 8 (1893): 803–814.

Morgan, Thomas Hunt. "The Development of Balanoglossus." *Journal of Morphology* 9 (1894): 1–86.

Morgan, Thomas Hunt. "Experimental Studies on Echinoderm Eggs." *Anatomischer Anzeiger* 9 (1894): 141–152.

Morgan, Thomas Hunt. "The Formation of the Embryo of the Frog." *Anatomischer Anzeiger* 9 (1894): 697–705.

Morgan, Thomas Hunt. "The Orientation of the Frog's Egg." *Quarterly Journal of Microscopical Science* 35 (1894): 373–405.

Morgan, Thomas Hunt. "The Formation of the Fish Embryo." *Journal of Morphology* 10 (1895): 419–472.

Morgan, Thomas Hunt. "Half-Embryos and Whole-Embryos from one of the first two Blastomeres of the Frog's Egg." *Anatomischer Anzeiger* 10 (1895): 623–628.

Morgan, Thomas Hunt. Review of William Sedgwick and E. B. Wilson, *General Biology,* 2d ed. *Science* 2 (1895): 740–741.

Morgan, Thomas Hunt. "Studies of the 'Partial' Larvae of Sphaerechinus." *Archiv für Entwickelungsmechanik der Organismen* 2 (1895): 81–126.

Morgan, Thomas Hunt. "A Study of Metamerism." *Quarterly Journal of Microscopical Science* 37 (1895): 395–476.

Morgan, Thomas Hunt. "Impressions of the Naples Zoölogical Station." *Science* 3 (1896): 16–18.

Morgan, Thomas Hunt. "The Number of Cells in Larvae from Isolated Blastomeres of Amphioxus." *Archiv für Entwickelungsmechanik der Organismen* 3 (1896): 269–294.

Morgan, Thomas Hunt. *The Development of the Frog's Egg.* New York: Macmillan, 1897.

Morgan, Thomas Hunt. "Regeneration in Allolobophora foetida." *Archiv für Entwickelungsmechanik der Organismen* 5 (1897): 570–586.

Morgan, Thomas Hunt. "Developmental Mechanics." *Science* 7 (1898): 156–158.

Morgan, Thomas Hunt. "Experimental Studies of the Regeneration of Planaria Maculata." *Archiv für Entwickelungsmechanik der Organismen* 7 (1898): 364–397.

Morgan, Thomas Hunt. "Some Problems of Regeneration." *Biological Lectures delivered at the Marine Biological Laboratory, 1898,* 1899, pp. 193–207.

Morgan, Thomas Hunt. "Regeneration in the Hydromedusa, Gonionemus Vertens." *American Naturalist* 33 (1899): 939–951.

Morgan, Thomas Hunt. "Regeneration: Old and New Interpretations." *Biological Lectures delivered at the Marine Biological Laboratory, 1898,* 1899, pp. 193–207; and *1899,* 1900, pp. 185–208.

Morgan, Thomas Hunt. "Regeneration in Teleosts." *Archiv für Entwickelungsmechanik der Organismen* 10 (1900): 120–134.

Morgan, Thomas Hunt. *Regeneration.* Columbia University Biological Series, No. 7. New York: Macmillan, 1901.

Morgan, Thomas Hunt. "The Relation between Normal and Abnormal Development of the Embryo of the Frog, as Determined by Injury to the Yolk-portion of the Egg." *Archiv für Entwickelungsmechanik der Organismen* 15 (1902): 238–313.

Morgan, Thomas Hunt. "Darwinism in the Light of Modern Criticism." *Harper's* 106 (1903): 476–479.

Morgan, Thomas Hunt. *Evolution and Adaptation.* New York: Macmillan, 1903.

Morgan, Thomas Hunt. "Recent Theories in Regard to the Determination of Sex." *Popular Science Monthly* 64 (1903): 97–116.

Morgan, Thomas Hunt. "Self-fertilization induced by Artificial Means." *Journal of Experimental Zoology* 1 (1904): 135–178.

Morgan, Thomas Hunt. "An Alternative Interpretation of the Origin of Gynandromorphous Insects." *Science* 21 (1905): 632–634.

Morgan, Thomas Hunt. "The Physiology of Regeneration." *Journal of Experimental Zoology* 3 (1906): 457–500.

Morgan, Thomas Hunt. *Experimental Zoology.* New York: Macmillan, 1907.

Morgan, Thomas Hunt. "Sex-determining Factors in Animals." *Science* 25 (1907): 382–384.

Morgan, Thomas Hunt. "The Dynamic Factor in Regeneration." *Biological Bulletin* 16 (1909): 265–276.
Morgan, Thomas Hunt. "For Darwin." *Popular Science Monthly* 74 (1909): 367–380.
Morgan, Thomas Hunt. "What are 'Factors' in Mendelian Explanations?" *American Breeders' Association* 5 (1909): 365–368.
Morgan, Thomas Hunt. "Chromosomes and Heredity." *American Naturalist* 64 (1910): 449–496.
Morgan, Thomas Hunt. "Cytological Studies of Centrifuged Eggs." *Journal of Experimental Zoology* 9 (1910): 593–655.
Morgan, Thomas Hunt. *Heredity and Sex.* New York: Columbia University Press, 1913.
Morgan, Thomas Hunt. "Edmund Beecher Wilson." *Science* 89 (1939): 258–259.
Morgan, Thomas Hunt. "Edmund Beecher Wilson 1856–1939." *National Academy of Sciences Biographical Memoirs* 21 (1941): 314–342.
Morgan, Thomas Hunt. "Edmund Beecher Wilson." *Science* 96 (1942): 239–242.
Morse, E. S. "Charles Otis Whitman 1842–1910." *National Academy of Sciences Biographical Memoirs* 7 (1912): 269–288.
Morse, E. S. "Agassiz and the School at Penikese." *Science* 58 (1923): 273–275.
Muller, H. J. "Edmund B. Wilson—An Appreciation." *American Naturalist* 77 (1943): 5–37, 142–172.
Müller, Irmgard. *Die Geschichte der Zoologischen Station Neapel von der Gründung durch Anton Dohrn (1872) bis zum ersten Weltkrieg und ihre Bedeutung für die Entwicklung der modernen biologischen Wissenschaften.* Habilitations-Schrift: Universität Düsseldorf, 1976.
Murray, John. "Alexander Agassiz." *Bulletin: Museum of Comparative Zoology* 54 (1911): 137–158.
"National Traits in Science." *Science* 2 (1883): 455–457.
"New Yale." *Fortune,* March 1934, pp. 70–81.
Nicholas, J. S. "Ross Granville Harrison." *Yale Journal of Biology and Medicine* 32 (1960): 407–412.
Nicholas, J. S. "Ross Granville Harrison." *National Academy of Sciences Biographical Memoirs* 35 (1961): 132–162.
Nunn, Emily. "The Naples Zoölogical Station." *Science* 1 (1883): 479–481, 507–510.
Nunn (Whitman), Emily. "The Zoölogical Station at Naples." *Century Illustrated,* 1886, pp. 791–799.
Nyhart, Lynn Keller. "Morphology and the German University, 1860–1900." Ph.D. dissertation, University of Pennsylvania, 1986.
Nyhart, Lynn Keller. "The Disciplinary Breakdown of German Morphology, 1870–1900." *Isis* 78 (1987): 365–389
Ogilvie, Marilyn Bailey, and Clifford J. Choquette. "Nettie Marie Stevens (1861–1912)." *Proceedings of the American Philosophical Society* 125 (1981): 292–311.

"Old-Fashioned." *Time*, 3 July 1939, pp. 39–42.

Oppenheimer, Jane M. "The Non-Specificity of the Germ-Layers." *Essays in the History of Embryology and Biology*, pp. 256–294. Cambridge, Mass.: MIT Press, 1967.

Oppenheimer, Jane M. "Ross Harrison's Contributions to Experimental Embryology." *Essays in the History of Embryology and Biology*, pp. 92–116. Cambridge, Mass.: MIT Press, 1967.

Oppenheimer, Jane M. "Ross Granville Harrison." *Dictionary of Scientific Biography*, 6:131–135. New York: Charles Scribner's Sons, 1970–80.

Oppenheimer, Jane M. "Historical Relationships between Tissue Culture and Transplantation Experiments." *Transactions and Studies. College of Physicians of Philadelphia* 39 (1971): 26–33.

Oppenheimer, Jane M. "Thomas Hunt Morgan as an Embryologist: The View from Bryn Mawr." *American Zoologist* 23 (1983): 845–854.

Oppenheimer, Jane M., ed. *Autobiography of Dr. Karl Ernst von Baer.* New Delhi: Amerind Publishing Co. for Science History Publications, 1986.

Owens, Larry. "Pure and Sound Government. Laboratories, Playing Fields, and Gymnasia in the Nineteenth-Century Search for Order." *Isis* 76 (1985): 182–194.

Packard, A. S. Jr., and E. D. Cope. "Editors' Table." *American Naturalist* 18 (1884): 392–395.

Parker, George Howard. *The World Expands.* Cambridge, Mass.: Harvard University Press, 1946.

Pauly, Philip J. "American Biologists in Wilhelmian Germany: Another Look at the Innocents Abroad." Paper presented at the History of Science Society meeting, Chicago, 1984.

Pauly, Philip J. "The Appearance of Academic Biology in Late Nineteenth-Century America." *Journal of the History of Biology* 17 (1984): 369–397.

Pauly, Philip J. *Controlling Life. Jacques Loeb and the Engineering Ideal in Biology.* New York: Oxford University Press, 1987.

"Penikese Island." *Frank Leslie's Illustrated Newspaper*, 23 August 1873, pp. 377–378.

Pflüger, Eduard. "Einige Beobachtungen zur Frage über die das Geschlecht bestimmenden Ursachen." *Pflüger's Archiv für die gesamte Physiologie des Menschen und der Tiere* 25 (1881): 243–258.

Pflüger, Eduard. "Die Bastardirung bei den Batrachien." *Pflüger's Archiv für die gesamte Physiologie des Menschen und der Tiere* 29 (1882): 48–75.

Pflüger, Eduard. "Hat die Concentration des Samens einen Einfluss auf des Geschlecht?" *Pflügers Archiv für die gesamte Physiologie des Menschen und der Tiere* 29 (1882): 1–12.

Pflüger, Eduard. "Über die das Geschlecht bestimmenden Ursachen und die Geschlechtsverhältnisse der Frösche." *Pflüger's Archiv für die gesamte Physiologie des Menschen und der Tiere* 29 (1882): 13–40.

Pflüger, Eduard. "Über den Einfluss der Schwerkraft auf die Theilung der Zellen und auf die Entwicklung des Embryo." *Pflüger's Archiv für die gesamte Physiologie des Menschen und der Tiere* 32 (1883): 1–79.

Pflüger, Eduard. "Über die Einwirkung der Schwerkraft und andere

Bedingungen auf die Richtung der Zelltheilung." *Pflüger's Archiv für die gesamte Physiologie des Menschen und der Tiere* 34 (1884): 607–616.

Rainger, Ronald. *An Agenda for Antiquity: Henry Fairfield Osborn, the American Museum, and American Vertebrate Paleontology, 1890–1936.* Tuscaloosa: University of Alabama Press, forthcoming.

Rainger, Ronald, Keith Benson, and Jane Maienschein, eds. *The American Development of Biology.* Philadelphia: University of Pennsylvania Press, 1988.

Ramon y Cajal, Santiago. *Textura del Sistema Nervioso del Hombre y de los Vertebrados.* Madrid: Moya, 1899.

Ramon y Cajal, Santiago. *Recollections of My Life.* Trans. E. Horne Craigie. Cambridge, Mass.: MIT Press, 1937.

Remak, Robert. *Untersuchungen über die Entwickelung der Wirbelthiere.* Berlin: G. Reimer, 1850–1855.

"Retrospect." *Journal of Experimental Zoology* 100 (1945): xi–xxxi.

Richards, A. "Edwin Grant Conklin." *Bios* 6 (1935): 187–211.

Richards, Robert J. *Darwin and the Emergence of Evolutionary Theories of Mind and Behavior.* Chicago: University of Chicago Press, 1987.

"The Role of Johns Hopkins University in the Development of Experimental and Quantitative Biology in America." Special section of *American Zoologist* 27 (1987): 748–817. Includes papers by Brother C. Edward Quinn, James Ebert, Keith Benson, Jane Maienschein, John A. Moore, Scott F. Gilbert, and Sharon Kingsland.

Roll-Hanson, Nils. "Drosophila Genetics: A Reductionist Research Program." *Journal of the History of Biology* 11 (1978): 159–210.

Rosenberg, Charles. "Henry Newell Martin." *Dictionary of Scientific Biography,* 9:142–143. New York: Charles Scribner's Sons, 1970–80.

Roux, Wilhelm. *Über die Bedeutung der Kerntheilungsfiguren. Eine hypothetische Erörterung.* Leipzig: Wilhelm Engelmann, 1883.

Roux, Wilhelm. *Über die Zeit der Bestimmung der Hauptrichtungen des Froschembryo.* Leipzig: Wilhelm Engelmann, 1883.

Roux, Wilhelm. "Beiträge zur Entwickelungsmechanik des Embryo No. 1." *Zeitschrift für Biologie* 21 (1885): 411–526.

Roux, Wilhelm. "Über die Bestimmung der Hauptrichtungen des Froschembryo im Ei und über die erste Theilung des Froscheies." *Breslauer ärtzlichen Zeitschrift* (1885): 1–54. Repaginated offprint.

Roux, Wilhelm. "Beiträge zur Entwickelungsmechanik des Embryo. No. 5. Über die künstliche Hervorbringung halber Embryonen durch Zerstörung einer der beiden ersten Furchungskugeln, sowie über die Nachentwickelung (Postgeneration) der fehlenden Köperhälfte." *Virchows Archiv für pathologisches Anatomie und Physiologie und klinische Medizin* 114 (1888): 113–153. Trans. as "Contributions to the Developmental Mechanics of the Embryo. On the Artificial Production of Half-Embryos by Destruction of One of the First Two Blastomeres, and the Later Development (Postgeneration) of the Missing Half of the Body." In Benjamin Willier and Jane Oppenheimer, eds., *Foundations of Experimental Embryology.* New York: Hafner Press, 1964. pp. 2–37.

Royce, Josiah. Contribution in Johns Hopkins University, *Celebration of the Twenty-fifth Anniversary,* pp. 112–118. Baltimore: The Johns Hopkins Press, 1902.

Rudnick, Dorothea. "Ross Harrison (1870–1959)." I. W. Haymaker and F. Schiller, eds. *Founders of Neurology,* pp. 123–128. Springfield, Ill.: Charles C. Thomas, 1970.

Russell, E. S. *Form and Function.* London: John Murray, 1916. Reprint. West Mead, Farnborough: Gregg International Publishers, 1972.

Sabin, Florence Rena. *Franklin Paine Mall.* Baltimore: The Johns Hopkins Press, 1934.

Sapp, Jan. *Beyond the Gene. Cytoplasmic Inheritance and the Struggle for Authority in Genetics.* New York: Oxford University Press, 1987.

Schleiden, Matthias. "Beiträge zur Phytogenesis." *Archiv für Anatomie, Physiologie, und wissenschaftliche Medizin,* 1838, 137–176.

Schultze, Oskar. "Über die multizelluläre Enstehung der peripheren sensiblen Nervenfaser und das Vorhandsein eines allgemeinen Endnetz sensibler Neuroblasten bei Amphibienlarven." *Archiv für mikroskopische Aatomie* 66 (1905): 41–111.

Schwann, Theodor. *Mikroskopische Untersuchungen über die Ubereinstimmung in der Struktur und dem Wachstum der Thiere und Pflanzen.* Berlin: G. Reimer, 1839.

Sedgwick, William, and E. B. Wilson. *An Introduction to General Biology.* New York: Henry Holt and Co., 1886 (2d ed. 1895).

Shine, Ian, and Sylvia Wrobel. *Thomas Hunt Morgan.* Lexington: University of Kentucky Press, 1976.

Stephenson, W. H. Article in *Filson Club Historical Quarterly* 20 (1946): 97–106.

Stevens, Nettie Marie. *Studies in Spermatogenesis with Especial Reference to the "Accessory Chromosome."* Carnegie Institution of Washington Publication, no. 36. Washington, D.C., 1905.

Sturtevant, A. H. "Thomas Hunt Morgan." *National Academy of Sciences Biographical Memoirs* 33 (1959): 283–325.

Sturtevant, A. H. "Reminiscences of T. H. Morgan." Transcribed from a talk at the Marine Biological Laboratory, 1967. MBL Director's Archives.

Sutton, Walter. "The Chromosome in Heredity." *Biological Bulletin* 4 (1902–3): 231–248.

Swanson, C. P. "A History of Biology at the Johns Hopkins University." *Bios* 22 (1951): 223–262.

Thom, Helen Hopkins. *Johns Hopkins: A Silhouette.* Baltimore: The Johns Hopkins Press, 1929.

Twitty, Victor Chandler. *Of Scientists and Salamanders.* San Francisco: W. H. Freeman and Co., 1966.

Uschmann, Georg, ed. *Ernst Haeckel. Biographie in Briefen.* Leipzig: Urania Verlag, 1983.

Vicedo, Marga. "T. H. Morgan, Neither an Epistemological Empiricist nor a 'Methodological' Empiricist." Paper presented at the International Society for History, Philosophy, and Social Studies of Science meeting, London, Ontario, 1989.

Virchow, Rudolf. *Die Cellularpathologie in ihrer Begründung auf physiologische und pathologische Gewebelehre.* Berlin: August Hirschwald, 1858. Trans. Frank Chance as *Cellular Pathology as based upon Physiological and Pathological Histology.* London: John Churchill, 1860.

Weismann, August. *Die Kontinuität des Keimplasmas als Grundlage einer Theorie der Vererbung.* Jena: Gustav Fischer, 1885. Trans. Selmar Schönland as "The Continuity of the Germ-Plasm as the Foundation of a Theory of Heredity." In Edward B. Poulton, Selmar Schönland, and Arthur Shipley, eds., *Essays Upon Heredity and Kindred Biological Problems,* pp. 163–255. Oxford: Clarendon Press, 1889.

Weismann, August. *Über die Zahl der Richtungskörper und über ihre Bedeutung für die Vererbung.* Jena: Gustav Fischer, 1887. Trans. Selmar Schönland as "On the Number of Polar Bodies and their Significance in Heredity." In Edward B. Poulton, Selmar Schönland, and Arthur Shipley, eds., *Essays Upon Heredity and Kindred Biological Problems,* pp. 343–396. Oxford: Clarendon Press, 1889.

Weismann, August. *Das Keimplasm. Eine Theorie der Vererbung.* Jcna: Gustav Fisher, 1892. Trans. W. Newton Parker and Harriet Rönnfeldt as *The Germ Plasm.* New York: Charles Scribner's Sons, 1893.

Weismann, August (subject). Sander, Klaus, ed. *August Weismann (1834–1914) und die theoretische Biologie des 19. Jahrhunderts.* Special issue of *Freiburger Universitätsblätter* 87/88 (1985): 1–203.

Wenrich, D. H. "Biology at the University of Pennsylvania." *Bios* 22 (1951): 151–190.

Werdinger, Jeffrey. "Embryology at Woods Hole. The Emergence of a New American Biology." Ph.D. dissertation, Indiana University, 1980.

Weygandt, Cornelius. *On the Edge of Evening.* New York: Putnam, 1946.

"What Is Life?" Pamphlet for Johns Hopkins University Fiftieth Anniversary. Baltimore: The Johns Hopkins Press, 1925.

Whitman, Charles Otis. "The Embryo of *Clepsine.*" *Quarterly Journal of Microscopical Science* 18 (1878): 215–315.

Whitman, Charles Otis. "The Advantages of Study at the Naples Zoölogical Station." *Science* 22 (1883): 93–97.

Whitman, Charles Otis. "The Germ Layers of *Clepsine.*" *Zoologischer Anzeiger* 9 (1886): 171–176.

Whitman, Charles Otis. "A Contribution to the History of the Germ-layers in Clepsine." *Journal of Morphology* 1 (1887): 105–182.

Whitman, Charles Otis. "The Kinetic Phenomena of the Egg During Maturation and Fecundation (Oökinesis)." *Journal of Morphology* 1 (1887): 227–252.

Whitman, Charles Otis. "The Inadequacy of the Cell Theory." *Journal of Morphology* 8 (1893): 639–658; also, with the title "The Inadequacy of the Cell-Theory of Development," *Biological Lectures delivered at the Marine Biological Laboratory, 1893,* 1894; pp. 105–124.

Whitman, Charles Otis. "Marine Biological Laboratory." *Biological Lectures delivered at the Marine Biological Laboratory, 1893,* 1894, pp. 235–242.

Whitman, Charles Otis. "Prefatory Note." *Biological Lectures delivered at the*

Marine Biological Laboratory, 1894, 1895, pp. iii–vii.
Whitman, Charles Otis. "Some of the Functions and Features of a Biological Station." *Science* 7 (1898): 37–44.
Whitman, Charles Otis. "Animal Behavior." *Biological Lectures delivered at the Marine Biological Laboratory, 1898,* 1899, pp. 285–338. Reprinted in Jane Maienschein, ed., *Defining Biology. Lectures from the 1890s.* pp. 217–272. Cambridge, Mass.: Harvard University Press, 1987.
Wilder, Burt. "What We Owe to Agassiz." *Popular Science Monthly* 71 (1907): 5–20.
Willier, Benjamin, and Jane M. Oppenheimer, eds. *Foundations of Experimental Embryology.* New York: Hafner Press, 1964.
Wilson, Edmund Beecher. "Notes on the Early Stages of some Polychaetous Annelids." *American Journal of Science* 20 (1880): 291–292.
Wilson, Edmund Beecher. "A Problem of Morphology as Illustrated by the Development of the Earthworm." *Johns Hopkins University Circulars,* May 1880, p. 66.
Wilson, Edmund Beecher. "The Development of Phoronis: A Contribution to the Study of Animal Metamorphosis." *Johns Hopkins University Circulars,* December 1880, p. 82.
Wilson, Edmund Beecher. "The Origin and Significance of the Metamorphosis of Actinotrocha." *Quarterly Journal of Microscopical Science* 21 (1881): 202–218.
Wilson, Edmund Beecher. "Observations upon the Structure and Development of Renilla and Leptogorgia." *Johns Hopkins University Circulars,* August 1882, p. 247.
Wilson, Edmund Beecher. "The Development of Renilla." *Proceedings of the Royal Society of London,* Series B, 34 (1883): 384–388.
Wilson, Edmund Beecher. "The Development of Renilla." *Royal Society of London Philosophical Transactions* 174 (1884): 723–815.
Wilson, Edmund Beecher. "The Mesenterial Filaments of the *Alcynoia.*" *Mittheilungen aus der Zoologischen Station zu Neapel* 5 (1884): 1–27.
Wilson, Edmund Beecher. "The Embryology of the Earthworm." *Journal of Morphology* 3 (1889): 387–462.
Wilson, Edmund Beecher. "The Origin of the Mesoblast-Bands in Annelids." *Journal of Morphology* 4 (1890): 205–219.
Wilson, Edmund Beecher. "Some Problems of Annelid Morphology." *Biological Lectures delivered at the Marine Biological Laboratory, 1890,* 1890, pp. 53–78.
Wilson, Edmund Beecher. "The Heliotropism of Hydra." *American Naturalist* 25 (1891): 413–433.
Wilson, Edmund Beecher. "The Cell-Lineage of *Nereis.*" *Journal of Morphology* 6 (1892): 361–463.
Wilson, Edmund Beecher. "On Multiple and Partial Development in Amphioxus." *Anatomischer Anzeiger* 7 (1892): 732–740.
Wilson, Edmund Beecher. "*Amphioxus,* and the Mosaic Theory of Development." *Journal of Morphology* 8 (1893): 579–630.
Wilson, Edmund Beecher. "The Mosaic Theory of Development." *Biological*

Lectures delivered at the Marine Biological Laboratory, 1893, 1894, pp. 1–14.

Wilson, Edmund Beecher. "Archoplasm, Centrosome, and Chromatin in the Sea-Urchin Egg." *Journal of Morphology* 1 (1895): 443–478.

Wilson, Edmund Beecher. *The Cell in Development and Inheritance.* New York: Macmillan Co., 1896 (2d ed. 1900).

Wilson, Edmund Beecher. "Embryological Criterion of Homology." *Biological Lectures delivered at the Marine Biological Laboratory, 1894,* 1895, pp. 101–124.

Wilson, Edmund Beecher. "On Cleavage and Mosaic-Work." *Archiv für Entwickelungsmechanik der Organismen* 3 (1896): 19–26.

Wilson, Edmund Beecher. "Considerations on Cell-Lineage and Ancestral Reminiscence." *Annals of the New York Academy of Sciences* 11 (1898): 1–27.

Wilson, Edmund Beecher. "The Structure of Protoplasm." *Biological Lectures delivered at the Marine Biological Laboratory, 1898,* 1899, pp. 1–20.

Wilson, Edmund Beecher. "Cell Lineage and Ancestral Reminiscence." *Biological Lectures delivered at the Marine Biological Laboratory, 1898,* 1899, pp. 21–42.

Wilson, Edmund Beecher. "The Chemical Fertilization of the Sea Urchin (eggs)." *Annals of the New York Academy of Sciences 13* (1900): 514–515.

Wilson, Edmund Beecher. "Some Aspects of Recent Biological Research." *International Monthly* 2 (1900): 74–93; reprinted and repaginated, 1900, pp. 3–22.

Wilson, Edmund Beecher. "Aims and Methods of Study of Natural History." *Science* 13 (1901): 14–23.

Wilson, Edmund Beecher. "Experimental Studies in Cytology. I. A Cytological Study of Artificial Parthenogenesis in Sea-urchin Eggs." *Archiv für Entwickelungsmechanik der Organismen* 12 (1901): 529–596.

Wilson, Edmund Beecher. "Experimental Studies in Cytology. II. Some Phenomena of Fertilization and Cell-division in Etherized Cells." *Archiv für Entwickelungsmechanik der Organismen* 13 (1901): 353–373.

Wilson, Edmund Beecher. "Experimental Studies in Cytology. III. The Effect on Cleavage of Artificial Obliteration of the First Cleavage-Furrow." *Archiv für Entwickelungsmechanik der Organismen* 13 (1901): 373–395.

Wilson, Edmund Beecher. "Mendel's Principles of Heredity and the Maturation of the Germ Cells." *Science* 16 (1902): 991–993.

Wilson, Edmund Beecher. "Experiments on Cleavage and Localization in the Nemertine-egg." *Archiv für Entwickelungsmechanik der Organismen* 16 (1903): 411–460.

Wilson, Edmund Beecher. "Mr. Cook on Evolution, Cytology, and Mendel's Laws." *Popular Science Monthly* 63 (1903): 88–89.

Wilson, Edmund Beecher. "Notes on Merogony and Regeneration in Renilla." *Biological Bulletin* 4 (1903): 215–226.

Wilson, Edmund Beecher. "Notes on the Reversal of Asymmetry in the Regeneration of the Chelae in Alpheus Heterochelis." *Biological Bulletin* 4 (1903): 197–210.

Wilson, Edmund Beecher. "Experimental Studies on Germinal Localization."

Journal of Experimental Zoology 1 (1904): 1–72, 197–268.

Wilson, Edmund Beecher. "Mosaic Development in the Annelid Egg." *Science* 20 (1904): 748–750.

Wilson, Edmund Beecher. "The Chromosomes in Relation to the Determination of Sex in Insects." *Science* 22 (1905): 500–502.

Wilson, Edmund Beecher. "The Problem of Development." *Science* 21 (1905): 281–294.

Wilson, Edmund Beecher. "Some Recent Studies on Heredity." *Harvey Lectures* (1906–7): 200–221.

Wilson, Edmund Beecher. "The Biological Significance of Sex: Sex-Determination in Relation to Fertilization and Parthenogenesis." *Science* 25 (1907): 376–379.

Wilson, Edmund Beecher. "Some Aspects of Progress in Modern Zoology." *Science* 4 (1915): 1–11.

Wilson, Edmund Beecher, and Edward Leaming. *An Atlas of the Fertilization and Karyokinesis of the Ovum.* New York: Macmillan, 1895.

Wilson, Edmund Beecher, and Albert P. Mathews. "Maturation, Fertilization, and Polarity in the Echinoderm Egg. New Light on the 'Quadrille of the Centers.'" *Journal of Morphology* 10 (1895): 319–342.

Wilson, Edmund Beecher, with H. L. Osborn and J. M. Wilson. "Variation in the yolk-cleavage of Renilla." *Zoologischer Anzeiger* 5 (1882): 545–548.

Witkowski, J. A. "Ross Harrison and the Experimental Analysis of Nerve Growth: The Origins of Tissue Culture." In T. J. Horder, J. A. Witkowski, and C. C. Wylie, *A History of Embryology*, pp. 149–177. Cambridge: Cambridge University Press, 1985.

Wright, Albert Hagen, and Anna Allen Wright. "Agassiz's Address at the Opening of Agassiz's Academy." *American Midland Naturalist* 43 (1950): 503–506.

Wunderlich, Klaus. *Rudolf Leuckart. Weg and Werk.* Jena: Gustav Fischer, 1978.

Index